Graphing Calculator Manual

Darryl Nester
Bluffton University

A Graphical Approach to Algebra and Trigonometry
Fourth Edition

and

A Graphical Approach to Precalculus
Fourth Edition

John Hornsby
University of New Orleans

Margaret L. Lial
American River College

Gary K. Rockswold
Minnesota State University, Mankato

Boston San Francisco New York
London Toronto Sydney Tokyo Singapore Madrid
Mexico City Munich Paris Cape Town Hong Kong Montreal

Reproduced by Pearson Addison-Wesley from electronic files provided by the author.

Publishing as Pearson Addison-Wesley, 75 Arlington Street, Boston, MA 02116.

ISBN 0-321-35801-5

2 3 4 5 6 OPM 09 08 07 06

PREFACE

This graphing calculator manual was written to help owners of Texas Instruments graphing calculators (TI-83/83+/84+, TI-86, and TI-89) use them to solve problems from *A Graphical Approach to Algebra and Trigonometry,* Fourth Edition, by Hornsby, Lial, and Rockswold. Owners of other TI calculators, like the TI-80, -81, -82, -85, and -92, as well as owners of other brands of graphers, may find the information useful as well, but certainly some of the specific details will not translate directly. In particular, TI-82 users should find many comments in the TI-83/83+/84+ chapter that apply to their calculator, TI-85 users can find useful information in the TI-86 chapter, and TI-92 users should read the chapter on the TI-89.

Please contact me with any questions or corrections concerning this material. My web site also contains additional resources for using graphing calculators; I welcome suggestions as to what else would be useful.

This manual was created with TI calculators (of course), the TI-Connect software, Adobe Photoshop, and TEX (Textures from Blue Sky Research).

Darryl Nester
Bluffton University
Bluffton, Ohio
`nesterd@bluffton.edu`
`http://www.bluffton.edu/~nesterd`
March 2006

TABLE OF CONTENTS

This manual contains three chapters, each devoted to one line of calculators. The table below lists, for each calculator, the topics covered in the introductory section (which contains a brief introduction to using the calculator), followed by the examples from *A Graphical Approach to Algebra and Trigonometry,* Fourth Edition, which are discussed in this manual. The page numbers in parentheses in the first column—e.g., "Section 1.1 Technology Note (page 3)"—refer to the text, while the page numbers in the other three columns (under "83/83+/84+," etc.) refer to this manual. For example, information about the TI-89 begins on page 97.

— *continued* —

			Page number		
			83/83+/84+	TI-86	TI-89
EXAMPLES (cont.)					
Section 1.5	Technology Note	(page 60)	15	64	114
	Example 11	(page 61)	15	65	114
Section 2.1	Technology Note	(page 96)	16	65	114
	Technology Note	(page 96)	16	65	114
	Technology Note	(page 98)	16	65	114
Section 2.2	Technology Note	(page 103)	16	65	115
Section 2.3	Technology Note	(page 113)	16	65	115
	Technology Note	(page 114)	16	65	115
	Technology Note	(page 115)	16	65	115
	Technology Note	(page 117)	16	65	115
Section 2.5	Example 1	(page 139)	17	66	115
	Example 2	(page 139)	17	66	115
	Example 4	(page 141)	18	67	116
Section 2.6	Example 1	(page 149)	18	67	116
	Example 2	(page 150)	18	67	116
	Example 4	(page 151)	–	–	116
	Example 5	(page 153)	18	68	117
	Example 6	(page 153)	18	68	117
Section 3.1	Technology Note	(page 174)	19	68	117
	Example 1	(page 175)	19	68	117
	Example 2	(page 176)	19	68	117
	Example 3	(page 176)	19	69	118
	Example 4	(page 177)	19	69	118
	Example 5	(page 178)	19	69	118
	Example 6	(page 179)	19	69	118
Section 3.2	Example 3	(page 185)	19	69	118
	Example 5	(page 186)	20	70	119
	Example 6	(page 187)	20	70	119
Section 3.3	Technology Note	(page 198)	20	70	119
	Example 4	(page 199)	–	–	119
	Example 5	(page 200)	–	–	119
	Example 6	(page 202)	–	–	119
	Example 7	(page 203)	–	–	119
Section 3.6	Example 2	(page 232)	–	–	120
Section 3.7	Example 2	(page 242)	21	71	120
Section 3.8	Example 7	(page 255)	21	71	120
Section 4.1	Example 2	(page 273)	21	71	120
Section 4.2	Example 8	(page 285)	21	71	121
Section 4.3	Example 1	(page 291)	–	–	121
	Example 2	(page 292)	–	–	121

— *continued* —

— continued —

			Page number		
			83/83+/84+	TI-86	TI-89
EXAMPLES (cont.)					
Section 7.4	Example 9	(page 508)	30	79	128
Section 7.5	Example 1	(page 514)	30	79	129
	Example 4	(page 517)	30	79	129
	Example 5	(page 518)	30	79	129
Section 7.6	Example 1	(page 525)	30	80	129
	Example 3	(page 528)	30	80	129
	Example 4	(page 530)	30	80	129
Section 7.7	Example 1	(page 538)	30	80	129
	Example 2	(page 539)	32	81	130
Section 7.8	Example 1	(page 548)	–	–	131
	Example 2	(page 549)	–	–	131
	Example 3	(page 550)	–	–	131
	Example 4	(page 551)	–	–	131
Section 8.1	Technology Note	(page 567)	32	82	132
	Example 2	(page 568)	32	82	132
	Example 3	(page 568)	33	82	132
	Example 5	(page 571)	33	83	133
	Example 6	(page 571)	33	83	133
Section 8.2	Example 3	(page 584)	33	83	133
	Example 4	(page 585)	34	83	133
	Technology Note	(page 589)	34	84	134
Section 8.3	Technology Note	(page 598)	34	84	134
	Example 7	(page 601)	34	84	134
	Technology Note	(page 602)	35	84	134
	Example 8	(page 602)	35	84	135
	Example 10	(page 603)	35	84	135
Section 8.4	Technology Note	(page 609)	35	85	135
	Example 7	(page 613)	35	85	135
Section 8.6	Technology Note	(page 630)	36	85	136
	Example 1	(page 631)	36	85	136
	Example 2	(page 633)	36	85	136
	Example 8	(page 638)	36	86	136
Section 8.7	Technology Note	(page 647)	36	86	136
	Example 1	(page 649)	36	86	136
	Example 2	(page 650)	36	86	136
Section 9.1	Example 2	(page 674)	37	86	137
	Example 3	(page 676)	37	86	137
	Example 4	(page 677)	37	86	137
	Example 5	(page 677)	37	86	137
	Example 6	(page 678)	37	86	137

— continued —

			Page number		
			83/83+/84+	TI-86	TI-89
EXAMPLES (cont.)					
Section 9.2	Example 1	(page 685)	38	87	138
	Example 3	(page 687)	38	87	138
Section 9.3	Example 2	(page 694)	–	–	138
	Example 3	(page 694)	–	–	138
	Example 4	(page 695)	38	87	138
	Example 6	(page 697)	–	–	138
	Example 7	(page 698)	–	–	138
	Example 8	(page 699)	–	–	138
Section 9.4	Example 1	(page 706)	38	88	139
	Example 2	(page 707)	38	88	139
	Example 4	(page 710)	39	88	139
	Example 7	(page 712)	–	–	140
Section 9.5	Example 1	(page 718)	–	–	140
Section 9.6	Example 8	(page 727)	39	88	140
Section 10.1	Technology Note	(page 743)	39	89	140
Section 10.2	Technology Note	(page 755)	39	89	140
Section 10.3	Example 1	(page 766)	40	89	141
	Example 2	(page 767)	40	89	141
	Example 3	(page 767)	–	90	142
	Example 5	(page 768)	40	90	142
	Example 6	(page 769)	40	90	142
Section 10.4	Example 2	(page 779)	41	90	142
	Example 3	(page 780)	41	90	143
	Technology Note	(page 782)	41	90	143
	Example 5	(page 782)	42	91	143
	Example 6	(page 784)	42	91	143
Section 10.5	Example 1	(page 787)	42	91	143
	Example 4	(page 790)	–	91	144
Section 10.6	Example 1	(page 794)	42	91	144
	Example 2	(page 795)	42	91	144
	Example 3	(page 796)	42	91	144
	Example 4	(page 796)	42	91	144
	Example 5	(page 797)	42	91	144
	Example 6	(page 798)	42	91	144
	Example 7	(page 798)	43	92	145
Section 10.7	Example 2	(page 805)	43	92	145
	Example 3	(page 805)	43	92	145
	Example 4	(page 806)	43	92	–
	Example 5	(page 807)	44	93	145
	Example 7	(page 808)	44	93	145

— continued —

			Page number		
			83/83+/84+	TI-86	TI-89
EXAMPLES (cont.)					
Section 11.1	Example 1	(page 821)	44	93	146
	Technology Note	(page 822)	44	93	146
	Example 2	(page 822)	45	94	147
	Example 4	(page 825)	46	94	148
Section 11.3	Example 7	(page 844)	–	–	148
Section 11.4	Technology Note	(page 851)	46	94	148
	Example 1	(page 852)	46	94	148
Section 11.6	Example 4	(page 865)	46	94	148
Section 11.7	Example 6	(page 878)	46	95	149
Section R.1	Example 4	(page 896)	–	–	149
	Example 5	(page 897)	–	–	149
Section R.2	Example 1	(page 900)	–	–	149
Section R.4	Example 4	(page 916)	46	95	149
	Example 5	(page 917)	47	95	150
Appendix B	Example 3	(page 934)	47	96	150
Appendix C	Example 1	(page 937)	47	96	150
	Example 4	(page 938)	47	96	151
Appendix: Simulating complex numbers with a TI-82			48	–	–

Chapter 1 The Texas Instruments TI-83/TI-83+/84+ Graphing Calculator

Introduction

The information in this section is essentially a summary of material that can be found in the TI-83 manual. Consult that manual for more details. **All references in this chapter to the TI-83 also apply to the TI-83+ and the TI-84+, including the "silver" editions of those calculators.**

While the TI-82 and TI-83 differ in some details, in most cases the instructions given in this chapter can be applied (perhaps with slight alteration) to a TI-82. The icon [82] is used to identify significant differences between the two, but some differences (e.g., a slight difference in keystrokes between the two calculators) are not noted. TI-82 users should watch for these comments. Also, see page 48 for information on computing with complex numbers on the TI-82.

1 Power

To power up the calculator, simply press the [ON] key. This should bring up the "home screen"—a flashing block cursor, and possibly the results of any previous computations that might have been done.

If the home screen does not appear, one may need to adjust the contrast (see the next section).

To turn the calculator off, press [2nd][ON] (note that the "second function" of [ON]—written in yellow type above the key—is "OFF"). The calculator will automatically shut off if no keys are pressed for several minutes.

2 Adjusting screen contrast

If the screen is too dark (all black), decrease the contrast by pressing [2nd] then pressing and holding [▼]. If the screen is too light, increase the contrast by pressing [2nd] and then press and hold [▲].

As one adjusts the contrast, the numbers 1 through 9 will appear in the upper right corner of the screen. If the contrast setting reaches 8 or 9, or if the screen never becomes dark enough to see, the batteries should be replaced.

3 Replacing batteries

To replace the four AAA batteries, first turn the calculator off ([2nd][ON]), then remove the back cover, remove and replace each battery, replace the back cover, then turn the calculator on again. (After replacing batteries, one may need to adjust the contrast down as described above.)

4 Basic operations

Simple computations are entered in essentially the same way they would be written. For example, to compute $2 + 17 \times 5$, press [2][+][1][7][×][5][ENTER] (the [ENTER] key tells the calculator to act on what has been typed). Standard order of operations (including parentheses) is followed.

```
2+17*5
                87
■
```

The result of the most recently entered expression is stored in `Ans`, which is typed by pressing [2nd][(-)] (the word "ANS" appears in yellow above this key). For example, [5][+][2nd][(-)][ENTER] will add 5 to the result of the previous computation.

```
2+17*5
                87
5+Ans
                92
■
```

After pressing [ENTER], the TI-83 automatically produces `Ans` if the first key pressed is one which requires a number before it; the most common of these are [+], [−], [×], [÷], [^], [x^{-1}], [x^2], and [STO▸]. For example, [+][5][ENTER] would accomplish the same thing as the keystrokes above (that is, it adds 5 to the previous result).

```
2+17*5
                87
5+Ans
                92
Ans+5
                97
■
```

Pressing [ENTER] by itself evaluates the previously typed expression again. This can be especially useful in conjunction with `Ans`. The screen on the right shows the result of pressing [ENTER] a second time.

```
2+17*5
                87
5+Ans
                92
Ans+5
                97
               102
■
```

Several expressions can be evaluated together by separating them with colons ([ALPHA][.]). When [ENTER] is pressed, the result of the *last* computation is displayed. The screen shown illustrates the computation $2(5 + 1)^2$.

```
3+2
                 5
Ans+1:Ans²:2Ans
                72
■
```

5 Cursors

When typing, the appearance of the cursor indicates the behavior of the next keypress. When the standard cursor (a flashing solid block, ■) is visible, the next keypress will produce its standard action—that is, the command or character printed on the key itself.

If [2nd][DEL] is pressed, the TI-83 is placed in INSERT mode and the standard cursor will appear as a flashing underscore. If the arrow keys ([▲], [▼], [▶], [◀]) are used to move the cursor around within the expression, and the TI-83 is placed in INSERT mode, subsequent characters and commands will be inserted in the line at the cursor's position. When the cursor appears as a block, the TI-83 is in DELETE (or OVERWRITE) mode, and subsequent keypresses will replace the character or command at the cursor's position. (When the cursor is at the end of the expression, this is irrelevant.)

The TI-83 will return to DELETE mode when any arrow key is pressed. It can also be returned to DELETE mode by pressing [2nd][DEL] a second time.

Pressing [2nd] causes an arrow to appear in the cursor: ⬆ (or an underscored arrow). The next keypress will produce its "second function"—the command or character printed in yellow above the key. (The cursor will then return to "standard.") If [2nd] is pressed by mistake, pressing it a second time will return the cursor to standard.

Pressing [ALPHA] places the letter "A" in the cursor: ▫ (or an underscored "A"). The next keypress will produce the letter or other character printed in green above that key (if any), and the cursor will then return to standard. Pressing [ALPHA] a second time cancels ALPHA mode. Pressing [2nd][ALPHA] "locks" the TI-83 in ALPHA mode, so that all of the following keypresses will produce characters until [ALPHA] is pressed again, or until some menu or second function is accessed.

6 Accessing previous entries ("deep recall")

By repeatedly pressing [2nd][ENTER] ("ENTRY"), previously typed expressions can be retrieved for editing and re-evaluation. Pressing [2nd][ENTER] once recalls the most recent entry; pressing [2nd][ENTER] again brings up the second most recent, etc. The number of previous entries thus displayed varies with the length of each expression (the TI-83 allocates 128 bytes to store previous expressions).

7 Menus

Keys such as [WINDOW], [MATH] and [VARS] bring up a menu screen with a variety of options. The top line of the menu screen gives a collection of submenus (if any), which can be selected with the [◄] and [►] keys. The lower lines list the available commands; these can be selected using the [▲] and [▼] keys and [ENTER], or by pressing the number (or letter) preceding the desired option. Shown is the menu produced by pressing [MATH]; the arrow next to the 7 in the bottom row indicates that there are more options available below.

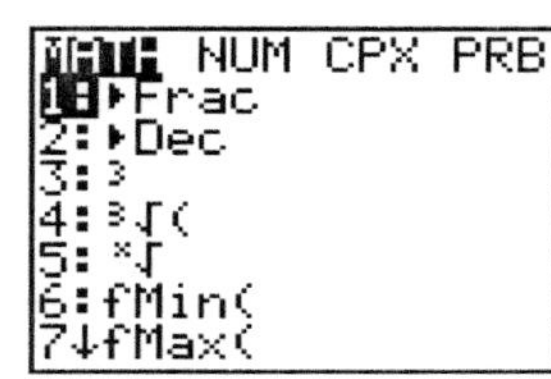

The various commands in these menus are too numerous to be listed here. They will be mentioned as needed in the examples.

8 Variables

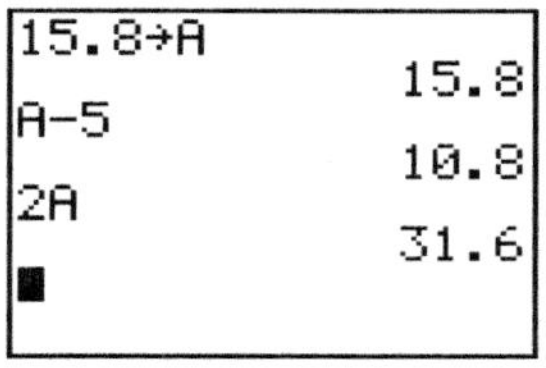

The letters A through Z can be used as variables (or "memory") to store numerical values. To store a value, type the number (or an expression) followed by [STO►], then a letter (preceded by [ALPHA] if necessary), then [ENTER]. That letter can then be used in the same way as a number, as demonstrated at right.

Note: The TI-83 interprets 2A as "2 times A"—the "*" symbol is not required (this is consistent with how we interpret mathematical notation). As for order of operations, this kind of multiplication is treated the same as "*" multiplication.

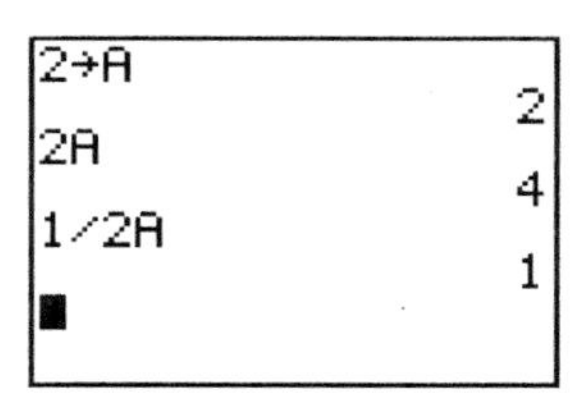

[82] *This latter comment is* **not** *true of the TI-82; on the TI-82, implied multiplication (such as* 2A*) is done before other multiplication and division, and even before some other operations, like the square root function* √*. Therefore, for example, the expression* 1/2A *is evaluated as 1/4 on the TI-82 (assuming that* A *is 2).*

9 Setting the modes

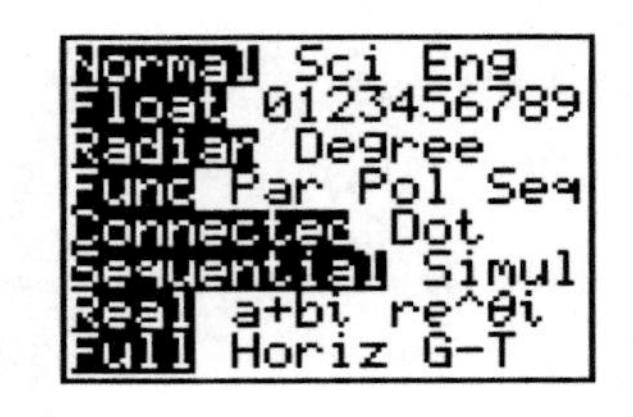

By pressing the MODE key, one can change many aspects of how the calculator behaves. For most of the examples in this manual, the "default" settings should be used; that is, the MODE screen should be as shown on the right. Each of the options is described below; consult the TI-83 manual for more details. Changes in the settings are made using the arrows keys and ENTER. (Note: The TI-84+ has essentially the same screen, but also includes a line for setting the clock.)

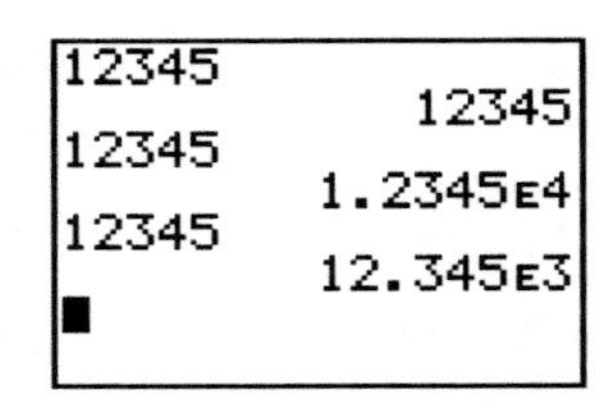

The `Normal Sci Eng` setting specifies how numbers should be displayed. The screen on the right shows the number 12345 displayed in `Normal` mode (which displays numbers in the range $\pm 9,999,999,999$ with no exponents), `Sci` mode (which displays all numbers in scientific notation), and `Eng` mode (which uses only exponents that are multiples of 3). Note: "E" is short for "times 10 to the power," so 1.2345E4 $= 1.2345 \times 10^4 = 1.2345 \times 10000 = 12345$.

The `Float 0123456789` setting specifies how many places after the decimal should be displayed. The default, `Float`, means that the TI-83 should display all non-zero digits (up to a maximum of 10).

`Radian Degree` indicates whether angle measurements should be assumed to be in radians or degrees. (A right angle measures $\frac{\pi}{2}$ radians, which is equivalent to 90°.) This text does not refer to angle measurement.

`Func Par Pol Seq` specifies whether formulas entered into the Y= screen are functions (specifically, y as a function of x), parametric equations (x and y as functions of t), polar equations (r as a function of θ), or sequences (u, v and w as functions of n). The text accompanying this manual uses all of these settings except for `Pol`.

When plotting a graph, the `Connected Dot` setting tells the TI-83 whether or not to connect the individually plotted points. `Sequential Simul` specifies whether individual expressions should be graphed one at a time (sequentially), or all at once (simultaneously).

`Real a+bi re^θi` specifies how to deal with complex numbers. `Real` means that only real results will be allowed (unless `i` is entered as part of a computation)—so that, for example, taking the square root of a negative number produces an error ("`NONREAL ANS`"). Selecting one of the other two options means that square roots of negative numbers are allowed, and will be displayed in "rectangular" ($a + bi$) or polar ($re^{i\theta}$) format. These two formats are essentially the same as the two used by the textbook (the term "trigonometric format" is used rather than "polar format"). More information about complex numbers can be found beginning on page 19 (Example 1 from Section 3.1) of this manual.

82 *The TI-82 does not support complex numbers, so it does not include this setting. However, some complex computations can be done with a TI-82; see the appendix at the end of this chapter, page 48.*

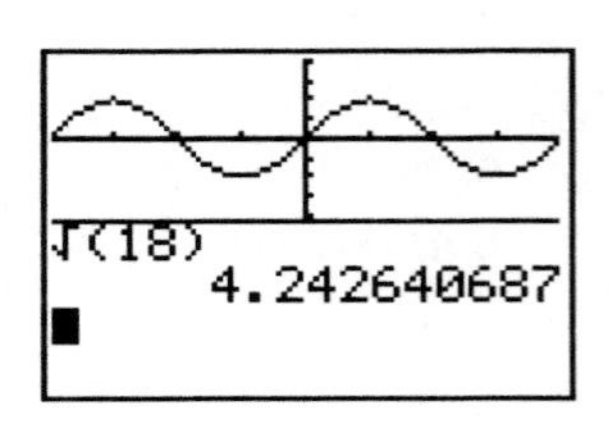

Finally, the `Full Horiz G-T` setting allows the option of, for example, showing both the graph and the home screen, as in the screen on the right (this shows a "horizontal split").

The third option, G-T, has no effect on the home screen display, but will show graphs and tables side by side when [GRAPH] is pressed.
[82] *The TI-82 supports only the horizontal split.*

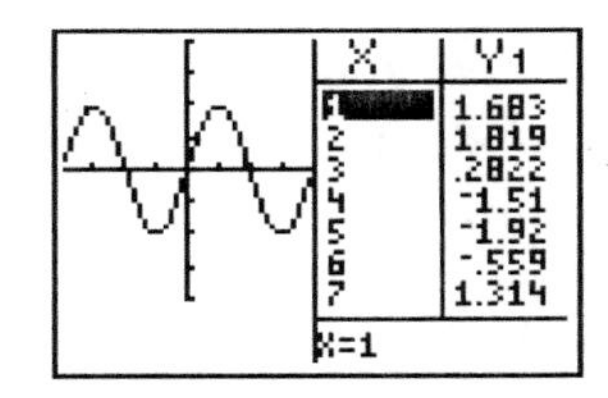

A related group of settings are found in the FORMAT menu ([2nd][ZOOM]). The default settings are shown in the screen on the right, and are generally the best choices for most examples in this book (although the last two settings could go either way).
[82] *The TI-82 does not include the* ExprOn ExprOff *option.*

RectGC PolarGC specifies whether graph coordinates should be displayed in rectangular (x, y) or polar (r, θ) format. Note that this choice is independent of the Func Par Pol Seq mode setting. The CoordOn CoordOff setting determines whether or not graph coordinates should be displayed. GridOff GridOn specifies whether or not to display a grid of dots on the graph screen, while AxesOn AxesOff and LabelOff LabelOn do the same thing for the axes and labels (y and x) on the axes. ExprOn ExprOff specifies whether or not to display the formula (expression) of the curves on the [GRAPH] screen when tracing. (This can be useful when more than one graph is displayed.)

10 Setting the graph window

The exact contents of the [WINDOW] menu vary depending on whether the calculator is in function, parametric, polar, or sequence mode; below are four examples showing the [WINDOW] menu in each of these modes.

```
WINDOW
 Xmin=-4.7
 Xmax=4.7
 Xscl=1
 Ymin=-3.1
 Ymax=3.1
 Yscl=1
 Xres=1
```
Function mode

```
WINDOW
 Tmin=0
 Tmax=6.2831853…
 Tstep=.1308996…
 Xmin=-4.7
 Xmax=4.7
 Xscl=1
↓Ymin=-3.1
```
Parametric mode

```
WINDOW
 θmin=0
 θmax=6.2831853…
 θstep=.1308996…
 Xmin=-4.7
 Xmax=4.7
 Xscl=1
↓Ymin=-3.1
```
Polar mode

```
WINDOW
 nMin=4
 nMax=10
 PlotStart=1
 PlotStep=1
 Xmin=-4.7
 Xmax=4.7
↓Xscl=1
```
Sequence mode

All these menus include the values Xmin, Xmax, Xscl, Ymin, Ymax, and Yscl. When the [GRAPH] key is pressed, the TI-83 will show a portion of the Cartesian (x-y) plane determined by these values. In function mode, this menu also includes Xres, the behavior of which is described in section 12 of this manual (page 7). The other settings in the [WINDOW] screen allow specification of the smallest, largest, and step values of t (for parametric mode) or θ (for polar mode), or initial conditions for sequence mode.

With settings as in the example screens shown above, the TI-83 would display the screen at right: x values from -4.7 to 4.7 (that is, from Xmin to Xmax), and y values between -3.1 to 3.1 (Ymin to Ymax). Since Xscl $=$ Yscl $= 1$, the TI-83 places tick marks on both axes every 1 unit; thus the x-axis ticks are at -4, $-3, \ldots, 3$, and 4, and the y-axis ticks fall on the integers from -3 to 3. This window is called the "decimal" window, and is most quickly set by pressing [ZOOM][4].

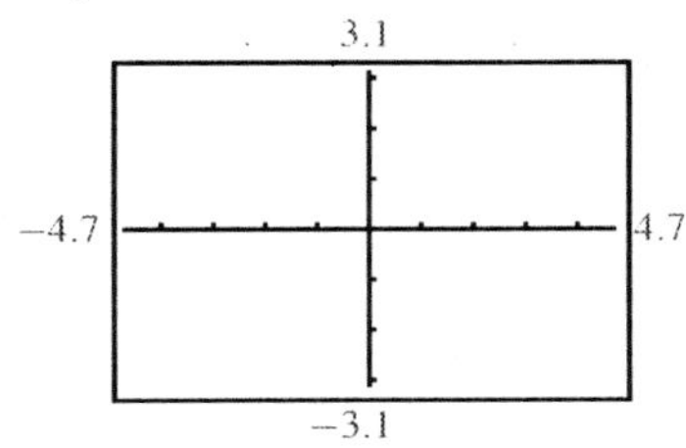

Below are four more sets of [WINDOW] settings, and the graph screens they produce. Note that the first graph on the left has tick marks every 10 units on both axes. The second window is called the "standard" viewing window, and is most quickly set by pressing [ZOOM][6]. The setting Yscl = 0 in the final graph means that no tick marks are placed on the y-axis.

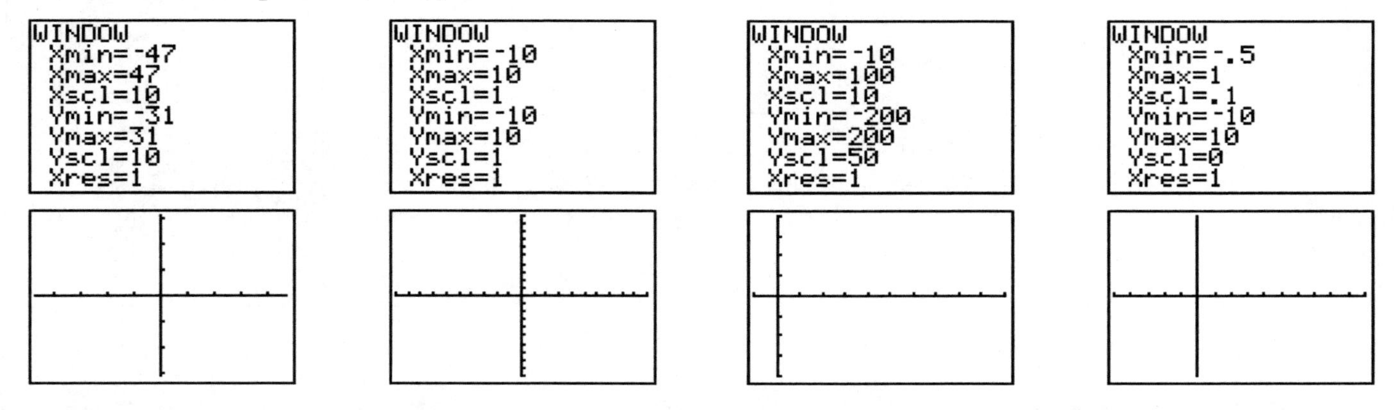

11 The graph screen

The TI-83 screen is made up of an array of square dots (pixels) with 63 rows and 95 columns. All the pixels in the leftmost column have x-coordinate Xmin, while those in the rightmost column have x-coordinate Xmax. The x-coordinate changes steadily across the screen from left to right, which means that the coordinate for the nth column (counting the leftmost column as column 0) must be Xmin + $n\Delta$X, where ΔX = (Xmax − Xmin)/94. Similarly, the nth row of the screen (counting up from the bottom row, which is row 0) has y-coordinate Ymin + $n\Delta$Y, where ΔY = (Ymax − Ymin)/62.

Note: In (horizontal) split screen mode, ΔY = (Ymax − Ymin)/30. In G-T (vertical split screen) mode, ΔX = (Xmax − Xmin)/46 and ΔY = (Ymax − Ymin)/50.

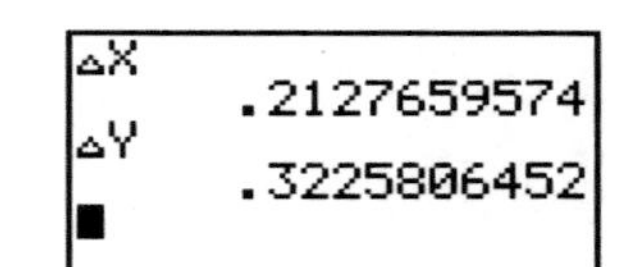

It is not necessary to memorize the formulas for ΔX and ΔY. Should they be needed, they can be determined by pressing [GRAPH] and then the arrow keys. When pressing [▶] or [◀] successively, the displayed x-coordinate changes by ΔX; meanwhile, when pressing [▲] or [▼], the y-coordinate changes by ΔY. Alternatively, the values can be found by pressing [VARS][1][8][ENTER] (for ΔX) or [VARS][1][9][ENTER] (for ΔY). This produces results like those shown on the right.

In the decimal window Xmin = −4.7, Xmax = 4.7, Ymin = −3.1, Ymax = 3.1, note that ΔX = 0.1 and ΔY = 0.1. Thus, the individual pixels on the screen represent x-coordinates −4.7, −4.6, −4.5, ..., 4.5, 4.6, 4.7 and y-coordinates −3.1, −3, −2.9, ..., 2.9, 3, 3.1. This is where the decimal window gets its name.

Windows for which ΔX = ΔY, such as the decimal window, are called square windows. Any window can be made square be pressing [ZOOM][5]. To see the effect of a square window, observe the two pairs of graphs below. In each pair, the first graph is on the standard window, and the second is on a square window (after pressing [ZOOM][5]). The first pair shows the lines $y = 2x - 3$ and $y = 3 - \frac{1}{2}x$; note that on the square window, these lines look perpendicular (as they should). The second pair shows a circle centered at the origin with a radius of 8. On the standard window, this looks like an oval since the screen is wider than it is tall. (The reason for the gaps in the circle will be addressed in the next section.)

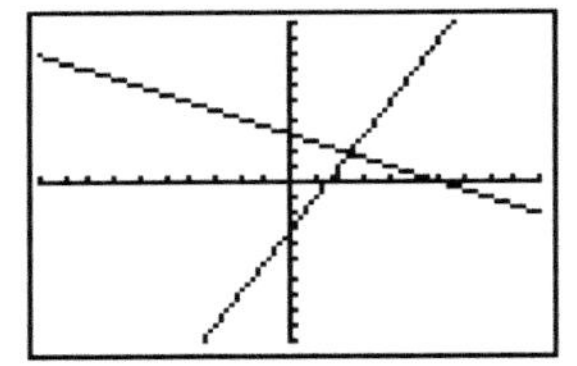
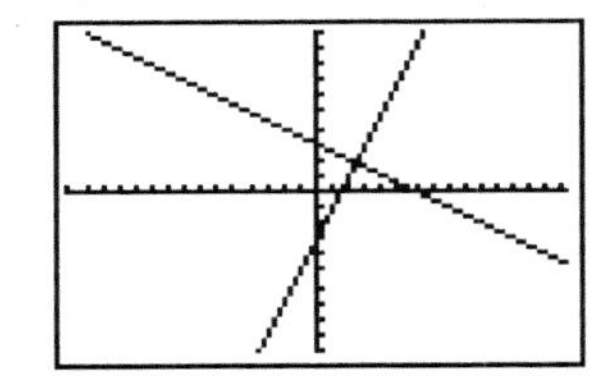
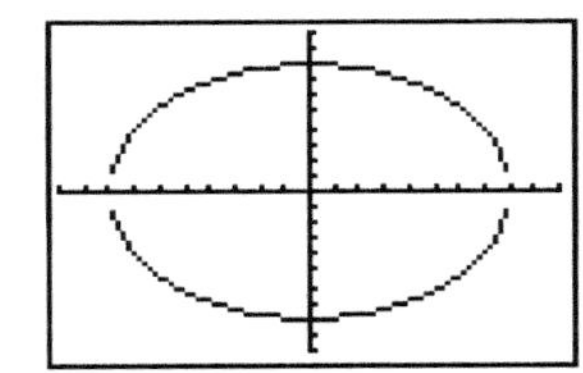
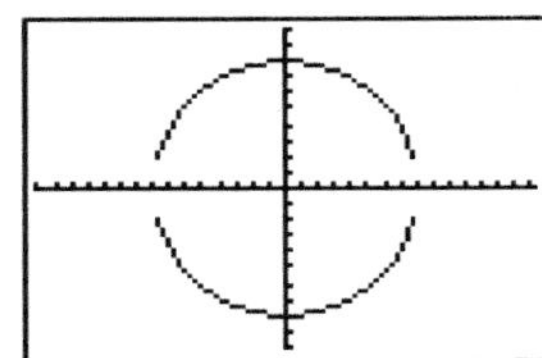

12 Graphing a function

This introductory section only addresses creating graphs in function mode. Procedures for creating parametric and polar graphs are very similar, and are described in this manual in the material related to the examples from the text.

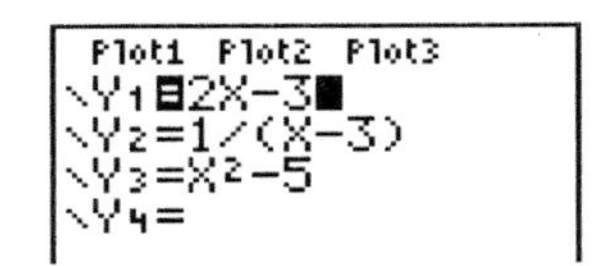

To see the graph of $y = 2x - 3$, begin by entering the formula into the calculator. This is done on the [Y=] screen of the calculator. Select one of the variables Y_1, Y_2, . . . , and enter the formula. If other y variables have formulas, either erase them (by positioning the cursor on that line and pressing [CLEAR]) or position the cursor on the equals sign "=" for that line and press [ENTER] (this has the effect of "unhighlighting" the equals sign, which tells the TI-83 not to graph that formula). Additionally, if any of `Plot1`, `Plot2` or `Plot3` are highlighted, move the cursor up until it is on that plot and press [ENTER]. In the screen on the right, only Y_1 will be graphed.

The next step is to choose a viewing window; see the previous section for more details on this. This example uses the standard window ([ZOOM][6]).

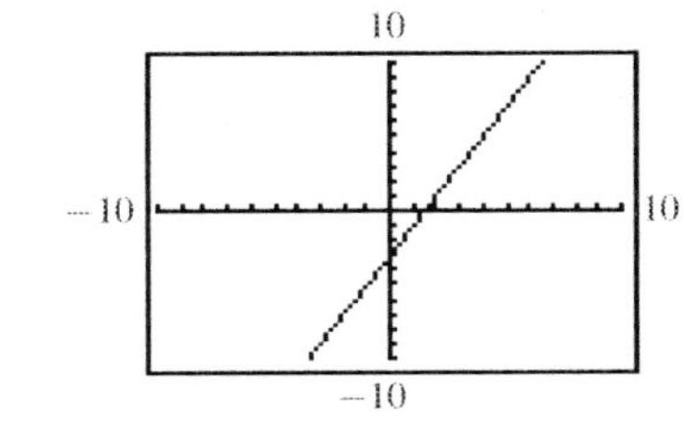

Finally, press [GRAPH], and the line should be drawn. In order to produce this graph, the TI-83 considers 95 values of x, ranging from `Xmin` to `Xmax` in steps of ΔX (assuming that `Xres` = 1; see below for other possibilities). For each value of x, it computes the corresponding value of y, then plots that point (x, y) and (if the calculator is in Connected mode) draws a line between this point and the previous one.

If `Xres` is set to 2, the TI-83 will only compute y for every other x value; that is, it uses a step size of 2ΔX. Similarly, if `Xres` is 3, the step size will be 3ΔX, and so on. Setting `Xres` higher causes graphs to appear faster (since fewer points are plotted), but for some functions, the graph may look "choppy" if `Xres` is too large, since detail is sacrificed for speed.

Note: If the line does not appear, or the TI-83 reports an error, double-check all the previous steps. Also, check the mode settings (discussed in section 9, page 4).

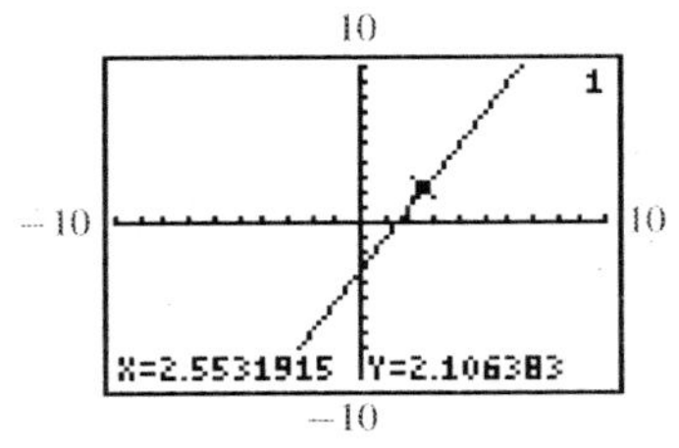

Once the graph is visible, the window can be changed using [WINDOW] or [ZOOM]. Pressing the [TRACE] key brings up the "trace cursor," and displays the x- and y-coordinates for various points on the line as the [◀] and [▶] keys are pressed. Tracing beyond the left or right columns causes the TI-83 to adjust the values of `Xmin` and `Xmax` and redraw the graph.

To graph the function

$$y = \frac{1}{x-3},$$

Plot1 Plot2 Plot3
\Y1=2X-3
\Y2=1/(X-3)
\Y3=X²-5
\Y4=

enter that formula into the [Y=] screen (note the use of parentheses). As before, this example uses the standard viewing window.

When [GRAPH] is pressed, the TI-83 produces the graph shown on the right. This illustrates one of the pitfalls of the connect-the-dots method used by the calculator: The nearly-vertical line segment drawn at $x = 3$ *should not be there*, but it is drawn because the calculator connects the points

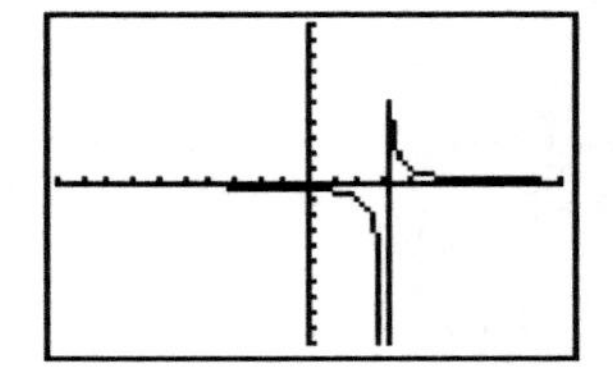

$x = 2.9787234$, $y = -46.99999$ and $x = 3.1914894$, $y = 5.2222223$.

Calculator users must learn to recognize these flaws in calculator-produced graphs. **Note:** Recent versions of the TI-84+ operating system "fix" this problem, so that this vertical line is not displayed.

The graph of a circle centered at the origin with radius 8 (shown on the square window [ZOOM][6][ZOOM][5]) shows another problem that arises from connecting the dots. When $x = -8.064516$, y is undefined, so no point is plotted (that is, there is no point on this circle that has x-coordinate less than -8, or greater than 8). The next point plotted on the upper half of the circle is $x = -7.741935$ and $y = 2.0155483$; since no point had been plotted for the previous x-coordinate, this is not connected to anything, so there appears to be a gap between the circle and the x-axis. The calculator is not "smart" enough to know that the graph should extend from -8 to 8.

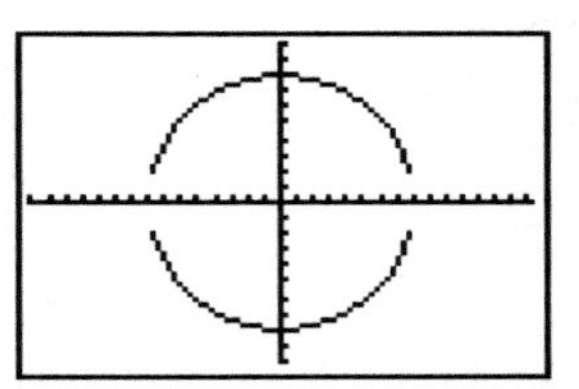

One additional feature of graphing with the TI-83 is that each function can have a "style" assigned to its graph. The symbol to the left of Y1, Y2, etc. indicates this style, which can be changed by pressing [◄] until the cursor is over the symbol, then pressing [ENTER] to cycle through the options. These options are shown on the right (with brief descriptive names); complete details are in the TI-83 manual. [82] *The TI-82 does not include graph-style features.*

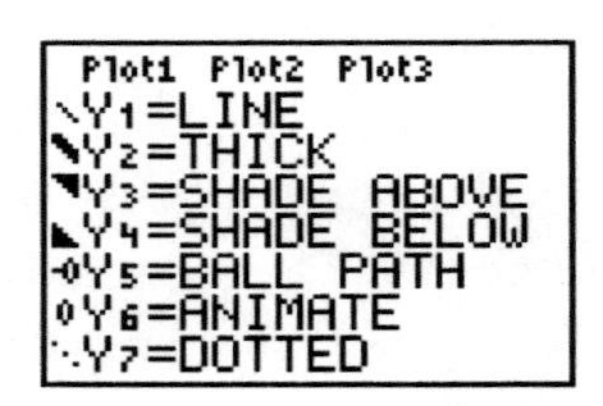

13 Adding programs to the TI-83

The TI-83's capabilities can be extended by downloading or entering programs into the calculator's memory. Instructions for writing a program are beyond the scope of this manual, but programs written by others and downloaded from the Internet (or obtained as printouts) can be transferred to the calculator in one of three ways:

1. If one TI-83 already has a program, it can be transferred to another using the calculator-to-calculator link cable. To do this, first make sure the cable is firmly inserted in both calculators. On the sending calculator, press [2nd][X,T,Θ,n] (LINK), then [3], and then select (by using the [▲] and [▼] keys and [ENTER]) the program(s) to be transferred. Now press the [►] to bring up the TRANSMIT submenu. *Before* pressing [ENTER] on the sending calculator, prepare the receiving calculator by pressing [2nd] [X,T,Θ,n][►][ENTER], and *then* press [ENTER] on the sending calculator.

2. If a computer with the TI-Graph Link is available, and the program file is on that computer (e.g., after having been downloaded from the Internet), the program can be transferred to the calculator using the TI Connect (or TI Graph Link) software. This transfer is done in a manner similar to the calculator-to-calculator transfer described above; specific instructions can be found in the documentation that

accompanies the software. (They are not given here because of slight differences between platforms and software versions.)

3. View a listing of the program and type it in manually. (**Note:** Even if the TI-Graph Link cable is not available, the software can be used to view program listings on a computer.) While this is the most tedious method, studying programs written by others can be a good way to learn programming. To enter a program, start by choosing [PRGM][◂][1] (“`Create New`”), then type a name for the new program (like “`QUADFORM`” or “`MIDPOINT`”)—note that the TI-83 is automatically put into ALPHA mode. Then type each command in the program, and press [2nd][MODE] (QUIT) to return to the home screen when finished.

To run the program, make sure there is nothing on the current line of the home screen, then press [PRGM], select the number or letter of the program (a sample screen is shown), and press [ENTER]. If the program was entered manually (option 3 above), errors may be reported; in that case, choose `GOTO`, correct the mistake and try again.

Programs can be found at many places on the Internet, including:

- `http://www.bluffton.edu/~nesterd`—the Web site of the author of this manual;
- `http://tifaq.calc.org`—A “Frequently Asked Questions” page maintained by Ray Kremer; and
- `http://www.ticalc.org`.

Additionally, the TI-83+ and TI-84+ calculators include a variety of “APPs”—applications (programs) which can extend the capabilities of the calculator in various ways. APPs can be viewed by selecting the [APPS] key; the number of installed APPs depends on the model. Shown is the list of applications available on the TI-84+ Silver Edition. Additional APPs can be downloaded from `education.ti.com`, then installed using a Graph Link cable.

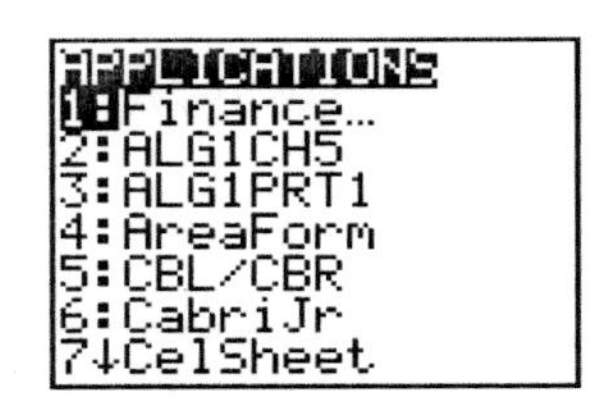

Examples

Here are the details for using the TI-83 for several of the examples from the textbook. Also given are the keystrokes necessary to produce some of the commands shown in the text's examples. In some cases, some suggestions are made for using the calculator more efficiently.

Throughout this section, it is assumed that the textbook is available for reference. The problems from the text are not restated here, and there are frequent references to the calculator screens shown in the text.

Owners of a TI-83 have an advantage over those who use other models, because the calculator screens shown in the text are from a TI-83.

Section 1.1 Technology Note (page 3) — Viewing Windows

Information about setting viewing windows is given in section 10 of the introduction, page 5.

Section 1.1 Example 1 (page 5) — Finding Roots on a Calculator

The three calculator screens shown in Figure 11 illustrate the main methods of computing roots with the TI-83. Aside from fractional exponents, we have the built-in square root function ([2nd][x^2]), the cube-root function $^3\sqrt{}$(and the "x-root" function $^x\sqrt{}$ ([MATH][4] and [MATH][5], respectively). It is worth noting that fourth roots can be typed in more efficiently using two square roots, since $\sqrt[4]{a} = a^{1/4} = (a^{1/2})^{1/2} = \sqrt{a^{1/2}} = \sqrt{\sqrt{a}}$.

Also, a quicker way to enter fractional exponents is to use the [x^{-1}] key, as the screen on the right shows. The TI-83's order of operations is such that the reciprocal takes place before the exponentiation.

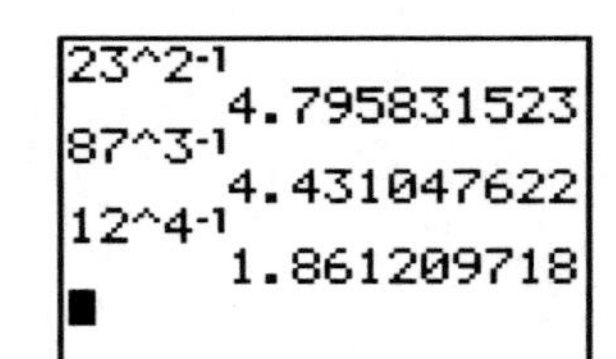

Section 1.2 Technology Note (page 18) — Producing Tables

To use the table features of the TI-83, begin by entering the formula ($y = 9x - 5$, in the example shown in Figure 26 of the text) on the [Y=] screen, as one would to create a graph. (The highlighted equals signs determine which formulas will be displayed in the table, just as they do for graphs.)

Next, press [2nd][WINDOW] to access the TABLE SETUP screen. The table will display y values for given values of x. The `TblStart` value sets the lowest value of x, while `ΔTbl` determines the "step size" for successive values of x. These two values are only used if the `Indpnt` option is set to `Auto`—this means, "automatically generate the values of the independent variable (x)." The effect of setting this option to `Ask` is illustrated at the end of this example. (The `Depend` option should almost always be set to `Auto`; if it is set to `Ask`, the y values are not displayed until [ENTER] is pressed.)

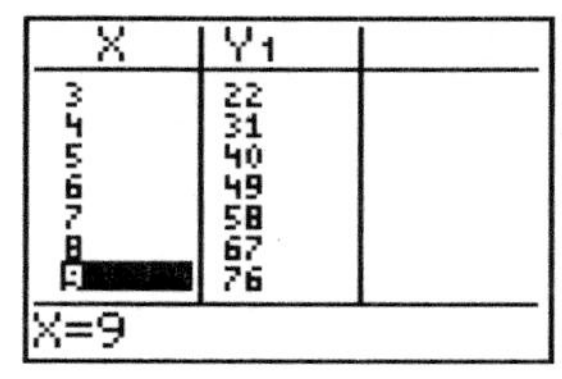

When the TABLE SETUP options are set satisfactorily, press [2nd][GRAPH] to produce the table. The first screen on the right shows the table generated based on the settings in the above screen (this is identical to Figure 26 of the text). By pressing [▼] repeatedly, the x values are increased, and the y values updated. After pressing [▼] nine times, the table looks like the second screen. Similarly, the x values can be decreased by pressing [▲].

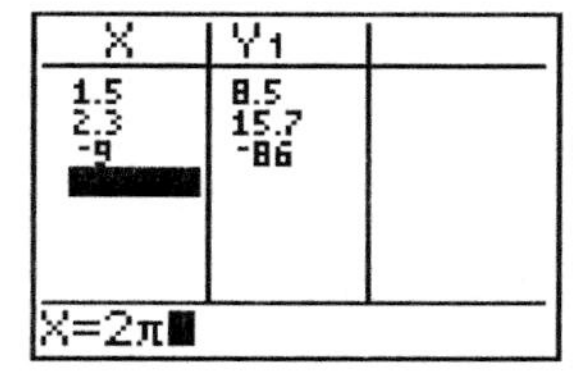

If `Indpnt` is set to `Ask` on the TABLE SETUP screen, the settings of `TblStart` and `ΔTbl` are ignored, and pressing [2nd][GRAPH] brings up a "blank" table. As values of x are entered, the y values are computed. Up to seven x values can be entered; if more are desired, the [DEL] key can be used to make room. Note that one can enter expressions (like 2π or $\sqrt{3}$) in addition to "simple" numbers.

Section 1.4 Example 5 (page 41) Using the Slope Relationship for Perpendicular Lines

See page 6 for information about setting square windows, including an illustration of perpendicular lines on square and non-square windows.

Section 1.4 Example 6 (page 42) Modeling Medicare Costs with a Linear Function

Given a set of data pairs (x, y), the TI-83 can produce a scatter diagram (as well as other types of statistical plots).

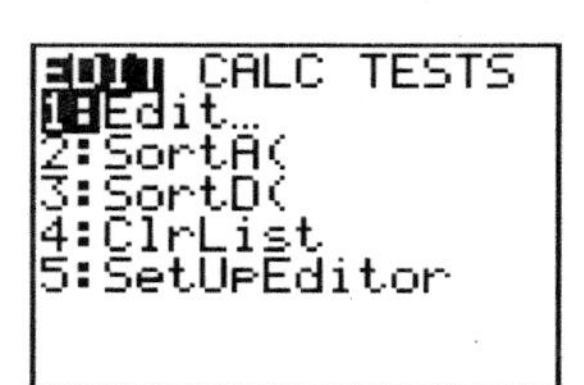

The first step is to enter the data into the TI-83. This is done by pressing [STAT] (which brings up the menu on the right), then choosing option 1 (`Edit`).

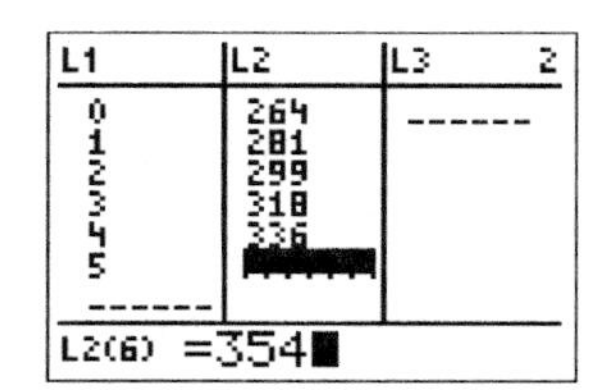

This brings up the list-editing screen. (If the column headings are not L_1, L_2, L_3 as shown on the right, press [STAT][5][ENTER] to reset the statistics [list] editor to its default. Then return to the list-editing screen.) Enter the x values (year) into the first column (L_1) and the y values (cost) into the second column (L_2). If either column already contains data, the [DEL] key can be used to delete numbers one at a time, or—to delete the whole column at once—press the [▲] key until the cursor is at the top of the column (on L_1 or L_2) and press [CLEAR][ENTER]. Make sure that both columns contain the same number of entries.

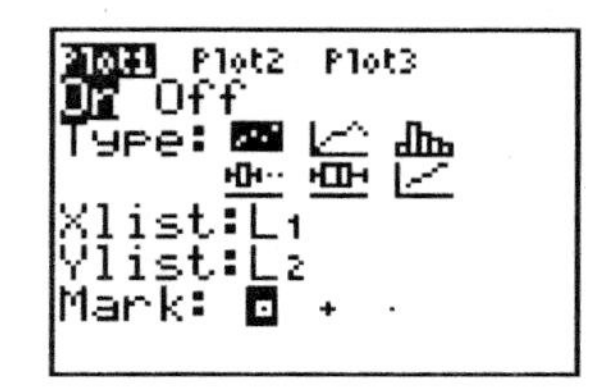

Next press [2nd][Y=][1] and make the settings for a STAT PLOT shown on the right. (If the data pairs had been stored somewhere other than L_1 and L_2, `Xlist` and `Ylist` can be set to those lists. Also, the `Mark` can be any of the three choices.)

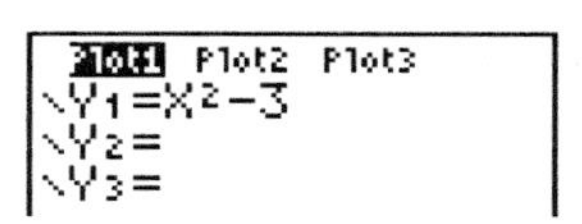

Now check that nothing else will be plotted: Press [Y=] and make sure that the only thing highlighted is `Plot1`. If `Plot2` or `Plot3` is highlighted, use the arrow keys to move the cursor up to that plot, then press [ENTER].

Finally, set up the viewing window as shown in Figure 58 of the text—or press [ZOOM][9] (`ZoomStat`), which automatically adjusts the window to show all the data in the plot. (The resulting window does not quite match the one shown in Figure 58.)

Note: When finished with a STAT PLOT, it is a good idea to turn all statistics plots off so that the TI-83 will not attempt to display them the next time [GRAPH] is pushed. This is most easily done by executing the `PlotsOff` command, using the key sequence [2nd][Y=][4][ENTER].

Section 1.4 Example 7 (page 43) Finding the Least-Squares Regression Line

Given a set of data pairs (x, y), the TI-83 can find various formulas (including linear, as well as more complex formulas) that approximate the relationship between x and y. These formulas are called "regression formulas."

The first step is to enter the data into the TI-83. See the previous example for a description of this process.

Once the data have been entered, press [STAT][▸][4] to choose `LinReg(ax+b)` from the CALC statistics submenu. This will place the command `LinReg(ax+b)` on the home screen.

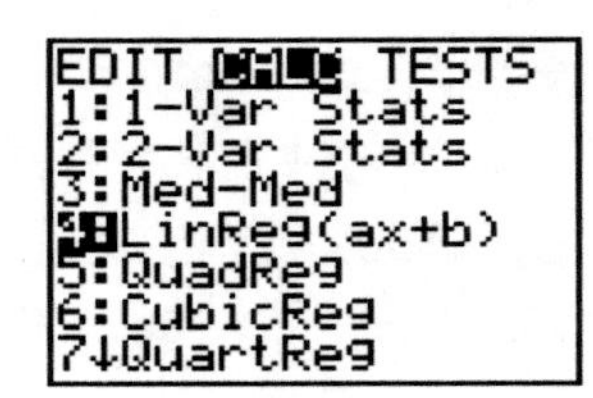

To tell the TI-83 where to find x and y, press [2nd][1][,][2nd][2], which adds "L_1,L_2" to the command. **Note:** This is not absolutely necessary, since the TI-83 will assume that x is in L_1 and y is in L_2 unless told otherwise. In other words, "`LinReg(ax+b)`" (by itself) works the same as "`LinReg(ax+b)` L_1,L_2". However, this should be done if, for example, x and y were stored in L_3 and L_4.

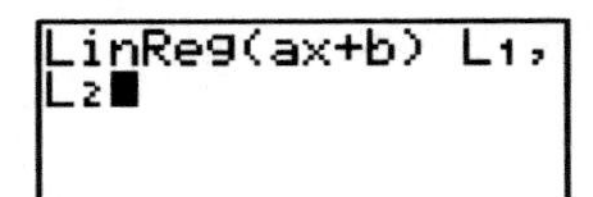

Pressing [ENTER] should then produce the screen shown in Figure 61(b) of the text.

If it produces an error message, it will likely either be a `DIM MISMATCH` (meaning that the two lists L_1 and L_2 have different numbers of entries) or a `SYNTAX` error, probably because the `LinReg` command was not on a line by itself. The screen on the right, for example, produces a syntax error.

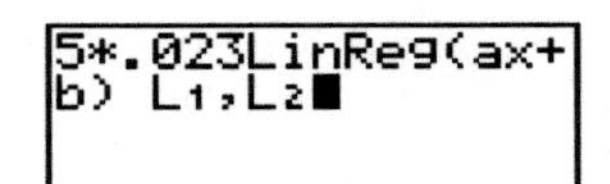

Following this example, the text introduces the idea of the correlation coefficient r. When the TI-83 finds a regression formula; it also computes r, but (in its default configuration) it does not display the correlation. One way to see the value of r is to press [VARS][5][▸][▸][7][ENTER] (which types the calculator variable `r` on the home screen, then displays its value). Alternatively, the TI-83 can be made to display `r` automatically every time a regression is performed by executing the `DiagnosticOn` command from the CATALOG ([2nd][0]), as described in the text's Technology Note. Once this command has been executed, the `LinReg(ax+b)` command will produce output like that show in Figure 62. The `DiagnosticOff` command can be used to turn this display off.

[82] *The TI-82 always displays r following a linear regression.*

Section 1.5 Example 4 (page 54) Applying the Intersection-of-Graphs Method

We need to solve the equation $f(x) = g(x)$, where $f(x) = 5.91x + 13.7$ and $g(x) = -4.71x + 64.7$. We are looking for an x value that will make the left and right sides of this equation equal to each other, which corresponds to the x-coordinate of the point of intersection of the graphs of $y = f(x)$ and $y = g(x)$.

In order to have the TI-83 locate this intersection, begin by setting up the TI-83 to graph the left side of the equation as Y1, and the right side as Y2.

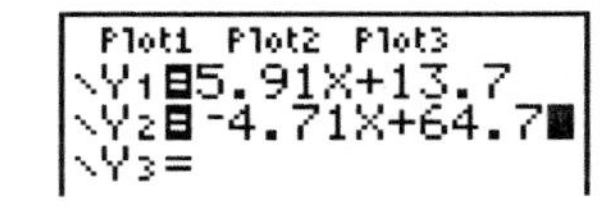

Next, select a viewing window which shows the point of intersection; we use the window shown in Figure 66(a) of the text: [0, 12] × [0, 100]. (The text also shows labels "CDS" and "TAPES" on the graphs. These are produced with the `Text` command; consult your calculator manual for more information.)

The TI-83 can automatically locate the point of intersection using the CALC menu ([2nd][TRACE]). Choose option 5 (intersect), use [▲], [▼] and [ENTER] to specify which two functions to use (in this case, the only two being displayed), and then use [◀] or [▶] to specify a guess. After pressing [ENTER], the TI-83 will try to find an intersection of the two graphs. The screens below illustrate these steps.

[2nd][TRACE][5] selects the CALC:intersect feature

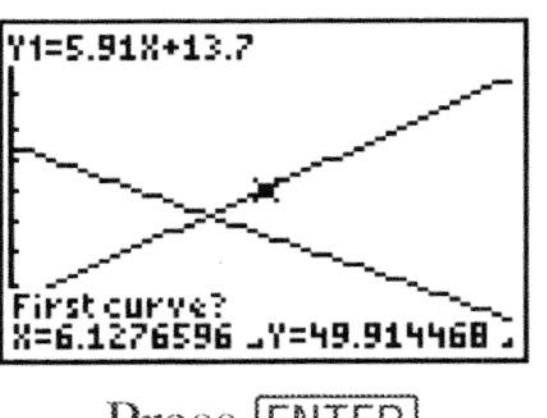

Press [ENTER] to choose Y1

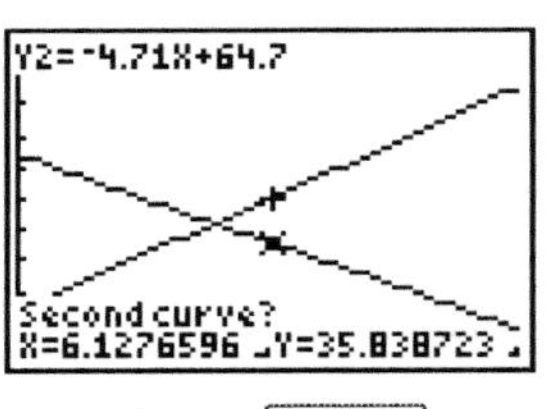

Press [ENTER] to choose Y2

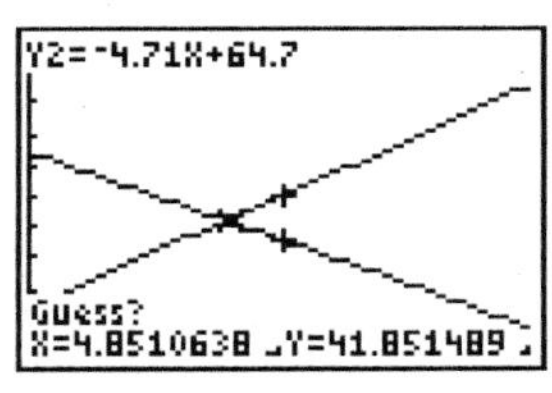

Move cursor to specify guess and then press [ENTER].

The final result of this process is the screen shown on the right, as well as in Figure 66(b). The x-coordinate of this point of intersection is calculated to many digits of accuracy.

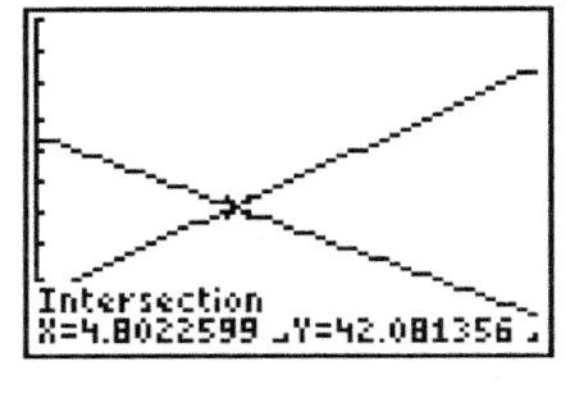

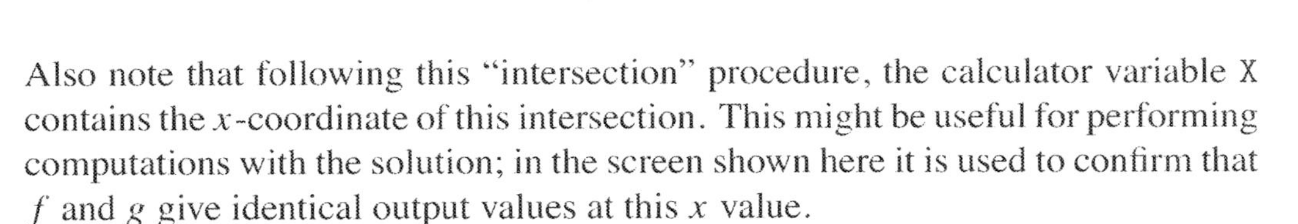

Also note that following this "intersection" procedure, the calculator variable X contains the x-coordinate of this intersection. This might be useful for performing computations with the solution; in the screen shown here it is used to confirm that f and g give identical output values at this x value.

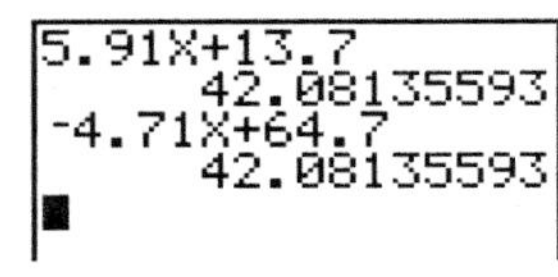

Note: An approximation for the point of intersection can be found simply by moving the TRACE cursor as near the intersection as possible. The amount of error can be minimized by "zooming in" on the graph. This is the only method available for graphing calculators such as the TI-81.

See the next example for another graphical approach to solving equations, as well as two other (non-graphical) approaches available on the TI-83.

Section 1.5 Example 5 (page 55) — Using the *x*-Intercept Method

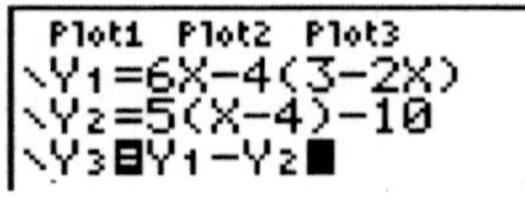

The text suggests graphing `Y1=6X-4(3-2X)-(5(X-4)-10)`, noting the importance of placing parentheses around the subtracted expression. An equivalent approach is illustrated on the right: Define `Y1` and `Y2` using the expressions on the left and right sides of the equation, and then set `Y3=Y1-Y2`. This method avoids the need for so many (potentially confusing) parentheses, at the cost of a few more keystrokes. To type `Y1` and `Y2`, press [VARS][▸][1] to access the Y-VARS:FUNCTION menu. Note that `Y1` and `Y2` have been "de-selected" so that they will not be graphed (see section 12 of the introduction, page 7).

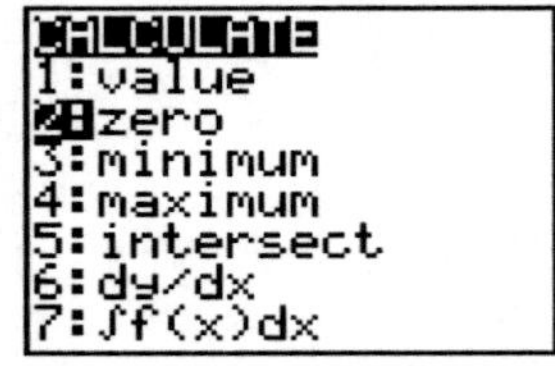

Next, select a viewing window which shows the x-intercept; we use the standard window, as shown in Figure 67(a) of the text. The TI-83 can automatically locate this point by choosing option 2 (zero) from the CALC ([2nd][TRACE]) menu. The TI-83 prompts for left and right bounds and a guess, then attempts to locate the zero between the given bounds. (Provided there is only one zero between the bounds, and the function is "well-behaved"—meaning it has some nice properties like continuity—the calculator will almost always find it.) The screens below illustrate these steps.

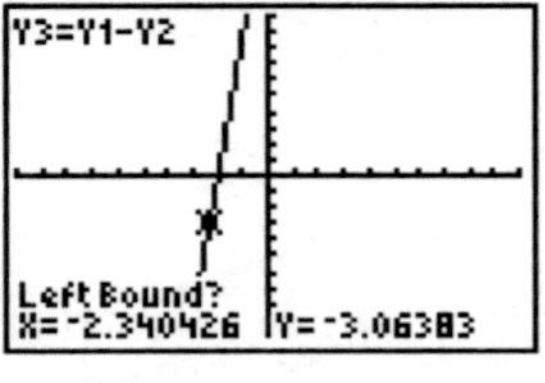

Move cursor to the left of the zero, press [ENTER]

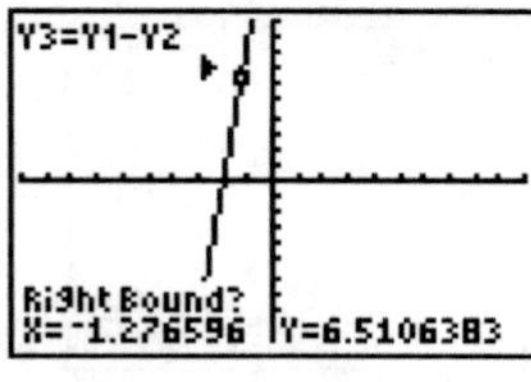

Move cursor to the right of the zero, press [ENTER]

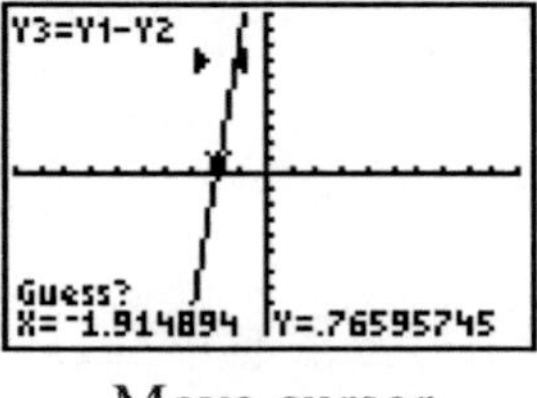

Move cursor close to the zero, press [ENTER]

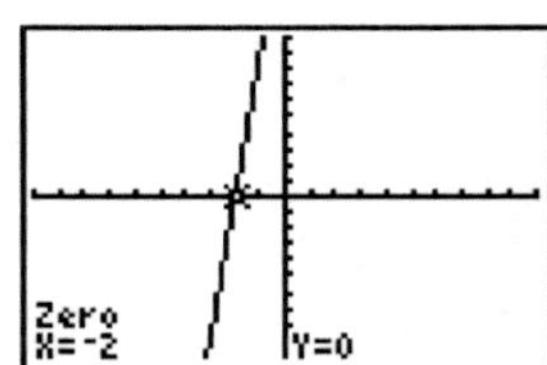

The TI-83 finds the zero.

As with the intersection method, after the TI-83 locates the x-intercept, the calculator variable `X` contains the x-coordinate.

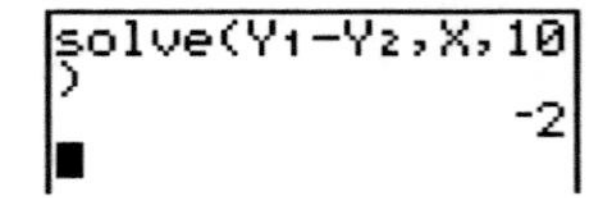

The TI-83 also offers some non-graphical approaches to solving this equation (or confirming a solution): As illustrated on the right, the TI-83's `solve` function attempts to find a value of `X` that makes the given expression equal to 0, given a guess (10, in this case). The entry shown makes use of the fact that `Y1` and `Y2` have been defined as the left and right sides of this equation; if that had not been the case, the same results could have been attained by entering (e.g.) `solve(6X-4(3-2X)-(5(X-4)-10),X,10)`. Full details on how to use this function (found in the CATALOG) can be found in the TI-83 manual.

Finally, the TI-83 includes an "interactive solver," accessed with [MATH][0]. This prompts for the equation to be solved, then allows the user to enter a guess for the solution (or a range or numbers between which a solution should be sought). To solve the equation, place the cursor on the line beginning with X= and press [ALPHA][ENTER] (SOLVE).

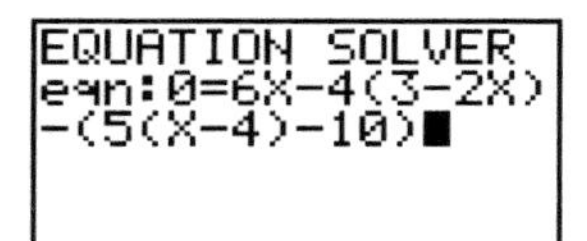

Enter the equation to be solved

6X-4(3-2X)-(5…=0
X=
bound={-1E99,1…

Specify a guess (optional), then press [ALPHA][ENTER]

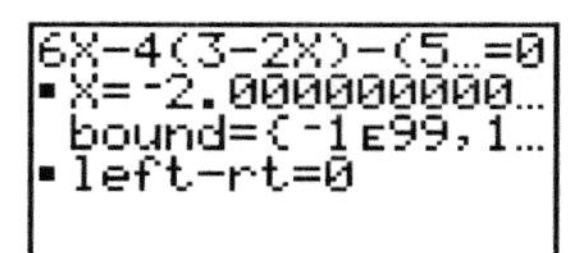

After a brief pause, the solution is found

The solver can also be used with equations containing more than one variable; simply provide values for all but one variable, then place the cursor on the line containing the variable for which a value is needed and press [ALPHA][ENTER].

[82] *The TI-82 does not include this interactive solver; the* solve *function is accessed with* [MATH][0].

Section 1.5 Example 8 (page 59) — Using the Intersection-of-Graphs Method

In Figure 71, the text illustrates using a graph to support the solution $[-3, \infty)$ to the inequality $3x - 2(2x + 4) \leq 2x + 1$, making the observation that solutions to this inequality correspond to those x-values for which the graph of $y = 3x - 2(2x + 4)$ *intersects or is below* the graph of $y = 2x + 1$. This connection between "$<$" and "below" (or "$>$" and "above") is an important one, and students should strive to understand it. However, it can sometimes be confusing, especially when one is just learning it, and the following graphical approach may be useful.

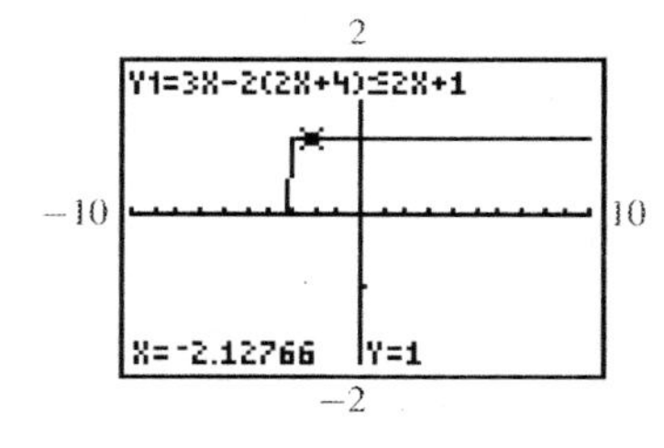

To solve (or confirm the solution of) an inequality like this, enter the formula Y1=3X-2(2X+4)≤2X+1, where the "≤" symbol is found in the TEST menu ([2nd] [MATH]). When one of these symbols is included in an expression, the TI-83 responds with 1 if the statement is true, and 0 otherwise. Therefore, Y1 will equal 1 for the values of X which satisfy the inequality, and 0 for all other values of X. With this understanding, one can observe the graph produced, and confirm that the solution is $x \geq -3$. (Care must be taken to determine whether or not -3 should be included in the solution set; the graph does not make that clear. This same observation is made in the Technology Note next to this example in the text.)

Section 1.5 Technology Note (page 60) — Typing Function Variables

For the technique of defining Y3=Y1−Y2, see the comments about Example 5 from Section 1.5 (page 14 of this manual).

Section 1.5 Example 11 (page 61) — Solving a Three-Part Inequality

As was mentioned in the discussion of Example 8 from Section 1.5 (page 15 of this manual), the understanding that "$<$" and "$>$" go with "below" and "above" (respectively) is very important. The technique mentioned in that discussion can be applied to three-part inequalities, but some modification is needed.

With single inequalities, one simply defines Y1 by typing in the inequality. For double inequalities, this approach does not work. The results are shown on the right; they would lead one to think that any value of x (or at least any x between -10 and 10) satisfies the inequality.

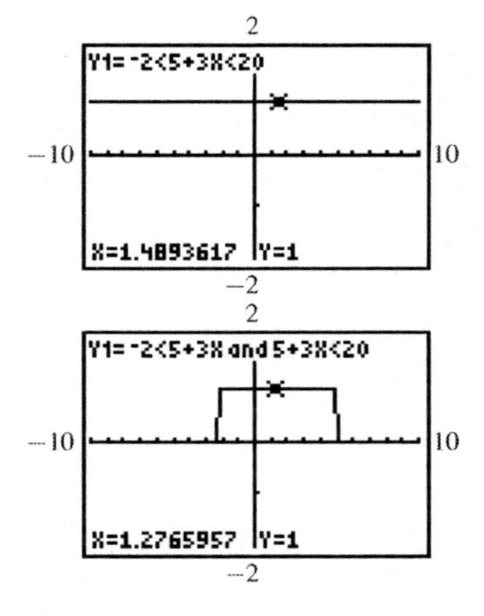

Instead, one must split the double inequality into two single inequalities, then use the conjunction and, found in the TEST:LOGIC menu ([2nd][MATH][▶]).

Section 2.1 Technology Note (page 96) — Rational Exponents

See page 10 of this manual for more information about roots and rational exponents.

Section 2.1 Technology Note (page 96) — Absolute Values

The absolute value function is typed using [MATH][▶][1]. To graph $y = |x|$, for example, define Y1=abs(X). Note that the TI-83 automatically supplies the opening parenthesis.

[82] *On the TI-82, the absolute value function is typed using* [2nd][x^{-1}], *and the opening parenthesis is* not *included. Parentheses are optional for expressions like* -abs -2, *but are needed for* abs (-8+2).

Section 2.1 Technology Note (page 98) — Function and Parametric Modes

For information about selecting function and parametric modes, see section 9 of the introduction, page 4.

Section 2.2 Technology Note (page 103) — Graphing Groups of Similar Functions
Section 2.3 Technology Note (page 113)
Section 2.3 Technology Note (page 114)
Section 2.3 Technology Note (page 115)
Section 2.3 Technology Note (page 117)

The screen shown on page 103 of the text illustrates one approach to graphing groups of similar functions: Set Y2=Y1+3, Y3=Y1-2, and Y4=Y1+5. This allows an entire "family" to be graphed by simply changing Y1.

An alternative is to define Y1=X²+{0,3,-2,5}. (The curly braces { and } are [2nd][(] and [2nd][)]). When a list (like {0,3,-2,5}) appears in a formula, it tells the TI-83 to graph this formula several times, using each value in the list; therefore, this one definition will graph the four functions $y = x^2$, $y = x^2 + 3$, $y = x^2 - 2$, and $y = x^2 + 5$. Different families can be produced simply by changing X² to another function.

This list approach translates nicely to other types of transformations. For horizontal shifts, use, e.g., Y1=(X+{0,-3,-5,4})². For vertical stretches and shrinks, use Y1={1,2,3,4}X²; the Technology Note on page 115 of the text illustrates a similar approach. For horizontal stretches and shrinks, use the approach shown in that Technology Note, or something like `Y2=Y1({2,0.5}X)`.

For reflections (page 117 of the text), use the approach shown in the text, or a variation of the one given above. For example, to graph $y = \sqrt{x}$ and $y = \sqrt{-x}$, for example, define `Y1=√({1,-1}X)`.

Note that the usefulness of using lists to accomplish horizontal transformations (horizontal shifts and reflection across the y axis) is limited. For example, the graph of $y = \sqrt{x^2 - 3x}$—or any expression in which x appears more than once—is not conveniently reflected across the y-axis or horizontally shifted by this method. An adaptation of the method shown in the text would work better: E.g., define `Y1=√(X2-3X)` and `Y2=Y1(X+{2,5,-3})` to see the graph of $f(x) = \sqrt{x^2 - 3x}$, $f(x+2)$, $f(x+5)$, and $f(x-3)$.

Section 2.5 Example 1 (page 139) Finding Function Values for a Piecewise-Defined Function

Section 2.5 Example 2 (page 139) Graphing a Piecewise-Defined Function

Recall that the inequality symbols $>$, $<$, $\geq$, $\leq$ are found in the TEST menu ([2nd][MATH]). The use of Dot mode (mentioned in the Technology Note) is not crucial to the graphing process, as long as one remembers that vertical line segments connecting the "pieces" of the graph (in this case, at $x = 0$) are not really part of the graph.

The discussion in the text for Example 2 shows how to enter a piecewise-defined function into the calculator, and also how the TI-83 evaluates such an expression. Note that the formula entered for `Y1` is a fairly simple translation of the written definition of the function. For Example 1, e.g., we have

$$f(x) = \begin{cases} x+2 & \text{if } x \leq 0 \\ \frac{1}{2}x^2 & \text{if } x > 0 \end{cases}$$ becomes `Y1=(X+2)(X≤0)+(1/2)X2(X>0)`.

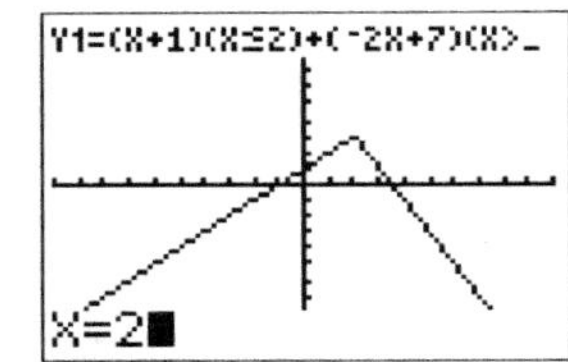

Figure 55 of the text shows that $y = 2$ when $x = 0$—a detail that is not evident from the calculator graph (since the TI-83 does not show open and closed circles as does the graph shown in text). Similarly, the screen in Figure 57(b) shows that $y = 3$ when $x = 2$, using the [TRACE] feature, but note that on the window shown, one cannot get to $x = 2$ by pressing [◄] and [►] to move the trace cursor. Instead, one must use the TI-83's "extended trace" feature, which allows one to trace to any x value between `Xmin` and `Xmax`. To do this, press [TRACE], then rather than pressing [◄] or [►], type a number and press [ENTER]. The screen on the right shows what happens when one presses [TRACE][2] (just before pressing [ENTER]). This same result can be achieved using option 1 (value) from the [2nd][TRACE] (CALC) menu.

[82] *This latter approach is also available on the TI-82.*

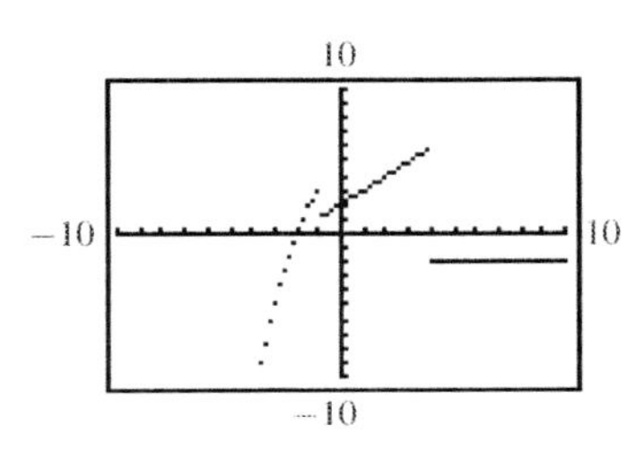

Extension: For more complicated piecewise-defined functions, a similar procedure can be used. Consider

$$f(x) = \begin{cases} 4 - x^2 & \text{if } x < -1 \\ 2 + x & \text{if } -1 \leq x \leq 4 \\ -2 & \text{if } x > 4 \end{cases}$$

shown on the right in Dot mode on the standard window.

It might be tempting to enter `Y1=(4-X2)(X<-1)+(2+X)(-1≤X≤4)+(-2)(X>4)`, but this does not work. Instead, use either

`Y1=(4-X2)(X<-1)+(2+X)(-1≤X and X≤4)+(-2)(X>4)`, or

`Y1=(4-X2)(X<-1)+(2+X)(-1≤X)(X≤4)+(-2)(X>4)`

(the "and" in the first formula is found using [2nd][MATH][▸][1]).

Section 2.5 Example 4 (page 141) — Evaluating [[x]]

The `int` function is [MATH][▸][5]. In Figure 59, note the use of lists to find $[\![x]\!]$ for all five values with one entry.

Do not confuse `int` with the `iPart` ("integer part") function, which is slightly different—specifically, `iPart(-6.5)` returns -6 while `int(-6.5)` gives -7.

Section 2.6 Example 1 (page 149) — Using the Operations on Functions

Section 2.6 Example 2 (page 150) — Using the Operations on Functions

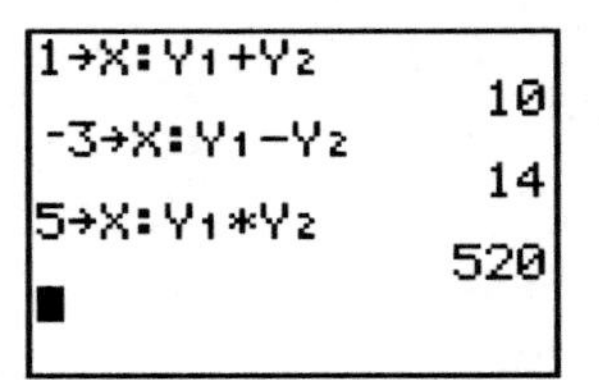

In Example 1, the screens in the text (Figure 64) show how to evaluate function sums, differences, products, and quotients using the TI-83's function notation. The screen on the right shows an alternative way to compute $(f+g)(1)$, etc.; these entries make use of the fact that, when not followed by a number in parentheses, `Y1` and `Y2` are evaluated using the current value of `X`. The approach shown here has a slight advantage over that shown in the text because fewer changes are required from one entry to the next. Recall that `Y1` and `Y2` (and other functions) are typed from the VARS:Y-VARS:Function ([VARS][▸][1]) menu.

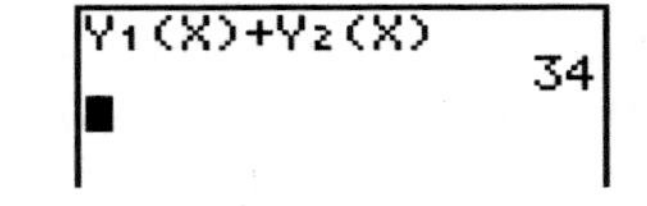

The TI-83 cannot give symbolic expressions like those required in Example 2. The entry `Y1(X)+Y2(X)` simply returns a number, using the current value stored in `X`. (In this case, `X` has the value 5.)

Section 2.6 Example 5 (page 153) — Evaluating Composite Functions

Section 2.6 Example 6 (page 153) — Finding Composite Functions

The screens shown in Figure 68 illustrate how to evaluate composite functions at specific input values. Note that compositions of three or more functions can be accomplished just as simply: If `Y1`, `Y2`, and `Y3` are defined as the functions f, g, and h, one can evaluate $(f \circ g \circ h)(3)$ by typing `Y1(Y2(Y3(3)))`.

```
Y2(Y1(X))
                987
```

The TI-83 cannot give symbolic expressions like those required in Example 6. The entry `Y2(Y1(X))` simply returns a number, using the current value stored in `X`. (In this case, `X` has the value 5.)

Section 3.1 Technology Note (page 174) — Complex Number Mode

See page 4 for more information about putting the TI-83 in `a+bi` mode.

[82] *The TI-82 has no complex number mode, since it does not have built-in support for complex numbers. See the appendix to this chapter (page 48) for information about doing computations with complex numbers on a TI-82.*

Section 3.1 Example 1 (page 175) — Writing $\sqrt{-a}$ as $i\sqrt{a}$

Section 3.1 Example 2 (page 176) — Finding Products and Quotients Involving $\sqrt{-a}$

If the TI-83 is in `Real` (rather than `a+bi`) mode, an entry like `√(-16)` results in a `NONREAL ANS` error. On the other hand, if the calculator is in `re^θi` mode, the output looks like one of the two results on the right (depending on whether the angle measurement mode is Radian or Degree). For the most part, `a+bi` mode is the best choice for problems like these.

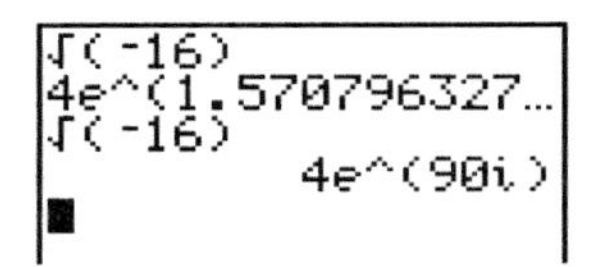

Section 3.1 Example 3 (page 176) — Adding and Subtracting Complex Numbers

Section 3.1 Example 4 (page 177) — Multiplying Complex Numbers

Section 3.1 Example 5 (page 178) — Simplifying Powers of *i*

Section 3.1 Example 6 (page 179) — Dividing Complex Numbers

The character "`i`" is [2nd][.]. Although not completely necessary, it is a good idea to put the TI-83 in `a+bi` mode; see page 4. (Even in Real mode, the TI-83 will display results for computations in which `i` is entered directly; it only complains if asked to find even roots of negative numbers.)

[82] *See the appendix to this chapter (page 48) for information about doing these computations on a TI-82.*

Section 3.2 Example 3 (page 185) — Using the Vertex Formula

The text identifies the vertex of the graph of $f(x) = -0.65x^2 + \sqrt{2}x + 4$ as a maximum at $(\sqrt{2}/1.3, 4 + 1/1.3) \approx (1.09, 4.77)$, and shows calculator screens that support those values. The TI-83 can automatically locate extreme values ("hills" and "valleys") in a graph using options 3 and 4 (minimum and maximum) in the CALC menu ([2nd][TRACE]).

Enter the function in `Y1`, and graph in a window that shows the extreme value (such as the window shown in the text: $[-2.4] \times [-2.5]$). Since the coefficient of x^2 is negative, this is a parabola that opens down, and the extreme point is a maximum value. Press [2nd][TRACE][4], then use the arrow keys and [ENTER] to define left and right bounds and specify a guess, as was done previously with option 2 (zero) and option 5 (intersect)—see the discussion on Examples 4 and 5 from Section 1.5 beginning on page 13 of this manual.

```
√(2)/1.3
      1.087856586
X
      1.087855765
■
```

```
Y1(√(2)/1.3)
      4.769230769
Y1(X)
      4.769230769
■
```

After going through the process of locating a maximum or minimum, the calculator variables X and Y contain the coordinates of the point. Depending on the window and the initial guess, the x value may be off a bit from the exact answer. This is the case in the screen shown in the text; observe that the value of X found by the TI-83 agrees with only the first six digits of the exact answer. A limitation of the technology is that the calculating algorithms are programmed to stop within a certain degree of accuracy. (In other words, the TI-83 looks for the vertex until it decides that it is "close enough"; it will not always find exact answers.) In this case, the value of Y_1 is nearly identical for both the exact answer and the TI-83's approximate value. It is important for the user to recognize this limitation for two reasons: First, do not report all digits displayed by the calculator, as they are not all reliable. Second, if the calculator reports a result of (say) 1.49999956, it is reasonable to guess that the exact answer might be 1.5.

Section 3.2 Example 5 (page 186) Identifying Extreme Points and Extreme Values

Figure 15 of the text illustrates the use of the `fMin` operation ([MATH][6]), which is similar to the `fMax` ([MATH][7]) operation. These two commands perform the same service as the graphical CALC:minimum and CALC:maximum features described in the previous example (except that the calculator variables X and Y are unaffected by `fMin` and `fMax`). The format is

`fMin`(*function*, *variable*, *low X*, *high X*)

Function can either be one of the function variables (e.g., Y_1) or an expression (like $4X^2-18X+3$). *Variable* is usually X, but can be any other variable used in *function*. The other two parameters specify the smallest and largest values of X (or whatever variable is used) between which a minimum function value is sought. (These are especially needed when the function has many high or low points.)

As was noted with the graphical minimum/maximum locating procedures, be aware that the values returned by these procedures are approximations of the exact answers. Note in the screen on the right (for which $Y_1=2X^2+4X-16$) that the answer reported can vary depending on the specified low and high values of X. In both cases shown, the answer only approximates the exact answer of -1.

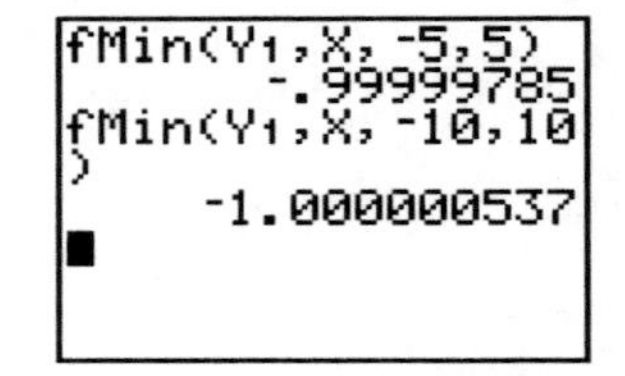

(These screens were produced on a TI-83, while the screen shown in the text was produced on a TI-83+; the latter has a slightly more sophisticated implementation of `fMin` and `fMax` that gives exact results more often.)

Section 3.2 Example 6 (page 187) Modeling Hospital Spending with a Quadratic Function

See page 11 for instructions on creating scatter diagrams on the TI-83. The TI-83's `QuadReg` (quadratic regression) feature, accessed with [STAT][▶][5], can be used to find a quadratic function to approximate a set of data. The procedures for doing this are similar to those for a linear regression, described on page 12 of this manual. (Note, though, that the quadratic regression formula is not the same as the approximating function given in the text.)

Section 3.3 Technology Note (page 198) Calculator Programs

See section 13 of the introduction (page 8) for information about installing programs in the TI-83.

Section 3.7 Example 2 (page 242) — Finding All Zeros of a Polynomial Function

If installed on your calculator (TI-83+ or TI-84+ only), the PolySmlt APP can find all zeros (real and complex) of a polynomial. The screens below illustrate the process. Note in the third screen, the top line contains a reminder that the expression must be equal to 0.

Option 1 finds polynomial roots (zeros).

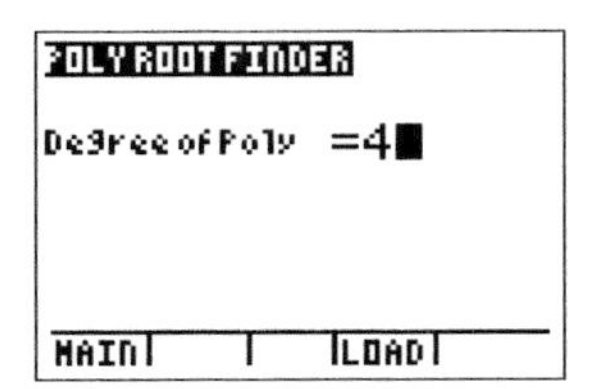

Give the degree of the polynomial.

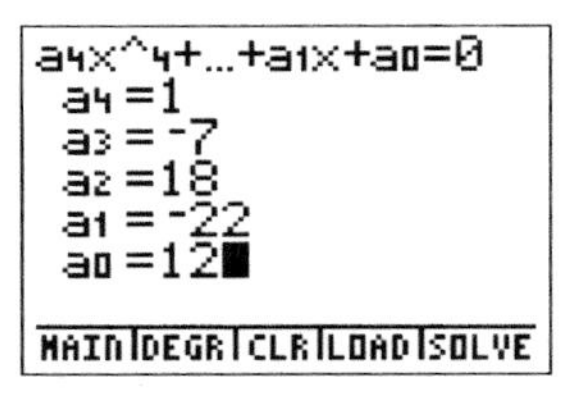

Enter coefficients on this screen.

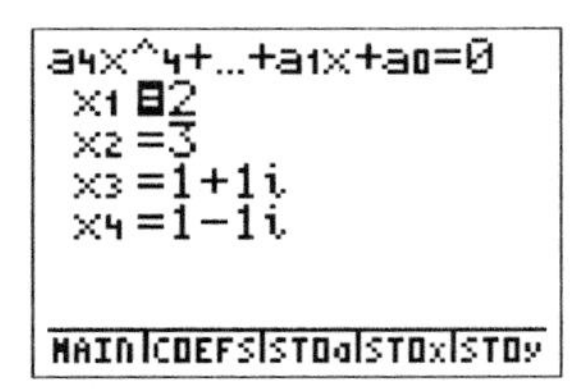

Press [GRAPH] to find the zeros.

Section 3.8 Example 7 (page 255) — Examining Polynomial Models for Debit Card Use

The procedures for creating scatter diagrams and performing regressions are described on pages 11 and 12 of this manual. Note that the first screen shown in Figure 88 includes the value "R²", which is only displayed after issuing the `DiagnosticOn` command (described on page 12).

Section 4.1 Example 2 (page 273) — Graphing a Rational Function

Note that this function is entered as `Y1=2/(X+1)`, **not** `Y1=2/X+1`.

The issue of incorrectly drawn asymptotes is addressed in section 12 of this chapter's introduction (page 7). This asymptote can also be eliminated by setting the window to `Xmin` $= -5$ and `Xmax` $= 3$ (or any choice of `Xmin` and `Xmax` which has -1 halfway between them). Since this function is not defined at -1, it cannot plot a point there, and as a result, it does not attempt to connect the dots across the "break" in the graph.

Section 4.2 Example 8 (page 285) — Graphing a Rational Function Defined by an Expression That Is Not in Lowest Terms

Figure 23 shows the graph of `Y1=(X²-4)/(X-2)` on a variation of the decimal window, on which the "hole" in the graph can be seen. There are many possible windows on which the hole is visible; any window for which $x = 2$ is halfway between `Xmin` and `Xmax` would work. Likewise, there are many windows for which the hole would not be visible.

It is important to realize that some holes cannot be made visible. For example, take the graph of the function `Y1=(X^4-4)/(X²-2)` —which looks like the function $y = x^2 + 2$, except at $x = \pm\sqrt{2}$. It is difficult (if not impossible) to find a window showing the holes at $x = \pm\sqrt{2}$.

Section 4.4 Example 5 (page 309) Modeling the Period of Planetary Orbits

Section 4.4 Example 6 (page 309) Modeling the Length of a Bird's Wing

The procedures for creating scatter diagrams are covered on page 11 of this manual. The "power regression" illustrated in Example 4 (Figures 47&48) is performed in a manner similar to linear regression (see page 12).

Here is the output for the power regression performed on the data in Example 3. Note that this gives further confirmation that the formula $f(x) = x^{1.5}$ does a good job of modeling the relationship between average distance from the sun x and period of revolution y.

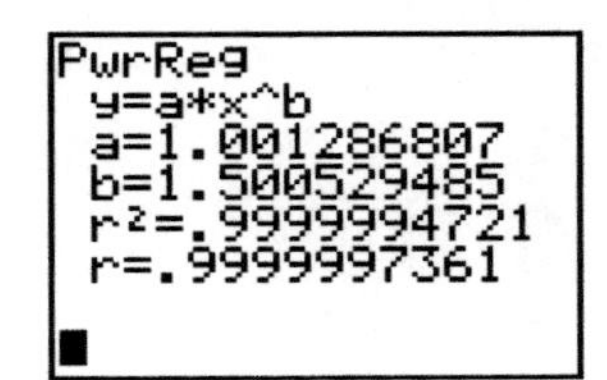

Section 4.4 Example 9 (page 313) Graphing a Circle

The text suggests graphing Y_1=√(4−X^2) and Y_2=−Y_1; the latter is typed with [(−)] [VARS] [▸] [1] [1]. Here are two alternative ways to graph this circle; these approaches might be useful in other situations where one wishes to graph to or more complicated (but similar) functions:

- After typing the formula in Y_1, move the cursor to Y_2, press [(−)], then press [2nd] [STO▸] (RCL) [VARS] [▸] [1] [1] [ENTER]. This will "recall" the formula of Y_1, placing that formula in Y_2. This takes a few additional keystrokes, but can be a useful approach in cases where the second function to be graphed is similar to the first, but cannot easily be written in terms of Y_1.
- Enter the single formula Y_1={−1,1}√(4−X^2). (The curly braces { and } are [2nd] [(] and [2nd] [)]). See page 16 for more information about this approach.

Section 5.1 Example 6 (page 345) Finding the Inverse of a Function with a Restricted Domain

The graph in Figure 8 shows the functions $f(x) = \sqrt{x+5}$ and $f^{-1}(x) = x^2 - 5,\ x \geq 0$. To produce a similar graph on the TI-83, enter the second function as a piecewise-defined function with only one "piece": Enter Y_2=(X^2−5)(X≥0), where ≥ is [2nd] [MATH] [4]. (See page 17 for more about piecewise-defined functions.)

An even more accurate graph can be created by entering Y_2=(X^2−5)/(X≥0). (Note the division symbol in the middle.) This function is undefined (because of division by 0) whenever $x < 0$.

Section 5.2 Example 3 (page 354) Comparing the Graphs of $f(x)=2^x$ and $g(x)=(1/2)^x$

The labels $(-2, 4)$ and $(2, 4)$ on the calculator screen shown were produced with the `Text(` command. Consult your TI-83 manual for more information.

Section 5.2 Example 5 (page 355) Using Graphs to Evaluate Exponential Expressions

The two calculator screens shown in Figure 20 were produced using the "extended trace" features of the TI-83, mentioned previously on page 17 of this manual. After pressing [TRACE], simply type a number or expression (like √(6) or −√(2)). The number appears at the bottom of the window in a larger font size than the TRACE coordinates. Pressing [ENTER] causes the TRACE cursor to jump to that x-coordinate. This same result can be achieved using option 1 (value) from the [2nd][TRACE] (CALC) menu.

[82] *This latter approach is also available on the TI-82.*

Section 5.3 Technology Note (page 365) Logarithms of Nonpositive Numbers

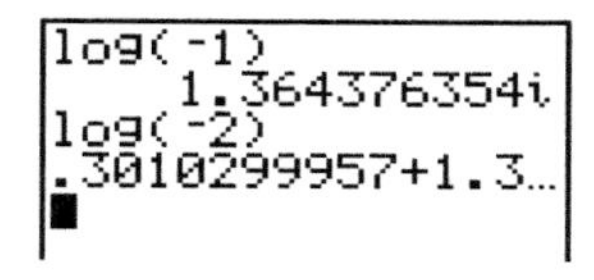

This Technology Note states that (under some circumstances) a calculator will give an error message when asked to compute (e.g.) `log(0)` or `log(-1)`. Specifically, `log(0)` gives a `DOMAIN` error (always), and `log(-1)` results in a `NONREAL ANS` error when the TI-83 is in `Real` mode. As the name of the latter error suggests, one can find a logarithm of a negative number if one allows for complex results. Shown on the right are the common logarithms of -1 and -2, when the TI-83 is in `a+bi` mode. In general, if $x > 0$, then $\log(-x) = \log(x) + \log(-1)$, as the product rule for logarithms would suggest.

[82] *The TI-82 does not support complex numbers, so it always reports an error when asked to compute the logarithm of a negative number.*

Section 5.3 Example 3 (page 365) Finding pH and $[H_3O^+]$

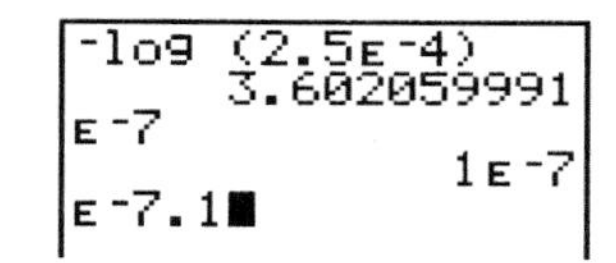

For (a), the text shows `-log(2.5*10^(-4))`, but this could also be entered as shown on the first line of the screen on the right, since "E" (produced with [2nd][,]) and "*10^" are nearly equivalent. The two are not completely interchangeable, however; in particular, in part (b), "10^" **cannot** be replaced with "E", because "E" is only valid when followed by an *integer*. That is, `E-7` produces the same result as `10^-7`, but the last line shown on the screen produces a syntax error.

(Incidentally, "10^" is [2nd][LOG], but [1][0][^] produces the same results.)

Section 5.4 Technology Note (page 375) Asymptotes in Logarithmic Graphs

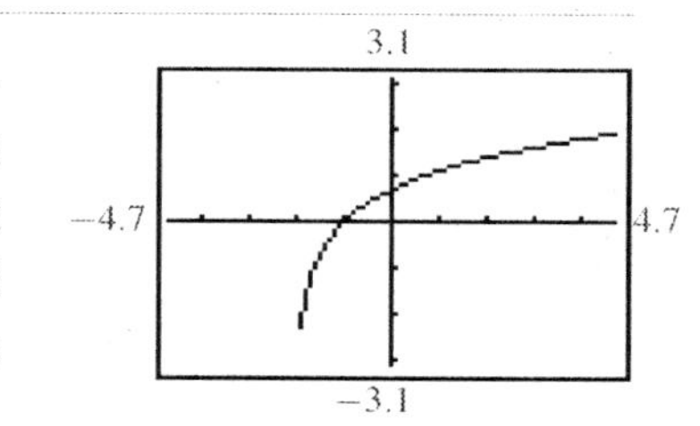

The calculator screen behavior described here is more evident when accompanied by a horizontal shift. Shown here (on the decimal window) is the graph of $y = \ln(x + 2)$; note that the graph seems to have an endpoint at about $(-1.9, -2.3)$. In reality, the graph has a vertical asymptote at $x = -2$, and that portion of the graph "goes down to $-\infty$"—that is, as $x \to -2$ from the right, $y \to -\infty$.

Section 5.6 Technology Note (page 398) Financial Calculations

To access the menu screen shown in the text, select the "Finance..." APP ([APPS][1]), or (on a "non-plus" TI-83, which has no [APPS] key) choose the FINANCE option, [2nd][x^{-1}]. The prefix `tvm` on many of these commands stands for "time value of money." See the next example for sample usages of this menu.

[82] *The TI-82 does not include these features.*

Section 5.6 Example 6 (page 399) Using Amortization to Finance an Automobile

To solve this problem with the TI-83's TVM features, access the TVM Solver by pressing ([APPS][1][1] or [2nd][x^{-1}][1]). Next, enter all of the given information, as the screen on the right shows; in the TI-83's notation, PV is the "present value" (initial loan balance), FV is the "future value," and P/Y and C/Y are the numbers of payments and compounding periods per year. The last line specifies that the payments are made at the end of each month (as the text specifies immediately before this example). Since we do not yet know the amount of the payments, that line is blank (although we could put any number on that line).

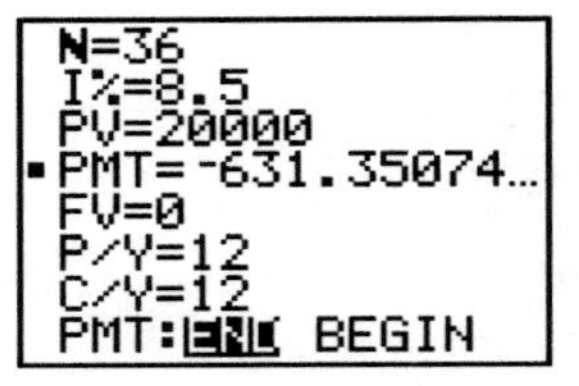

Once everything is specified (except for the payment), move the cursor to the PMT= line and press [ALPHA][ENTER] (SOLVE). The results are shown on the right; note the black square next to PMT=, indicating that this value was computed as a function of the other supplied values. The fact that the PMT value is negative indicates that it will cause the balance to decrease (from \$20,000 to \$0).

The real power of the TVM Solver is in the ability to experiment with how changes in one variable affect the others. For example, one can examine the effect of paying over four years rather than three, or making slightly higher payments, or changing the interest rate.

```
tvm_Pmt
            -631.3507485
tvm_Pmt(48,8)
            -488.2584468
■
```

The payment amount can also be computed at the home screen by using the `tvm_Pmt` function ([APPS][1][2] or [2nd][x^{-1}][2]). The format is

```
tvm_Pmt(N,I%,PV,FV,P/Y,C/Y)
```

If any of these values are omitted, the value supplied on the TVM Solver screen is used. Note that this means one must ensure that the payments are set to occur at the end of each period, since there is no way to change that in the `tvm_Pmt` function. (The FINANCE:CALC menu does include functions `Pmt_End` and `Pmt_Bgn` to set this.) The screen on the right shows the output of `tvm_Pmt` with no additional parameters (assuming that the TVM Solver screen contains the values shown above), and the payment needed at an 8% interest rate spread over 4 years.

[82] *The TI-82 does not have built-in support for these financial computations.*

Section 5.6 Example 8 (page 400) Modeling Atmospheric CO_2 Concentrations

Section 5.6 Example 9 (page 401) Modeling Interest Rates

The procedures for creating scatter diagrams and performing regressions are described on pages 11 and 12 of this manual. Note that the screens in Figures 55(a) and 57(b) include the values `r²` and `r`, which are only displayed after issuing the `DiagnosticOn` command (described on page 12).

Section 6.1 Example 3 (page 420) Graphing a Circle

For part (b), the text suggests graphing `Y1=4+√(36-(X+3)²)` and `Y2=4-√(36-(X+3)²)`. Here are three options to speed up entering these formulas (see also page 22):

- After typing the formula in Y1, move the cursor to Y2 and press [2nd][STO▸] (RCL), then [VARS][▸][1] [1][ENTER]. This will "recall" the formula of Y1, placing that formula in Y2. Now edit this formula, changing the first "+" to a "−."
- After typing the formula in Y1, enter Y2=8-Y1. (To type Y1, press [VARS][▸][1][1].) This produces the desired results, since $8 - \text{Y1} = 8 - \left(4 + \sqrt{36-(x+3)^2}\right) = 8 - 4 - \sqrt{36-(x+3)^2} = 4 - \sqrt{36-(x+3)^2}$.
- Enter the single formula Y1=4+{-1,1}√(36-(X+3)²). (The curly braces { and } are [2nd][(] and [2nd] [)]). When a list (like {-1,1}) appears in a formula, it tells the TI-83 to graph this formula several times, using each value in the list.

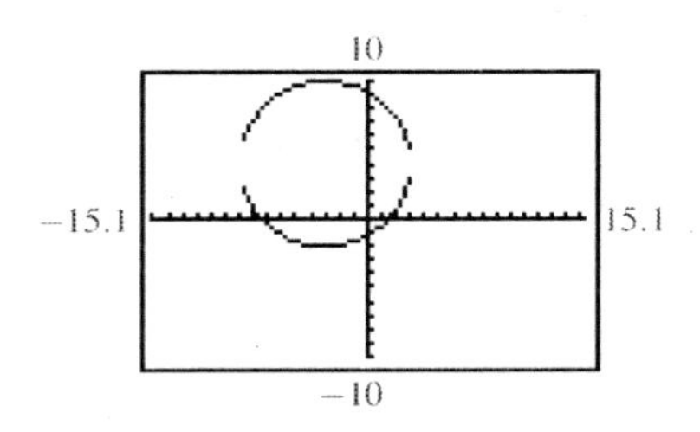

The window chosen in Figure 6 of the text is a square window (see section 11 of this chapter's introduction), so that the graph looks like a circle. (On a non-square window, the graph would look like an ellipse—that is, a squashed circle.) Other square windows would also produce a "true" circle, but some will leave gaps similar to those circles shown in sections 11 and 12 of the introduction (pages 6–7). Shown is the same circle on the square window produced by pressing [ZOOM][6][ZOOM][5]. The Technology Note next to this example shows a similar graph for the circle in part (a).

Section 6.1 Example 7 (page 426) Graphing a Parabola with vertex (h,k)

Section 6.2 Example 1 (page 434) Graphing an Ellipse Centered at the Origin

Section 6.2 Example 3 (page 436) Graphing an Ellipse Translated from the Origin

Section 6.2 Example 6 (page 439) Using Asymptotes to Graph a Hyperbola

Section 6.2 Example 8 (page 440) Graphing a Hyperbola Translated from the Origin

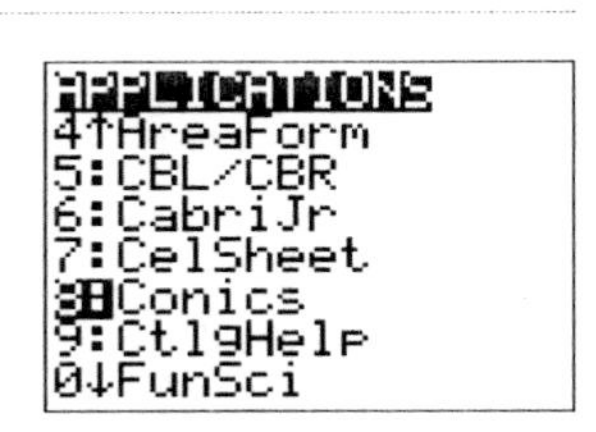

If installed on your calculator (TI-83+ or TI-84+ only), the Conics APP will graph conic sections, given the appropriate information. Shown below is the process of graphing the parabola in Example 7; the other graphs are accomplished in a similar manner. Press [GRAPH] to produce the graph, [TRACE] to see points on the graph, and [Y=] to change the conic parameters. The APP chooses the viewing window for you.

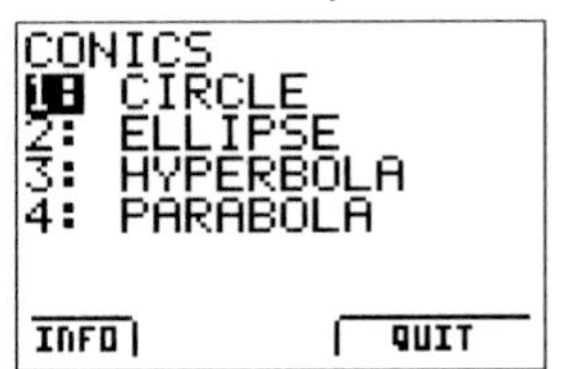

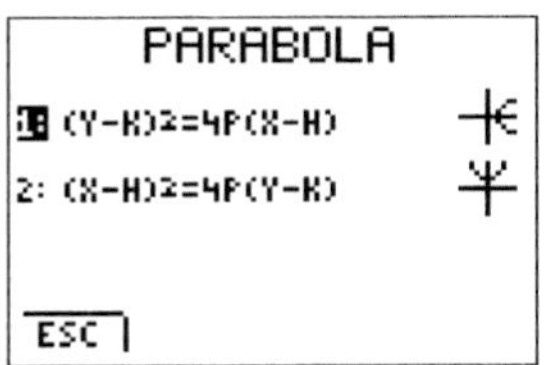

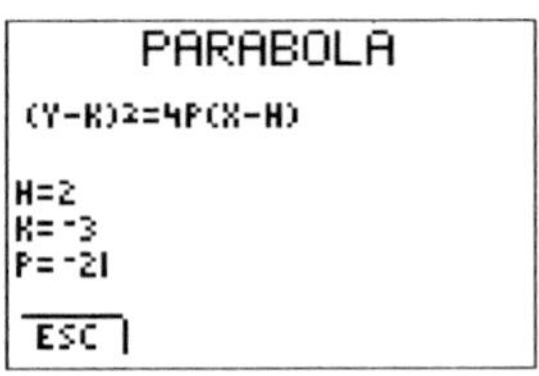

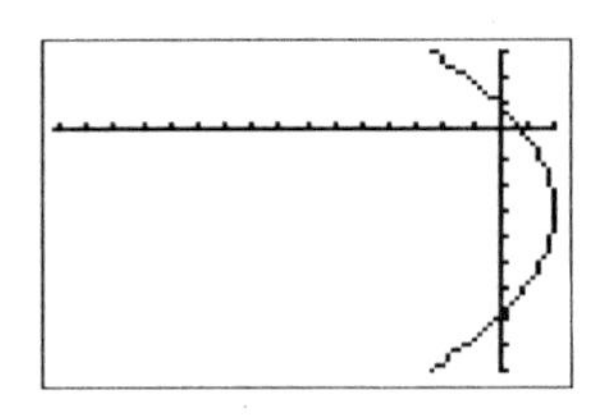

Section 6.4 Technology Note (page 454) Parametric Mode

See the next example, as well as section 9 of the introduction (page 4), for information about selecting Parametric mode. The process of creating a graph in this mode is described in the next example.

Section 6.4 Example 1 (page 454) Graphing a Plane Curve Defined Parametrically

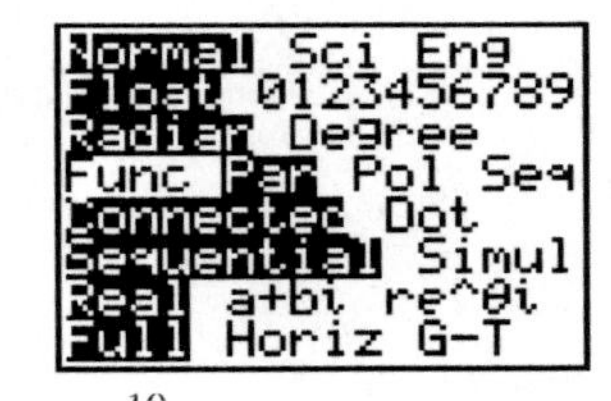

Place the TI-83 in Parametric mode, as the screen on the right shows. In this mode, the Y= key allows entry of up to six pairs of parametric equations (x and y as functions of t). No graph is produced unless both functions in the pair are entered and selected (that is, both equals signs are highlighted).

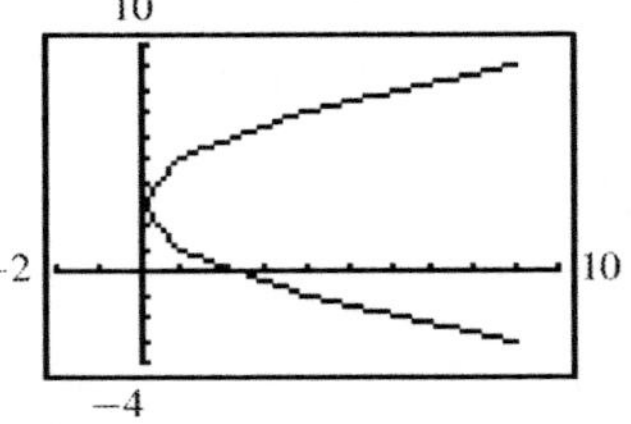

The text shows the WINDOW settings with `Tmin=-3`, `Tmax=3`, and `Tstep=0.05`. The first two of these are specified in the example, but the value of `Tstep` does not need to be 0.05, although that choice works well for this graph. Too large a choice of `Tstep` produces a less-smooth graph, like the one shown on the right (drawn with `Tstep=1`); note the angularity of the parabola near its vertex. Setting `Tstep` too small, on the other hand, produces a smooth graph, but it is drawn very slowly. Sometimes it may be necessary to try different values of `Tstep` to choose a good one.

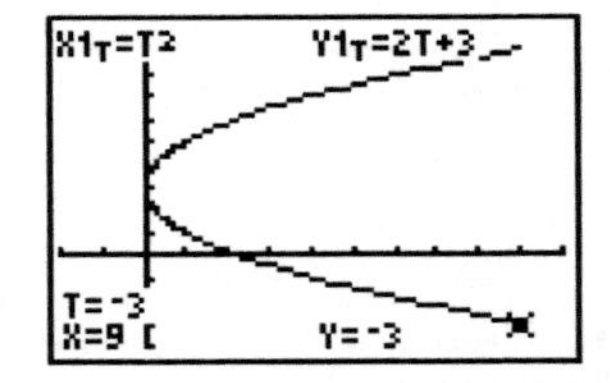

One can trace on a parametric graph, just as on a function-mode graph. The screen shown here is what appears when TRACE is first pressed: The trace cursor begins at the (x, y) coordinate corresponding to `Tmin`, and pressing ▶ increases the value of t (and likewise, ◀ decreases t). This can be somewhat disorienting, since for this graph, pressing ▶ moves the cursor to the *left*.

Section 7.1 Example 1 (page 468) Solving a System by Substitution

Section 7.1 Example 2 (page 468) Solving a System by Elimination

See page 13 for a description of the TI-83's intersection-locating procedure.

Section 7.1 Example 6 (page 472) Solving a Nonlinear System by Elimination

See the discussion related to Example 9 from Section 4.4 (page 22 of this manual) for tips on entering formulas like these.

In Figure 7, the text shows the intersections as found by the procedure built in to the calculator (described on page 13 of this manual). However, that is somewhat misleading; for these equations, the TI-83 can only find these intersections if the "guesses" supplied by the user are the exact x coordinates of the intersections (that is, -2 and 2). This is because the two circle equations are only valid for $-2 \leq x \leq 2$, while the hyperbola equations are only valid for $x \leq -2$ and $x \geq 2$. Since only ± 2 fall in both of these domains, any guess other than these two values results in a `BAD GUESS` error.

Section 7.2 Example 1 (page 481) Solving a System of Three Equations in Three Variables

If installed on your calculator (TI-83+ or TI-84+ only), the PolySmlt APP allows you to solve systems of linear equations. The screens below illustrate the process. When all coefficients have been entered (in the form of an *augmented matrix*, discussed in Section 7.3 of the text), pressing GRAPH solves for the three unknowns (which the TI-83 calls x1, x2, and x3, rather than x, y, and z).

MAIN MENU
1:PolyRootFinder
2:SimultEqnSolver
3:About
4:PolyHelp
5:SimultHelp
6:QuitPolySmlt

Option 2 solves systems of equations.

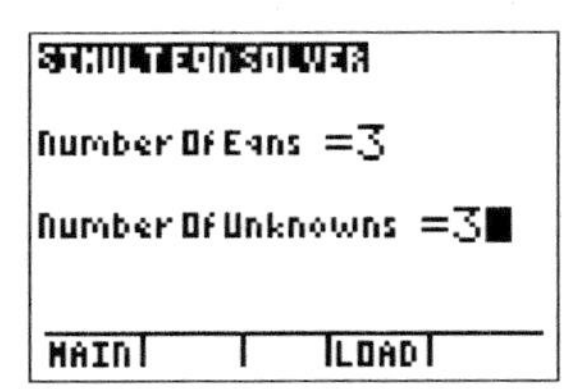

Give the number of equations and unknowns.

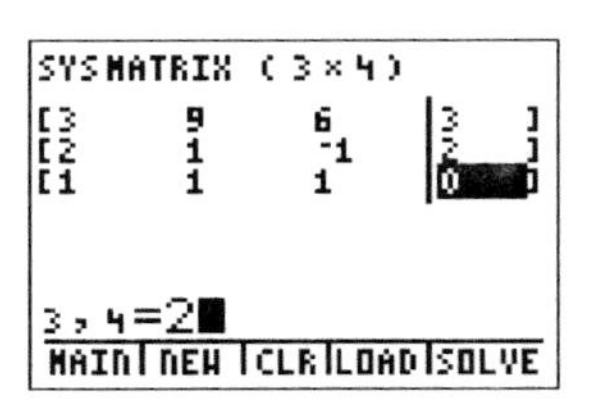

Enter coefficients on this screen.

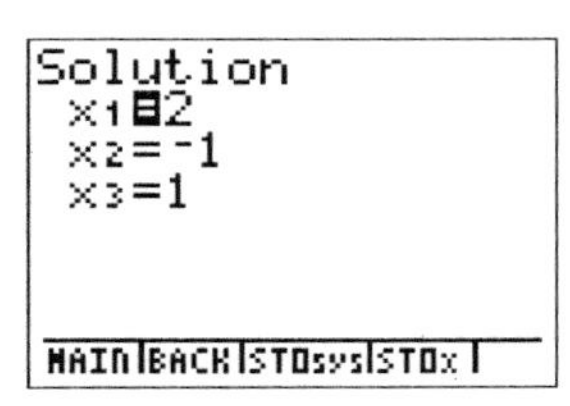

Press GRAPH to solve the system.

Section 7.2 Example 5 (page 484) Using a System to Fit a Parabola to Three Data Points

The text mentions that quadratic regression can be used to find the quadratic function fitting the three given data points. Shown on the right is the output of the TI-83's `QuadReg` procedure, with the diagnostic result `R²=1` indicating an exact match with the data points.

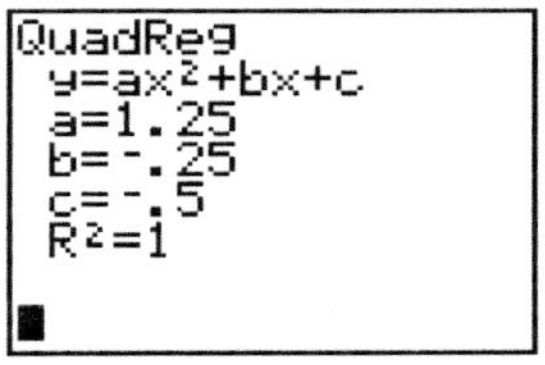

QuadReg
y=ax²+bx+c
a=1.25
b=-.25
c=-.5
R²=1

It is worth noting that the `CubicReg` and `QuartReg` procedures could likewise be used to find functions to exactly fit sets of four or five data points (much more easily than solving the related systems of equations). Furthermore, the `LinReg` procedure can be used with a pair of points to find the equation of the line through those points.

Section 7.3 Technology Note (page 488) Entering Matrices

Note that the TI-82 and TI-83 have a MATRX key, while on the TI-83+ and TI-84+, 2nd x^{-1} accesses the matrix commands.

The TI-83 has ten matrices, named `[A]` through `[J]`. One way to enter a matrix is by "storing" the contents on the home screen; an example is shown on the right. "`[B]`" is typed by pressing MATRX 2 (or 2nd x^{-1} 2), while the other square brackets `[` and `]` are 2nd × and 2nd −.

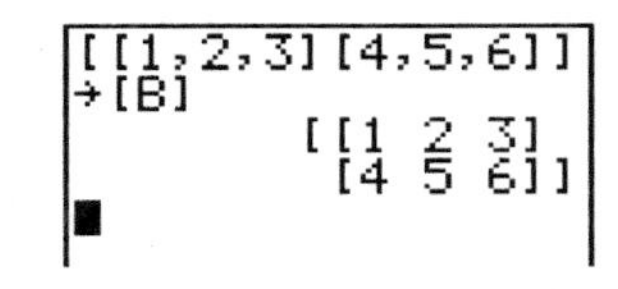

Alternatively, press MATRX ◄ (or 2nd x^{-1} ◄), then choose one of the matrices, specify the number of rows and columns, and type in the entries, moving around by pressing the arrow keys and ENTER. The screen on the right shows the process of editing matrix `[A]`. Press 2nd MODE (QUIT) when finished.

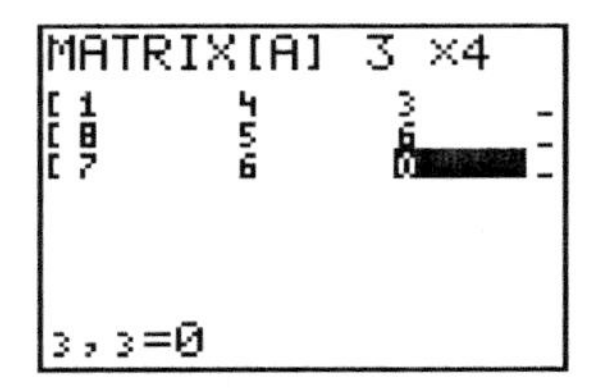

These two methods are illustrated in Figure 13 of the text.

Section 7.3 Technology Note (page 489) — Matrix Row Transformations
Section 7.3 Example 1 (page 489) — Using Row Transformations

The formats for the matrix row-operation commands, found in the MATRX:MATH ([MATRX][▸]) menu, are:

- `rowSwap(`*matrix*`,`*A*`,`*B*`)` produces a new matrix that has row *A* and row *B* swapped.
- `row+(`*matrix*`,`*A*`,`*B*`)` produces a new matrix with row *A* added to row *B*.
- `*row(`*number*`,`*matrix*`,`*A*`)` produces a new matrix with row *A* multiplied by *number*.
- `*row+(`*number*`,`*matrix*`,`*A*`,`*B*`)` produces a new matrix with row *A* multiplied by *number* and added to row *B* (row *A* is unchanged).

Shown below are examples of each of these operations on the matrix $[E]=\begin{bmatrix}1&4&7\\2&5&8\\3&6&9\end{bmatrix}$.

```
rowSwap([E],1,2
        [[2 5 8]
         [1 4 7]
         [3 6 9]]
```

```
row+([E],1,2
       [[1 4 7 ]
        [3 9 15]
        [3 6 9 ]]
```

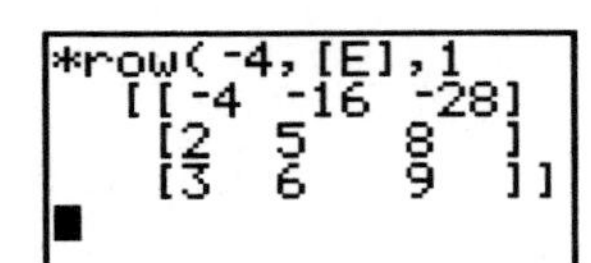

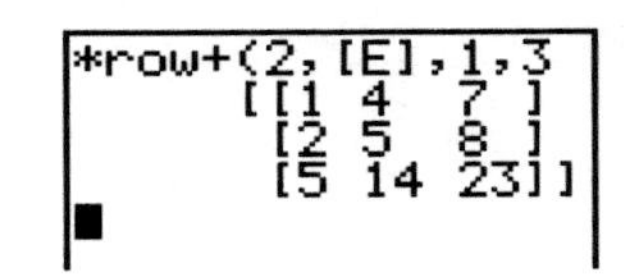

Keep in mind that these row operations leave the matrix [E] untouched. To perform a sequence of row operations, each result must either be stored in a matrix, or use the result variable Ans as the matrix in successive steps.

For example, with [A] equal to the augmented matrix given in the text just before this Technology Note and Example, the following screens illustrate the initial steps in the solving the system. Note the use of Ans in the last three screens.

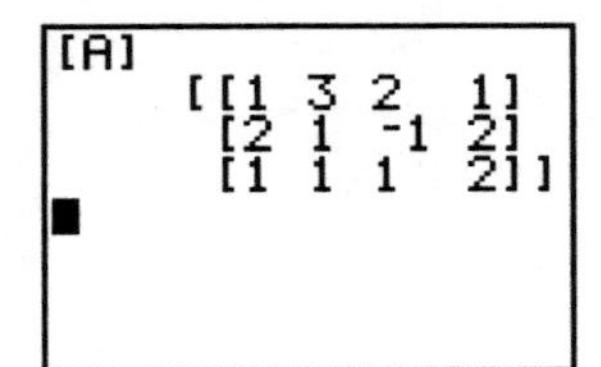

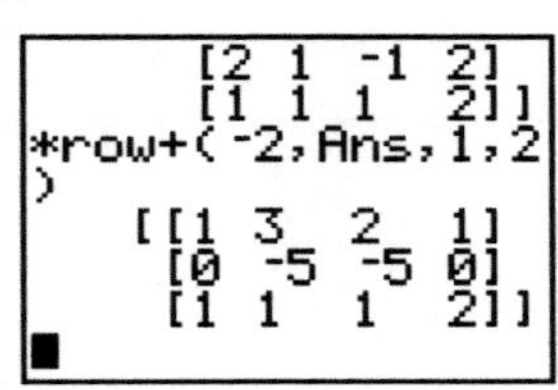

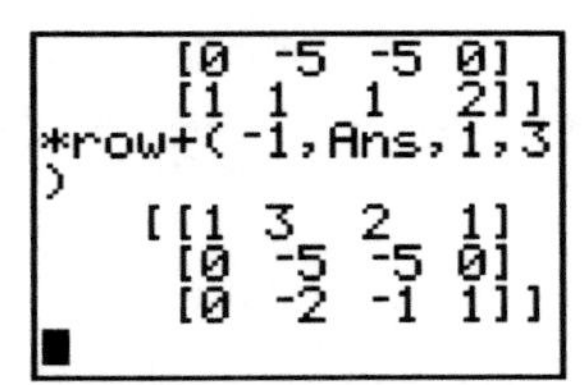

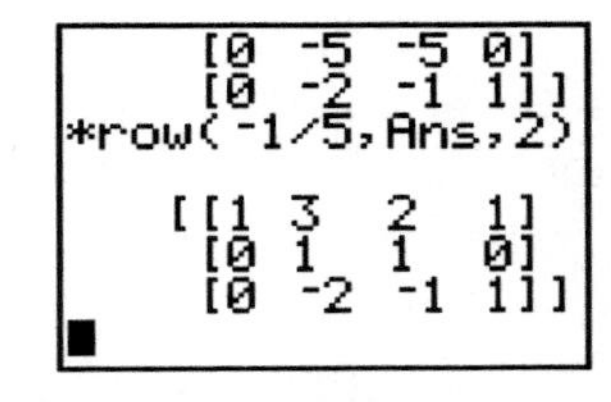

Section 7.3 Example 2 (page 490) — Solving a System by the Row Echelon Method
Section 7.3 Example 5 (page 493) — Solving a System by the Reduced Row Echelon Method

The `ref` and `rref` commands are options A and B in the MATRX:MATH menu; press [MATRX][▸][ALPHA] (or [2nd][x^{-1}][▸][ALPHA]), then [MATH] for `ref`, or [MATRX] (or [APPS]) for `rref`. These will do all the necessary row operations at once, making these individual steps seem tedious. However, doing the whole process step-by-step can be helpful in understanding how it works.

[82] *The TI-82 has the same simple row operations as the TI-83, but does not have the* `ref` *and* `rref` *commands.*

Section 7.4 Example 1 (page 501) Classifying Matrices by Dimension

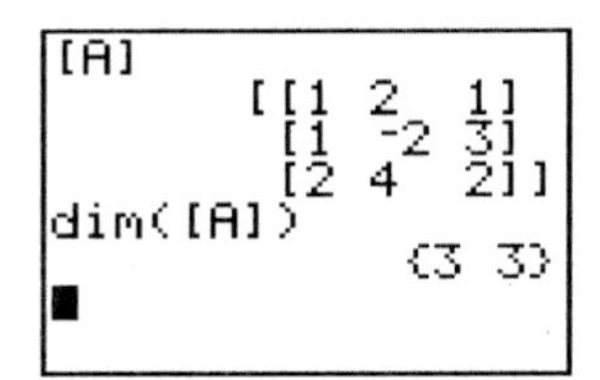

If a matrix has been entered in the TI-83, the dimensions can be found using the dim command—[MATRX][▸][3] (or [2nd][x⁻¹][▸][3])—which returns a list containing two numbers: { row count, column count }. Of course, counting rows and columns is arguably much simpler than using the dim command.

Section 7.4 Example 2 (page 501) Determining Equality of Matrices

If two matrices have been stored in the TI-83's matrix variables (say, [A] and [B]), one can test for equality by entering the command [A]=[B] on the home screen; the equals sign is found in the TEST menu ([2nd] [MATH]). This will give 1 if the matrices are equal, and 0 otherwise.

This test will produce a DIM MISMATCH error if the two matrices do not have the same dimensions—but of course, they are obviously not equal in that case!

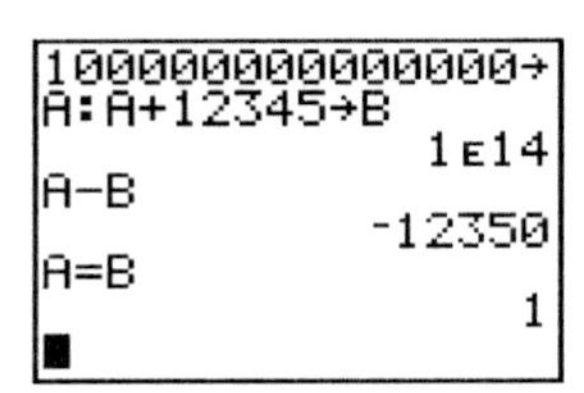

A word about the TI-83's testing of equality: The TI-83 keeps track of the first 14 digits of any number entered into it, so it cannot distinguish between numbers that differ beyond the 14th digit. In fact, the TI-83 only compares numbers out to the 10th digit. This may yield some surprising results, such as that shown on the screen on the right: A and B differ by over 12000, but since the first 10 digits are the same, the TI-83 reports that they are equal. (Also note that B has been rounded to the nearest 10, since the TI-83 stores only 14 digits, and the ones digit is the 15th.)

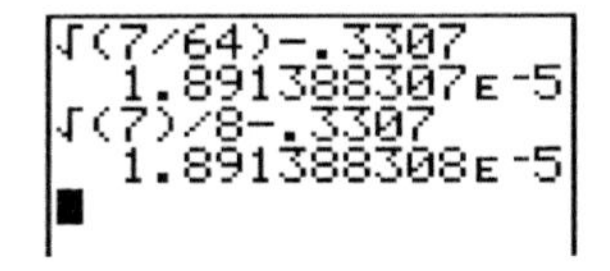

The TI-83 takes this approach because, when the calculator does a fairly complex computation, unavoidable numerical errors (due to rounding and other causes) can creep into the result. Such errors can affect the last digit (or in extreme cases, the last several digits), so two supposedly equal numbers might differ in that 14th digit. For example, although $\sqrt{7/64}$ and $\sqrt{7}/8$ are equal, they do not compute to exactly the same value; the screen on the right shows they differ in the 14th digit. By considering only the first 10 digits in testing for equality, the TI-83 avoids being tripped up by these numerical errors (at the risk of calling two close-but-different numbers equal).

Section 7.4 Example 3 (page 502) Adding Matrices
Section 7.4 Example 4 (page 504) Subtracting Matrices
Section 7.4 Example 5 (page 505) Multiplying Matrices by Scalars
Section 7.4 Example 8 (page 508) Multiplying Matrices

As the screens in the text indicate, basic arithmetic with matrices is relatively straightforward. When multiplying one matrix by another, the multiplication symbol * is not needed; that is, the entry [B][A] would produce the same results as those shown in Figure 32.

Section 7.4 Example 9 (page 508) — Multiplying Square Matrices

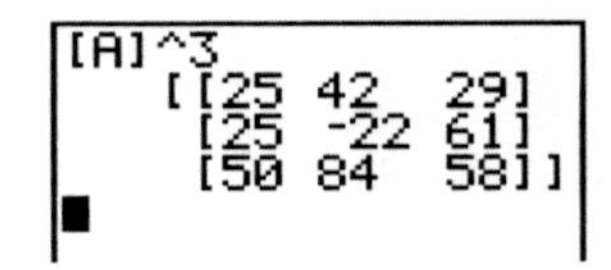

One additional observation about multiplying square matrices: One can raise a square matrix to any non-negative integer power using the [^] key (or [x^2] or [MATH] [3]). Negative or non-integer powers result in a `DOMAIN` error.

Section 7.5 Example 1 (page 514) — Evaluating the Determinant of a 2×2 Matrix

Section 7.5 Example 4 (page 517) — Evaluating the Determinant of a 3×3 Matrix

Section 7.5 Example 5 (page 518) — Evaluating the Determinant of a 4×4 Matrix

The `det` command is found in the MATRX:MATH submenu. To type "`det([A])`," press [MATRX][▸][1] (or [2nd] [x^{-1}][▸][1]), then [MATRX][1] (or [2nd][x^{-1}][1]), then [)] (although this command will work correctly without the closing parenthesis).

The determinant of a 3 × 3—or larger—matrix is as easy to find with a calculator as that of a 2 × 2 matrix. (At least, it is as easy for the user; the calculator is doing all the work!) Note, however, that trying to find the determinant of a non-square matrix (for example, a 3 × 4 matrix) results in an `INVALID DIM` error.

Section 7.6 Example 1 (page 525) — Using the 2×2 Identity Matrix

The `identity` command is option 5 in the MATRX:MATH menu: [MATRX][▸][5] (or [2nd][x^{-1}][▸][5]).

Section 7.6 Example 3 (page 528) — Finding the Inverse of a 3×3 Matrix

Section 7.6 Example 4 (page 530) — Finding the Inverse of a 2×2 Matrix

An inverse matrix is found using [x^{-1}]. (Do not try to use [^][(-)][1] in place of [x^{-1}].) For example, to find the inverse of matrix `[A]`, press [MATRX][1] (or [2nd][x^{-1}][1]), then [x^{-1}][ENTER]. An error will occur if the matrix is not square, or if it is a singular matrix, as in Example 4(b).

Section 7.7 Example 1 (page 538) — Graphing a Linear Inequality

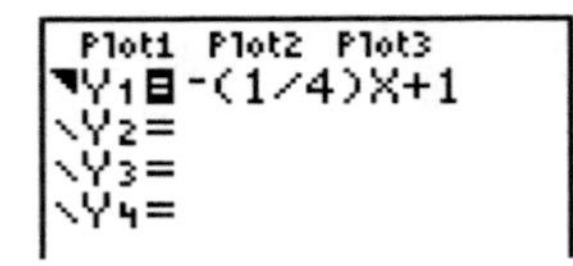

With the TI-83, there are two ways to shade above or below a function. The simpler way is to use the "shade above/below" graph style (see page 8). The screen on the right shows the "shade above" symbol next to Y1, which produces the graph shown in the text (Figure 52). Note that the TI-83 is not capable of showing the detail that the line is "dashed."

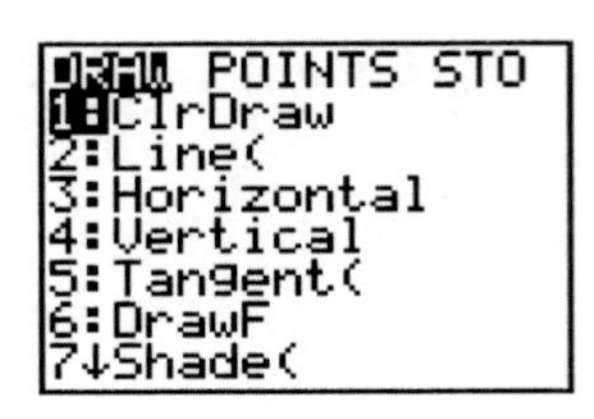

The other way to shade is the `Shade(` command, found as option 7 in the [2nd] [PRGM] (DRAW) menu. The format is

`Shade(`*lower*`,`*upper*`,`*min X*`,`*max X*`,`*pattern*`,`*resolution*`)`

Here *lower* and *upper* are the functions between which the TI-83 will draw the shading (above *lower* and below *upper*).

The last four options can be omitted. *min X* and *max X* specify the starting and ending x values for the shading. If omitted, the TI-83 uses Xmin and Xmax.

[82] *The TI-82 does not have graph styles, so the* Shade *command is the only method available.*

The last two options specify how the shading should look. *pattern* determines the direction of the shading: 1 (vertical—the default), 2 (horizontal), 3 (negative-slope 45°—that is, upper left to lower right), or 4 (positive-slope 45°—that is, lower left to upper right). *resolution* is a positive integer (1,2,3,. . .) which specifies how dense the shading should be (1 = shade every column of pixels, 2 = shade every other column, 3 = shade every third column, etc.). If omitted, the TI-83 shades every column; i.e., it uses *resolution* = 1.

To produce the graph shown in Figure 52 of the text, the appropriate command (typed on the home screen) would be something like

```
Shade(-(1/4)X+1,10,-2,6,1,2).
```

The use of 10 for the *upper* function simply tells the TI-83 to shade up as high as necessary; this could be replaced by any number greater than 4 (the value of Ymax for the viewing window shown). Also, if the function Y_1 had previously been defined as -(1/4)X+1, this command could be shortened to Shade(Y_1,10,-2,6,1,2).

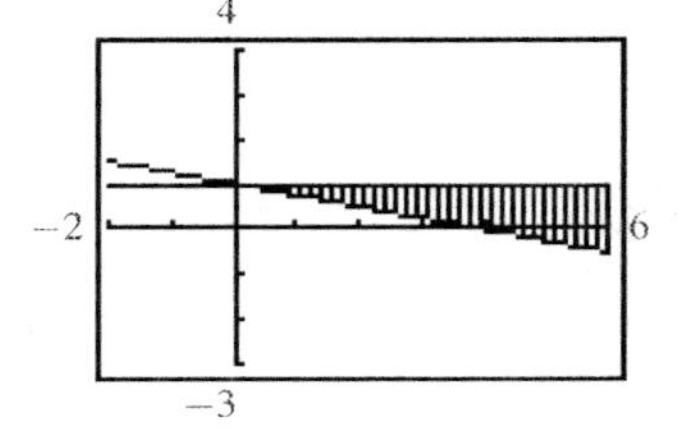

One more useful piece of information: Suppose one makes a mistake in typing the Shade(command (e.g., switching *upper* and *lower*, or using the wrong value of *resolution*), resulting in the wrong shading. The screen on the right, for example, arose from typing Shade(-(1/4)X+1,1,-2,6,1,2). In order to achieve the desired results, the mistake must be erased using the Clr-Draw command ([2nd][PRGM][1][ENTER]). Then—perhaps using deep recall (see page 3)—correct the mistake in the Shade(command and try again.

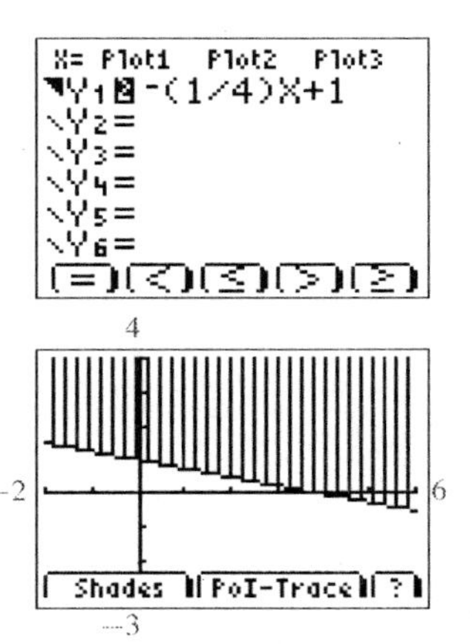

One more approach is available on the TI-83+ and TI-84+ calculators: The Inequalz APP (see section 13 of the introduction, page 8) changes the [Y=] screen so that one can enter inequalities rather than equations. After entering the formula (Y_1=-(1/4)X+1), place the cursor on the equals sign and press [ALPHA][ZOOM] to choose the ≤ graphing option.

The chief advantage of using this APP is that the calculator attempts to show more detail—specifically, it draws either a solid or a dashed line to indicate that the region satisfying the inequality does or does not include the line. (In this case, the line is solid.)

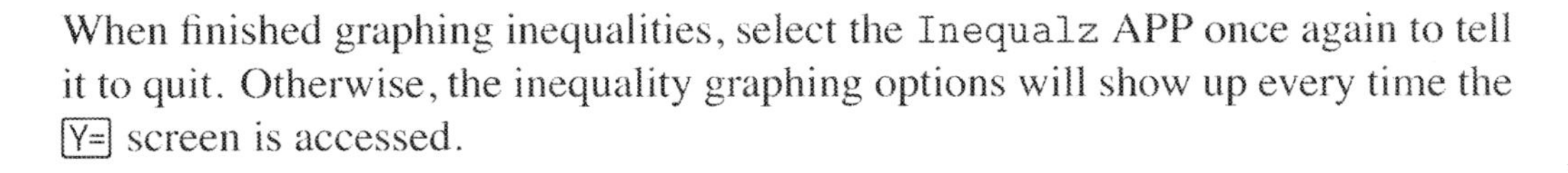

When finished graphing inequalities, select the Inequalz APP once again to tell it to quit. Otherwise, the inequality graphing options will show up every time the [Y=] screen is accessed.

Section 7.7 Example 2 (page 539) Graphing a System of Two Inequalities

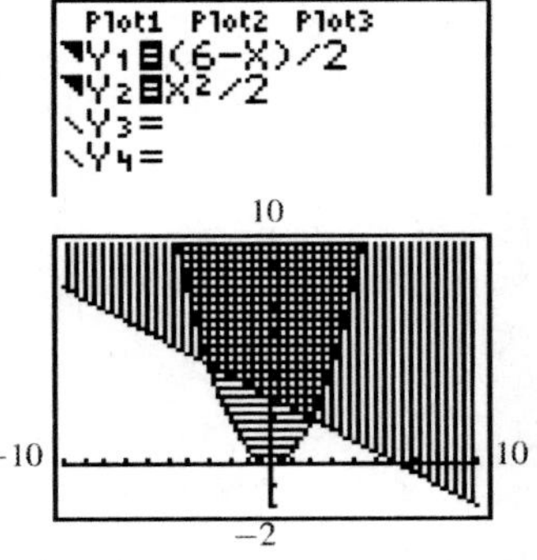

The easiest way to produce (essentially) the same graph as that shown in the text is to use the "shade above" graph style (see page 8). The screen on the right (above) shows the "shade above" symbol next to Y1 and Y2, with the results shown on the graph below (the same as that shown in Figure 52 of the text). When more than one function is graphed with shading, the TI-83 rotates through the four shading patterns (see page 31); that is, it graphs the first with vertical shading, the second with horizontal, and so on. All shading is done with a resolution of 2 (every other pixel).

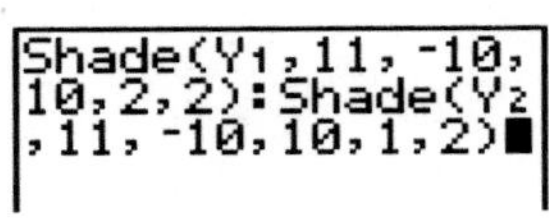

The `Shade` command can be used to produce this from the home screen. If Y1=(6-X)/2 and Y2=X²/2, the commands at right produce the graph shown in the text.

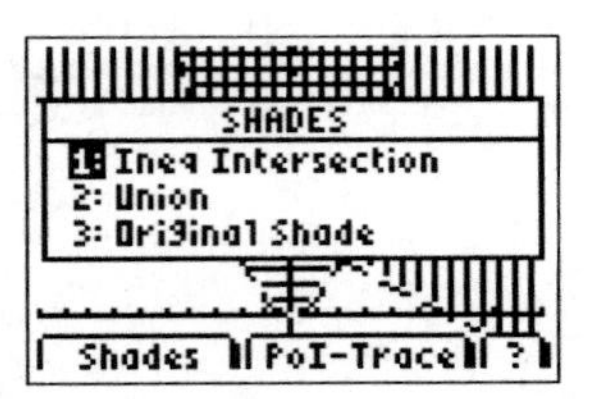

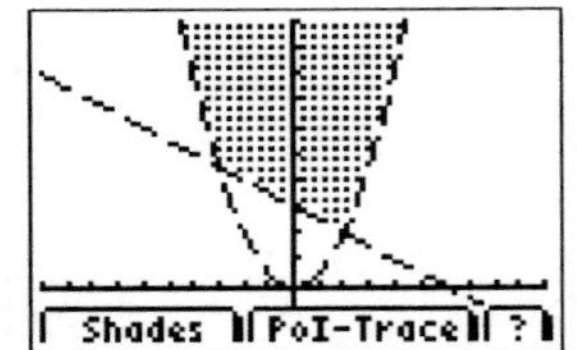

Using the `Inequalz` APP on the TI-83+ or TI-84+ (see the previous example), a nice picture can be made fairly easily. First, enter Y1>(6-X)/2 and Y2> X²/2, which produces a graph much like the first one shown above. Now, pressing [ALPHA][Y=] or [ALPHA][WINDOW] will bring up the Shades options; if we select `Ineq Intersection`, we get the desired region.

Section 8.1 Technology Note (page 567) Radian and Degree modes

See section 9 of the introduction (page 4) for information about selecting Degree and Radian modes.

Section 8.1 Example 2 (page 568) Calculating with Degrees and Minutes

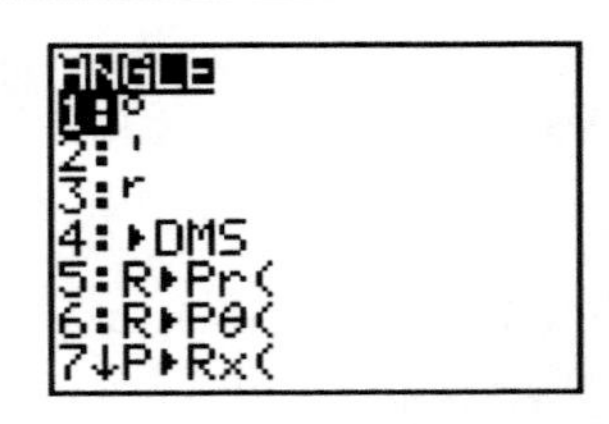

The degrees and minutes symbols, and the ▸DMS operator (which causes an angle to be displayed in degrees, minutes, and seconds, rather than as a decimal), are all found in the ANGLE menu ([2nd][MATRX] on the 82/83, or [2nd][APPS] on the 83+/84+), shown on the right.

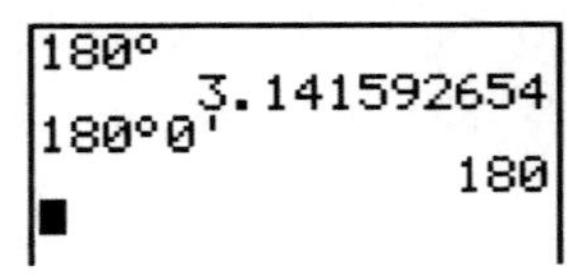

See section 9 of the introduction (page 4) for information about selecting Degree mode. Note that the results are displayed correctly regardless of the mode, *provided* the angle measurement includes a "minutes" portion. The two entries and outputs shown here were both produced with the TI-83 in Radian mode; the first entry is converted to radians, but the second is not.

[82] *The TI-82 uses only the minutes symbol "'" for entering angles in degrees, minutes, and seconds (output uses all the symbols). For example, the computation of part (a) would be entered on a TI-82 as* `51'29'+32'46'▸DMS`.

Section 8.1 Example 3 (page 568)

Converting between Decimal Degrees and Degrees, Minutes, Seconds

As was noted in the previous example, the degrees and minutes symbols are found as options 1 and 2 in the ANGLE menu; the "seconds" symbol is simply the double quotes, entered as [ALPHA][+].

[82] *On the TI-82, the conversion for part (a) would be entered as* `74'8'14'`.

Section 8.1 Example 5 (page 571)

Converting Degrees to Radians

The number π is available as [2nd][^], and the degree symbol is [2nd][MATRX][1] or [2nd][APPS][1]. With the calculator in Radian mode (see page 4), entering 45° causes the TI-83 to automatically convert to radians.

A useful technique to aid in recognizing when an angle is a multiple of π is to divide the result by π. This approach is illustrated in the screen on the right, showing that 45° is $\pi/4$ radians, and 30° is $\pi/6$ radians. This screen also makes use of the ▸`Frac` command ([MATH][1]), which simply means "display the result of this computation as a fraction, if possible."

```
45°
          .7853981634
Ans/π▸Frac
                  1/4
30°/π▸Frac
                  1/6
```

An alternative to using the degree symbol is to store $\pi/180$ in the calculator variable `D` (see page 3). Then typing, for example, `45D` [ENTER] will multiply 45 by $\pi/180$. This approach will work regardless of whether the calculator is in Degree or Radian mode. (A value stored in a variable will remain there until it is replaced by a new value.)

```
π/180→D
          .0174532925
45D
          .7853981634
249.8D
          4.359832471
```

Section 8.1 Example 6 (page 571)

Converting Radians to Degrees

With the TI-83 in Degree mode (see page 4), the radian symbol, produced with [2nd][MATRX][3] or [2nd][APPS][3], will automatically change a radian angle measurement to degrees.

Alternatively, with the value $180/\pi$ stored in the calculator variable `R` (see page 3), typing `(9π/4)R` [ENTER] will convert from radians to degrees regardless of whether the calculator is in Degree or Radian mode. (The same result can be achieved by *dividing by* the calculator variable `D` as defined in the previous example.)

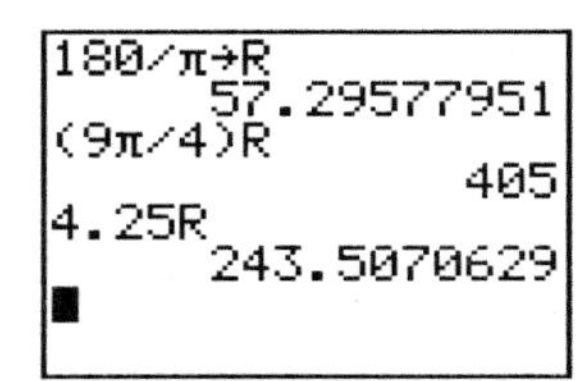

Section 8.2 Example 3 (page 584)

Finding Function Values of an Angle

To enter a restriction (like $x \geq 0$) on the TI-83, enter `Y1=(-1/2)X(X≥0)`—as is shown in the screen in the text. The greater-than-or-equal-to symbol is found in TEST menu ([2nd][MATH]). The TI-83 evaluates an expression like `X≥0` as 1 if it is true, and 0 if it is false; therefore, `Y1` has the value 0 for all $x < 0$, and has the value $-0.5x$ for $x \geq 0$.

Section 8.2 Example 4 (page 585) — Finding Function Values of Quadrantal Angles

The alternative to putting the calculator in Degree mode is to use the degree symbol ([2nd][MATRX][1] or [2nd] [APPS][1]) following each angle measure.

Since the cotangent, secant, and cosecant functions are the reciprocals of the tangent, cosine, and sine, they can be entered as (e.g.) `1/sin(90)`. Note, though, that this will not properly compute $\cot 90^\circ$, since `1/tan(90)` produces a domain error. Entering $\cot x$ as $\cos x / \sin x$ will produce the correct result at 90°.

One might guess that the other three trigonometric functions are accessed with [2nd] followed by [SIN], [COS], or [TAN] (which produce, e.g., $\sin^{-1}$). This is **not** what these functions do; in this case, the exponent -1 does not mean "reciprocal," but instead indicates that these are inverse functions (which are discussed in detail in Chapter 9 of the text).

Section 8.2 Technology Note (page 589) — Powers of trigonometric functions

Because the TI-83 automatically includes an opening parenthesis on the trigonometric functions, one cannot enter (e.g.) `sin²30`. The screen on the right shows two correct ways to enter this, one incorrect way (which computes the sine of 900°), and one method which—not too surprisingly—produces a syntax error.

```
sin(30)²
                .25
(sin(30))²
                .25
sin(30²)
                  0
sin(²30)■
```

[82] *On the TI-82, which does not automatically include an opening parenthesis, it is possible to type something like* `sin²30`*, but this produces a syntax error. One must instead type* `(sin 30)²`.

Section 8.3 Technology Note (page 598) — Decimal approximations and exact values

As this note suggests, exact values cannot always be found. Sometimes, one can recognize exact values by squaring the result, as the screen on the right illustrates: We can see that $\cos 30^\circ$ is the square root of $0.75 = \frac{3}{4}$ and that $\sin 45^\circ = \sqrt{0.5} = \frac{1}{\sqrt{2}} = \frac{\sqrt{2}}{2}$.

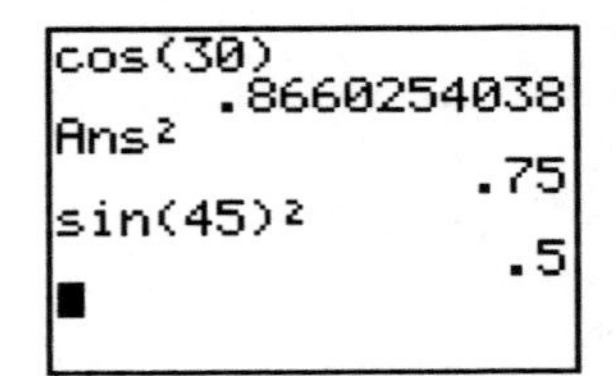

Section 8.3 Example 7 (page 601) — Approximating Trigonometric Function Values with a Calculator

The screen on the right (with computations done in Radian mode) illustrates a somewhat unexpected behavior: Even if an angle is entered in DMS format, the TI-83 assumes that the angle is in radians. (See the related discussion on page 32.) In order to remedy this, either put the calculator in Degree mode, or use the degree symbol ([2nd][MATRX][1] or [2nd][APPS][1]) at the very end of the angle measurement, as was done in the second entry.

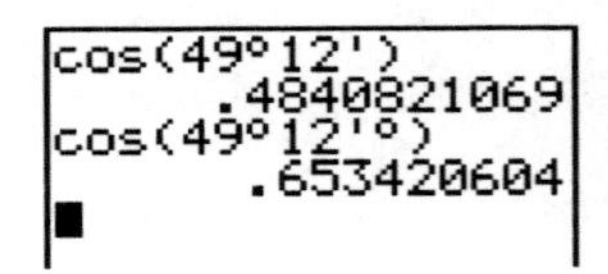

[82] *On a TI-82, the computation in part (a) would be entered* `cos 49'12'`.

Section 8.3 Technology Note (page 602)

Inverse versus reciprocal functions

The inverse trigonometric functions were mentioned earlier, and are covered in the next example; more details are given in Chapter 9 of the text. The screen shown here illustrates the two ways to compute csc 197.977° on the TI-83.

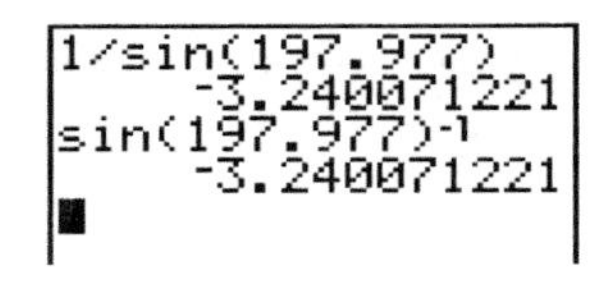

Section 8.3 Example 8 (page 602)

Using Inverse Trigonometric Functions

Section 8.3 Example 10 (page 603)

Finding Angle Measures

The inverse trigonometric functions are accessed with [2nd][SIN], [2nd][COS], and [2nd][TAN]; they *cannot* be entered as [SIN][x^{-1}], etc.

Note that the first output shown in text Figure 55 was produced in Degree mode, and the second was produced in Radian mode. None of the options available in the ANGLE menu ([2nd][MATRX] or [2nd][APPS]) can be used to avoid changing the mode; for example, when in Degree mode, the inverse trigonometric functions will always give an angle measure in degrees.

Section 8.4 Technology Note (page 609)

Programs to solve right triangles

See section 13 of the introduction (page 8) for information about installing and running programs on the TI-83. The text notes that one must "consider the various cases"; there are five such cases: two legs, leg and hypotenuse, angle and hypotenuse, angle and adjacent leg, angle and opposite leg.

Section 8.4 Example 7 (page 613)

Solving a Problem Involving the Angle of Elevation

The TI-83 can automatically locate the intersection of two graphs using the CALC menu ([2nd][TRACE]). This feature was previously illustrated on page 13, but we repeat the description here: Choose option 5 (intersect), use [▲], [▼] and [ENTER] to specify which two functions to use (in this case, the only two being displayed), and then use [◄] or [►] to specify a guess. After pressing [ENTER], the TI-83 will try to find an intersection of the two graphs. The screens below illustrate these steps; the final result is the screen shown as Figure 71 of the text. The guessing step in the fourth screen below is not crucial in this case, since the calculator would locate the intersection even if a very poor guess was given.

CALCULATE
1:value
2:zero
3:minimum
4:maximum
5:intersect
6:dy/dx
7:∫f(x)dx

[2nd][TRACE][5] selects the CALC:intersect feature

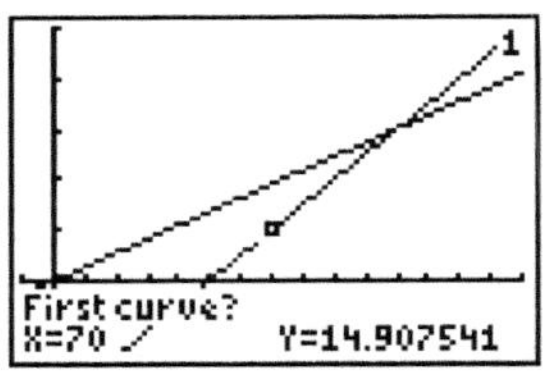

Press [ENTER] to choose Y1

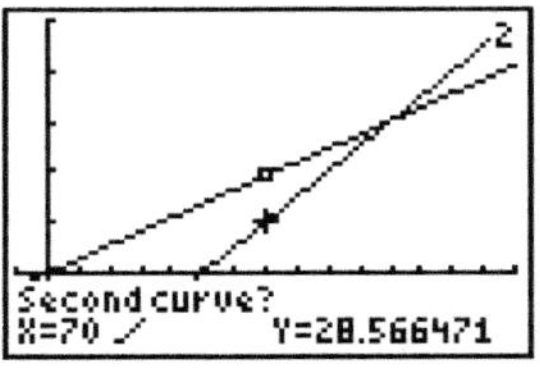

Press [ENTER] to choose Y2

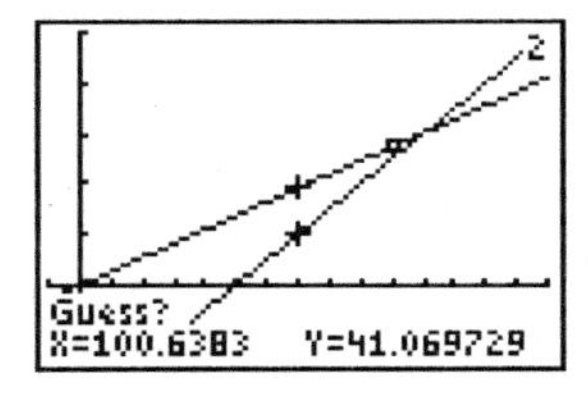

Move cursor to specify guess and then press [ENTER].

Section 8.6 Technology Note (page 630) — The trig viewing window

The trig viewing window is option 7 (`ZTrig`) in the [ZOOM] menu. The values of `Xmin` and `Xmax` are not quite $\pm 2\pi$; they are actually $\pm\frac{47}{24}\pi$, chosen so that ΔX (see page 4) equals $\pi/24$.

If the TI-83 is in Degree mode, choosing `ZTrig` sets `Xmin` and `Xmax` to about ± 360 instead of $\pm 2\pi$. (There are actually $\pm 352\frac{1}{2}$, so $\Delta\text{X} = 7.5$.)

Section 8.6 Example 1 (page 631) — Graphing $y = a \sin x$

Section 8.6 Example 2 (page 633) — Graphing $y = \sin bx$

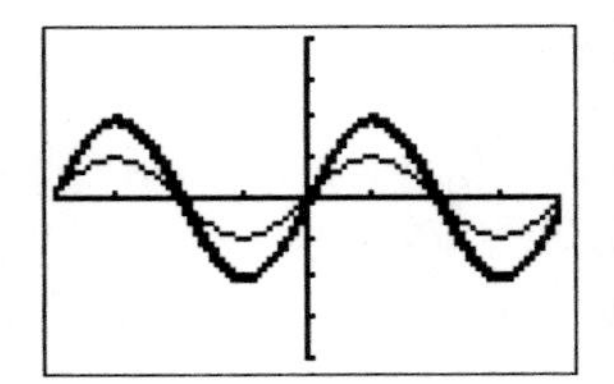

The TI-83's graph styles can produce screens like the one shown in Figure 89 or the graph screen accompanying Example 2. See page 8 for information on setting the thickness of a graph. Note that the TI-83 must be in Radian mode in order to produce the correct graph.

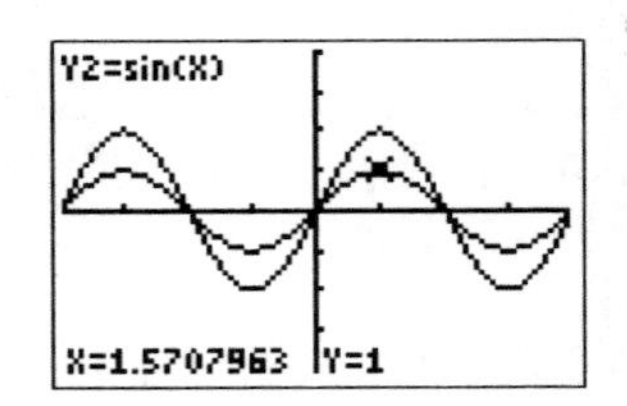

It is not absolutely necessary to vary the graphs' thicknesses, since pressing [TRACE] causes the formula to be displayed in the upper left corner of the screen (assuming the `ExprOn` has been chosen; see page 5).

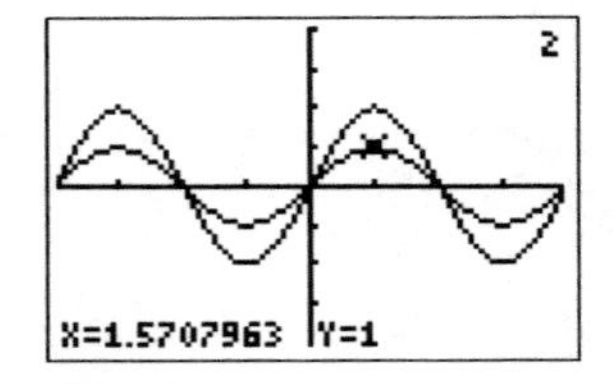

[82] *Since the TI-82 has neither graph styles nor an* `ExprOn` *option, one must rely on the number in the upper right corner of the screen to know which function is being traced. Here the number "2" means that the trace cursor is currently on the graph of* `Y2`.

Section 8.6 Example 8 (page 638) — Modeling Temperature with a Sine Function

The "sine regression" illustrated at the end of this example is a built-in feature of the TI-83. The procedures for creating scatter diagrams and performing regressions are described on pages 11 and 12 of this manual.

[82] *The TI-82 does not support sine regression. It might be possible to find a program to do this; see section 13 of this chapter's introduction, on page 8.*

Section 8.7 Technology Note (page 647) — Cosecant and secant functions

Section 8.7 Example 1 (page 649) — Graphing $y = a \sec bx$

Section 8.7 Example 2 (page 650) — Graphing $y = a \csc (x - d)$

See section 9 of the introduction (page 4) for information about Connected versus Dot mode. To graph the cosecant function, the actual entry on a TI-83 would be `Y1=1/sin(X)` or `Y1=sin(X)-1`. The function in Example 1 can be entered as `Y1=2cos(X/2)-1` (or as shown in the text). The function in Example 2 can be entered as `Y1=(3/2)(1/sin(X-π/2))` or `Y1=3/(2sin(X-π/2))`. A reminder: `sin-1` ([2nd][SIN]) is *not* the cosecant function.

82 *On a TI-82, parentheses are not automatically included with the sine and cosine functions, so in some cases, they must be added. In other cases, they can be omitted. For example, the cosecant function can be entered as either* `Y1=1/sin X` *or* `Y1=(sin X)-1`.

Section 9.1	Example 2	(page 674)	Rewriting an Expression in Terms of Sine and Cosine
Section 9.1	Example 3	(page 676)	Verifying an Identity (Working with One Side)
Section 9.1	Example 4	(page 677)	Verifying an Identity (Working with One Side)
Section 9.1	Example 5	(page 677)	Verifying an Identity (Working with One Side)
Section 9.1	Example 6	(page 678)	Verifying an Identity (Working with Both Sides)

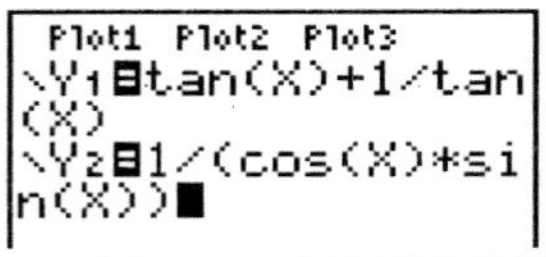

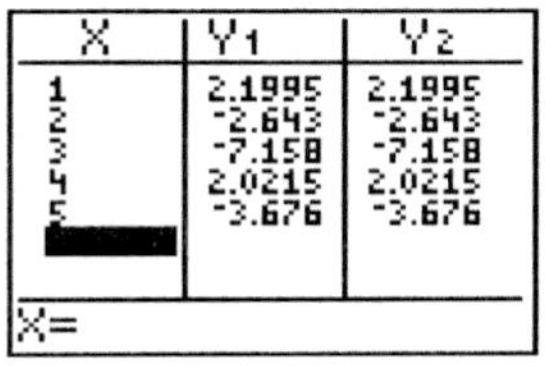

The top screen on the right shows how the expressions for Example 2 are entered on the [Y=] screen, so that they can be graphed and compared, as is illustrated in the text. As an alternative to graphing these two functions, the TI-83's table feature (see page 10 of this manual) can be used: If the y values are the same for a reasonably large sample of x values, one can be fairly sure (though not certain) that the two expressions are equal. To make this approach more reliable, be sure to choose x values that are not, for example, all multiples of π. This method is used in Example 4.

```
Plot1 Plot2 Plot3
\Y1=tan(X)+1/tan
(X)
\Y2=1/(cos(X)*si
n(X))
```

X	Y1	Y2
1	2.1995	2.1995
2	-2.643	-2.643
3	-7.158	-7.158
4	2.0215	2.0215
5	-3.676	-3.676

X=

The fact that the two graphs are identical on the calculator screen does provide strong support for the identity, especially when confirmed by tracing, as described in the Technology Note next to Example 3 in the text. The tables shown in Figures 3 and 4 also show that the two expressions produce identical output—at least to the number of digits visible in the table. The exception is Figure 3 when X equals $\pi/2 \approx 1.5708$, for which Y1 shows ERROR while Y2 equals 1. In fact, $\cot(\pi/2) + 1 = 1$, but the value of Y1 is reported as ERROR since $\tan(\pi/2)$ is undefined.

The footnote on page 676 points out that the graph (or the table) cannot be used to prove the identity. In particular, the graph in Figure 2 only plots points for values of x that are $\Delta X = \pi/24$ units apart, while the table in Figure 3 shows outputs for input values spaced $\pi/8$ units apart. For example, the function `Y3=cos(X)/sin(X)+cos(48X)` would look identical to Y1 and Y2 on both the graph and the table, although this function is different from these two functions at any point other than those shown in the graph.

Section 9.2 Example 1 (page 685) Finding Exact Cosine Function Values
Section 9.2 Example 3 (page 687) Finding Exact Sine and Tangent Function Values

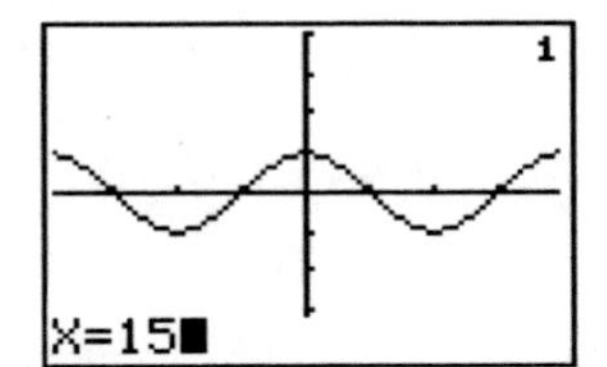

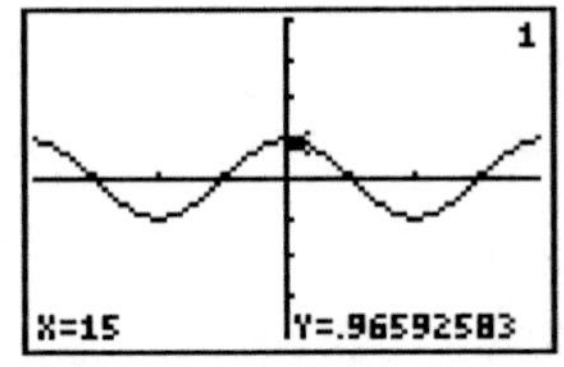

The TI-83 can graphically and numerically support exact value computations such as $\cos 15^\circ = \frac{\sqrt{6}+\sqrt{2}}{4}$. Starting with a graph of Y1=cos(X), the TI-83 makes it possible to trace to any real number value for x between Xmin and Xmax. Simply type a number or expression (like 1/π or √(2)) while in TRACE mode. The number appears at the bottom of the window in a larger font size than the TRACE coordinates. Pressing [ENTER] causes the TRACE cursor to jump to that x-coordinate. This same result can be achieved using option 1 (value) from the [2nd][TRACE] (CALC) menu. (This feature was previously discussed on page 17.)

[82] *This latter approach is also available on the TI-82.*

Alternatively, a table of values like those shown here can be used to find the value of $\cos 15^\circ$. Of course, the screen shown in the text in support of Example 1(b) shows the other part of this process: Computing the decimal value of $(\sqrt{6}+\sqrt{2})/2$ and observing that it agrees with those found here.

X	Y1
9	.98769
10	.98481
11	.98163
12	.97815
13	.97437
14	.9703
15	[illegible]

Y1=.965925826289

Note that Example 8 in Section 9.3 (page 699) shows that $\cos 15^\circ = \cos \frac{\pi}{12}$ can also be written as $\frac{\sqrt{2+\sqrt{3}}}{2}$.

Section 9.3 Example 4 (page 695) Deriving a Multiple-Number Identity

The table in Figure 9 shows Y1=sin(3X) and Y2=3sin(X)-4sin(X)³—recall the proper way to enter this second expression—with ΔTbl $= \pi/8$ and TblStart= $-3\pi/8$. Note that choosing x values which are multiples of a fraction of π is somewhat risky, since the periods of $\sin 3x$ and $\sin x$ involve fractions of π. Stronger support can be obtained by trying input values that are not multiples of π—say, $x = 1$, or $x = \sqrt{2}$.

Section 9.4 Example 1 (page 706) Finding Inverse Sine Values
Section 9.4 Example 2 (page 707) Finding Inverse Cosine Values

Of course, it is not necessary to graph $y = \sin^{-1} x$ or $y = \cos^{-1} x$ to find these values; one can simply enter, e.g., sin⁻¹(1/2) on the home screen. The first entry of the screen on the right shows what happens when the calculator is in Degree mode; note that the result is not in $[-\pi/2, \pi/2]$. With the calculator in Radian mode, results similar to those in the text are found, and the method employed on page 33 (Example 5 from Section 8.1) confirms that these values are $\pi/6$ and $3\pi/4$.

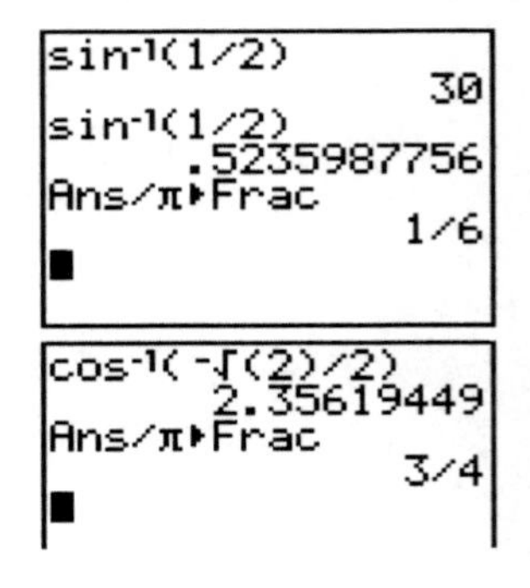

Section 9.4 Example 4 (page 710) Finding Inverse Function Values with a Calculator

Note that the answer given for (b), 109.499054°, overrepresents the accuracy of that value. A typical rule for doing computations involving decimal values (like −0.3541) is to report only as many digits in the result as were present in the original number—in this case, four. This means the reported answer should be "about 109.5°," and in fact, any angle θ between about 109.496° and 109.501° has a cotangent which rounds to −0.3541. (See also the discussion in the text on page 608.)

Section 9.6 Example 8 (page 727) Describing a Musical Tone from a Graph

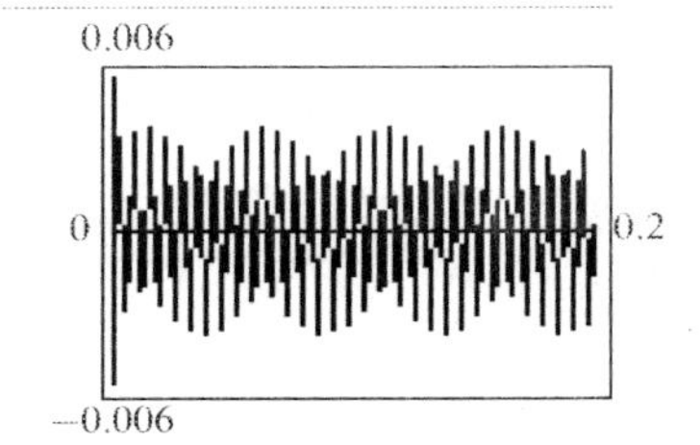

Note that the calculator screen shown in Figure 40 illustrates the importance of choosing a "good" viewing window. If we choose the wrong vertical scale (Ymin and Ymax), we might not be able to see the graph at all—it might be squashed against the x-axis. If we make the window too wide—that is, if Xmax minus Xmin is too large—we might see the "wrong" picture, like the one on the right: We see a periodic function in this view, but not the one we want. There are actually 30 periods in this window, but the TI-83's limited resolution cannot show them all.

This observation—that a periodic function, viewed at fixed intervals, can appear to be a *different* periodic function—is the same effect that causes wagon wheels to appear to run backwards in old movies.

Section 10.1 Technology Note (page 743) Programs to solve triangles
Section 10.2 Technology Note (page 756)

See section 13 of the introduction (page 8) for information about installing and running programs on the TI-83. If you wish to type these in on your own, here are two programs that give the same output as those shown in the text. (The first line of the second program ensures that the TI-83 is in Degree mode. The first program will operate correctly regardless of the mode.)

```
PROGRAM:ASA
Input "1ST ANGLE:",A
Input "2ND ANGLE:",B
Input "COMMON SIDE:",D
180-A-B→C
Dsin(A°)/sin(C°)→E
Dsin(B°)/sin(C°)→F
Disp "OTHER ANGLE",C
Disp "OTHER SIDES",E,F
```

```
PROGRAM:SSS
Degree
Input "1ST SIDE:",D
Input "2ND SIDE:",E
Input "3RD SIDE:",F
cos⁻¹((D²+E²-F²)/(2DE))→C
cos⁻¹((D²+F²-E²)/(2DF))→B
180-B-C→A
Disp "ANGLES ARE",A,B,C
```

Section 10.3 Example 1 (page 766) — Finding Magnitude and Direction Angle

The `R▸Pr` and `R▸Pθ` conversion functions (illustrated in Figure 28) are in the TI-83's ANGLE menu ([2nd][MATRX] or [2nd][APPS]) as options 5 and 6. `R▸Pθ(x,y)` returns an angle in degrees if the TI-83 is in Degree mode.

"R" and "P" stand for "rectangular" and "polar" coordinates. Rectangular coordinates are the familiar x and y values. Polar coordinates, described in Section 10.6 of the text, are r (which corresponds to the magnitude of a vector) and θ (the direction angle).

Section 10.3 Example 2 (page 767) — Finding Vertical and Horizontal Components

```
P▸Rx(25,41.7)
         18.66595456
P▸Ry(25,41.7)
         16.63075887
```

The `P▸Rx` and `P▸Ry` conversion functions are in the ANGLE menu ([2nd][MATRX]) or [2nd][APPS]) as options 7 and 8. Note that the screen shown in Figure 30 was created with the calculator set to display only one digit after the decimal, and in Degree mode. If the TI-83 is in Radian mode, the degree symbol (also in the ANGLE menu) can specify that the angle is in degrees. Also note that these functions yield exactly the same results as, e.g., `25cos(41.7)`; which to use is a matter of personal preference.

"R" and "P" stand for "rectangular" and "polar" coordinates. Rectangular coordinates are the familiar x and y values. Polar coordinates, described in Section 10.6 of the text, are r (which corresponds to the magnitude of a vector) and θ (the direction angle).

Section 10.3 Example 5 (page 768) — Performing Vector Operations
Section 10.3 Example 6 (page 769) — Finding the Dot Product

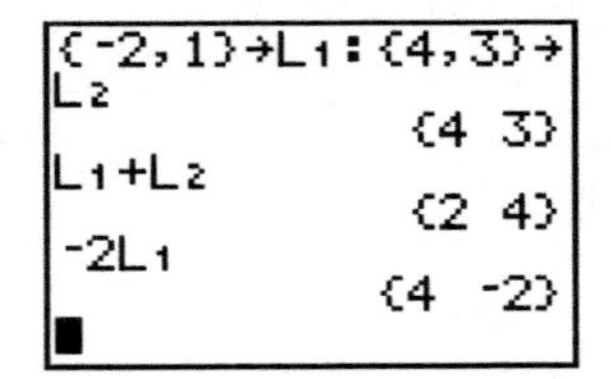

The TI-83 does not have built-in support for vector mathematics, but as Figure 35 shows, some vectors operations can be done using the TI-83's lists. If one must do several computations with the same vectors, it may be useful to store them in the list variables L_1, L_2, . . . , and perform the computations as shown in the screen on the right.

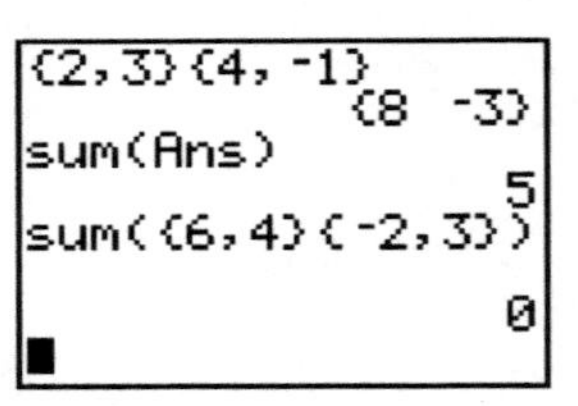

In order to compute dot products, use list multiplication combined with the TI-83's `sum` command from the LIST:MATH menu ([2nd][STAT][▸][5]).

Alternatively, programs might be found to perform tasks like this; see section 13 of the introduction (page 8) for information about installing programs in the TI-83.

Section 10.4 Example 2 (page 779) Converting from Trigonometric Form to Rectangular Form

While the TI-83 must be in `a+bi` or `re^θi` mode in order to compute (e.g.) square roots of negative numbers, it will perform computations involving `i` ([2nd][.]) even while in `Real` mode (see page 4).

```
2e^(300°i)
  1-1.732050808i
2e^(300°i)
-.0441932386-1.…
```

A shorter way to enter a complex number in trigonometric form is to use the TI-83's "polar" form: For modulus r and argument θ, enter $re^{i\theta}$. **Note:** The angle θ must be in radians regardless of whether the TI-83 is in Degree or Radian mode. For the first output in screen on the right, the TI-83 was in Radian mode (so 300° was converted to $\frac{5\pi}{6}$ radians); for the second, it was in Degree mode, so 300° was unchanged, producing an incorrect result.

[82] *The TI-82 does not support complex numbers. However, some complex computations can be done with a TI-82; see the appendix at the end of this chapter, page 48.*

Section 10.4 Example 3 (page 780) Converting from Rectangular Form to Trigonometric Form

Section 10.4 Technology Note (page 782) The angle and abs commands

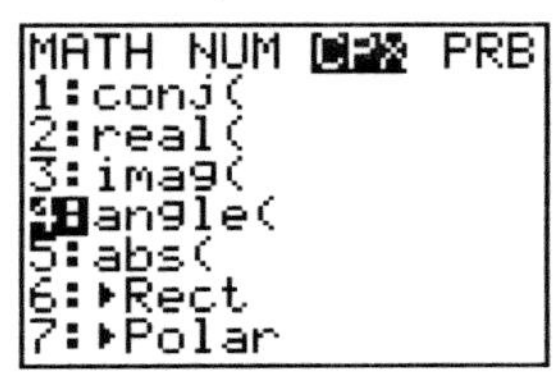

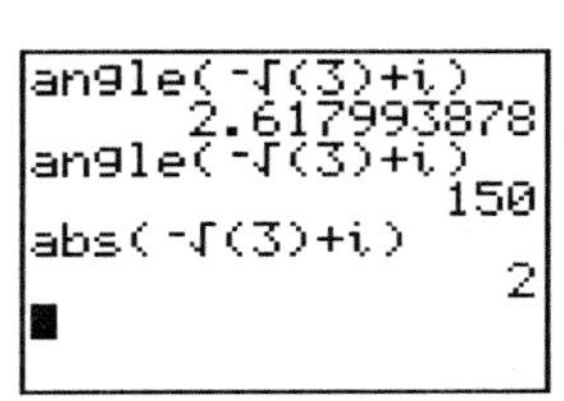

The TI-83 can convert from rectangular to trigonometric form in several ways. Aside from the commands `R▸Pr` and `R▸Pθ` shown in the text—found in the ANGLE menu ([2nd][MATRX]) or [2nd][APPS]), and previously discussed on page 40—one can use the MATH:CPX menu commands `angle(` and `abs(`. The `angle(` command gives the angle θ in radians or degrees (depending on the mode setting), and the `abs(` function (also found in the MATH:NUM menu) gives the modulus r. Note that `angle(-3i)` and `R▸Pθ(-3i)` give the result −90°, which is coterminal with the answer given in the text, 270°.

```
-√(3)+i
        2e^(150i)
-3i
        3e^(-90i)
```

Furthermore, with the calculator placed in `re^θi` mode (see page 4), the conversion can be done all at once, as the screen on the right shows. As the name of the mode suggests, the values of r and θ can be read directly from the output: r is the leading coefficient, and θ is in the exponent. (The computations here were done in Degree mode. In Radian mode, these inputs would produce the outputs

`2e^(2.617993878i)` and `3e^( -1.570796327i)`,

respectively. However, as was noted in the previous example, if a complex number is entered in this polar form, the angle θ must be given in radians.)

```
-√(3)+i▸Polar
        2e^(150i)
-3i▸Polar
        3e^(-90i)
```

Additionally, using the `▸Polar` command (option 7 in the MATH CPX menu; [MATH] [▸][▸][7]), and with the TI-83 in **any** mode (`Real`, `a+bi`, or `re^θi`), the output is displayed in this polar format, as the screen on the right shows. Note: Polar format is equivalent to trigonometric format because of the (fairly deep) mathematical fact that $e^{i\theta} = \cos\theta + i\sin\theta = \text{cis}\,\theta$.

[82] *Since the TI-82 has no* MATH:CPX *menu, the* `R▸Pθ` *and* `R▸Pr` *commands are the only option available.*

Section 10.4 Example 5 (page 783) Using the Product Theorem

Section 10.4 Example 6 (page 784) Using the Quotient Theorem

If doing this computation on a calculator, the TI-83's polar format can save a lot of typing. Shown on the right are the appropriate input, and the two possible outputs that can result: The first output arises from `Real` or `a+bi` mode, and the second from `re^θi` mode; note that this is $6e^{\pi i} = -6i$.

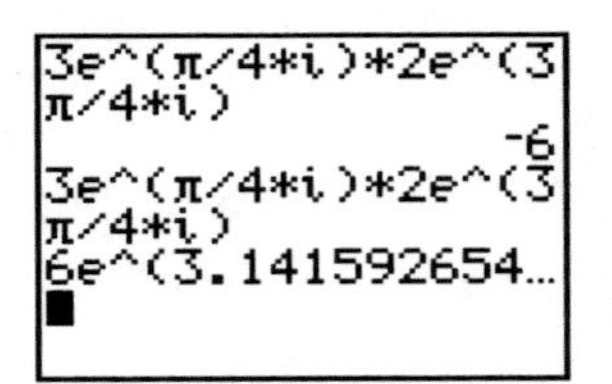

Section 10.5 Example 1 (page 787) Finding a Power of a Complex Number

The screen on the right shows several options for computing $(1 + i\sqrt{3})^8$ with the TI-83 (in `a+bi` mode). The first is fairly straightforward, but the reported result shows the complex part of the answer $(128i\sqrt{3})$ given in decimal form. For an "exact" answer, the TI-83's polar format (`re^θi`) can be used. The TI-83 was in Degree mode for the first of the two polar-format answers, and in Radian mode for the second.

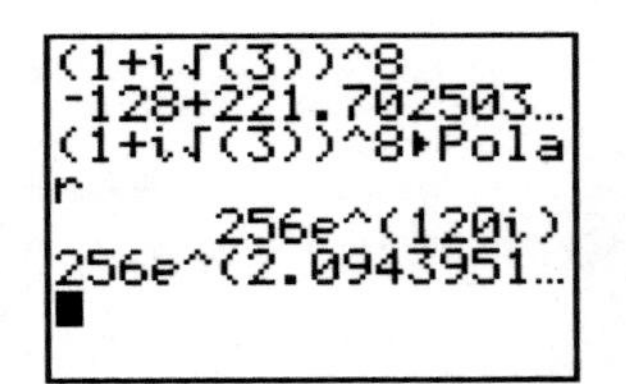

Section 10.6 Example 1 (page 794) Plotting Points with Polar Coordinates

Section 10.6 Example 2 (page 795) Giving Alternative Forms for Coordinates of a Point

Converting between polar and rectangular coordinates can be done using any of the conversion methods for vectors and complex numbers (covered beginning on page 40).

Section 10.6 Example 3 (page 796) Examining Polar and Rectangular Equations of Lines and Circles

Section 10.6 Example 4 (page 796) Graphing a Polar Equation (Cardioid)

Section 10.6 Example 5 (page 797) Graphing a Polar Equation (Rose)

Section 10.6 Example 6 (page 798) Graphing a Polar Equation (Lemniscate)

To produce these polar graphs, the TI-83 should be set to Degree and Polar modes (see the screen on the right). In this mode, the [Y=] key allows entry of up to six polar equations (r as a function of θ). One could also use Radian mode, adjusting the values of `θmin`, `θmax`, and `θstep` accordingly (e.g., use 0, 2π, and $\pi/30$ instead of 0, 360, and 5).

For the cardioid, rose, and lemniscate, the window settings shown in the text show these graphs on "square" windows (see section 11 of the introduction, page 6), so one can see how their proportions compare to those of a circle.

For the cardioid, the value of `θstep` does not need to be 5, although that choice works well for this graph. Too large a choice of `θstep` produces a graph with lots of sharp "corners," like the one shown on the right (drawn with `θstep=30`). Setting `θstep` too small, on the other hand, produces a smooth graph, but it is drawn very slowly. Sometimes it may be necessary to try different values of `θstep` to choose a good one.

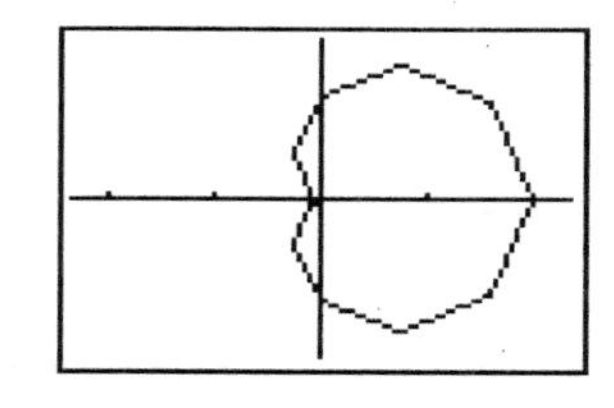

The lemniscate can be drawn by setting `θmin=0` and `θmax=180`, or `θmin=-45` and `θmax=45`. In fact, with θ ranging from -45 to 225, the graph of `r1=√(cos(2θ))` (alone) will produce the entire lemniscate. (`θstep` should be about 5.) The rose can be produced by setting `θmin=0` and `θmax=360`, or using any 360°-range of θ values (with `θstep` about 5).

Section 10.6 Example 7 (page 798) Graphing a Polar Equation (Spiral of Archimedes)

To produce this graph on the viewing window shown in the text, the TI-83 must be in Radian mode. (In Degree mode, it produces the same shape, but magnified by a factor of $180/\pi$ — meaning that the viewing window needs to be larger by that same factor.)

Section 10.7 Example 2 (page 805) Graphing an Ellipse with Parametric Equations

See page 26 for information about using parametric mode. This curve can be graphed in Degree mode with `Tmin=0` and `Tmax=360`, or in Radian mode with `Tmax=2π`. In order to see the proportions of this ellipse, it might be good to graph it on a square window. This can be done most easily by pressing [ZOOM] [5]. On a TI-83, initially with the window settings shown in the text, this would result in the window $[-6, 6] \times [-4, 4]$.

Section 10.7 Example 3 (page 805) Graphing a Cycloid

The TI-83 *must* be in Radian mode in order to produce this graph.

Section 10.7 Example 4 (page 806) Creating a Drawing With Parametric Equations

To turn off the display of the coordinate axes, set `AxesOff` on the [2nd][ZOOM] (FORMAT) screen. It is a good idea to restore this setting to `AxesOn` when finished, since for most uses, the absence of the axes can be confusing.

Section 10.7 Example 5 (page 807) — Simulating Motion With Parametric Equations

Section 10.7 Example 7 (page 808) — Analyzing the Path of a Projectile

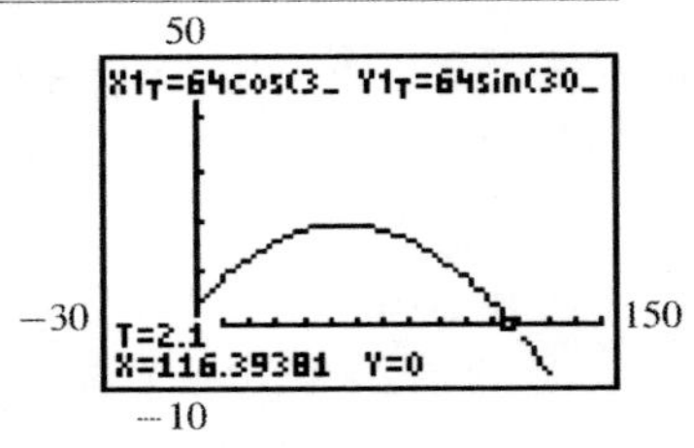

Parametric mode is particularly nice for analyzing motion, because one can picture the motion by watching the calculator create the graph, or by pressing [TRACE] and watching the motion of the trace cursor. (When tracing in parametric mode, the [▶] and [◀] keys increase and decrease the value of t, and the trace cursor shows the location (x, y) at time t.) The screen on the right is essentially the same as Figure 82, and illustrates tracing on the projectile path in Example 7. Note that the value of t changes by ±`Tstep` each time [▶] or [◀] is pressed, so obviously the choice of `Tstep` affects which points can be traced. This graph was produced by setting `Tmin=0`, `Tmax=3`, and `Tstep=0.1`.

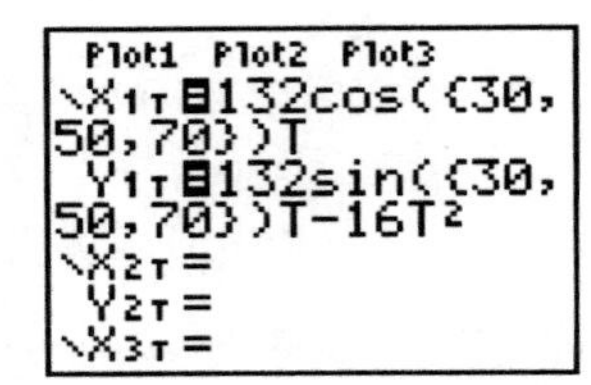

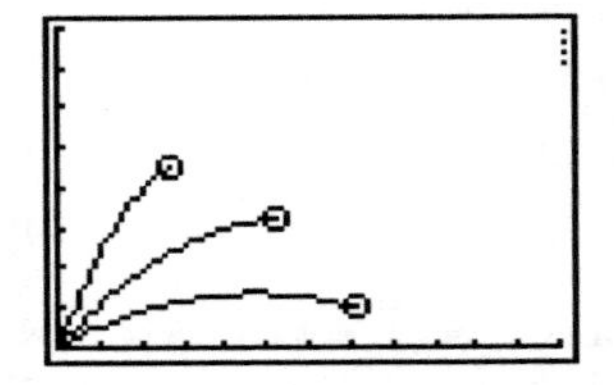

Rather than entering three separate pairs of equations for Example 5, the TI-83's list features can be used to graph all three curves with a single pair of equations. The TI-83's graph styles (see page 8) can be useful, too. The screen on the far right shows the three paths for Example 5 being plotted in "ball path" style.

Section 11.1 Example 1 (page 821) — Finding Terms of Sequences

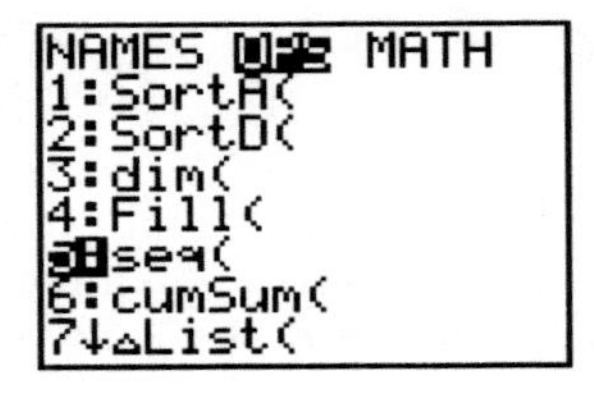

The `seq(` command is item 5 in the [2nd][STAT][▶] (LIST:OPS) menu. Given a formula a_n for the nth term in a sequence, the command

`seq(`*formula*`,`*variable*`,`*start*`,`*end*`,`*step*`)`

produces the list $\{a_{start}, a_{start+step}, \ldots, a_{end}\}$. In most uses, the value of *step* is 1, which is the assumed value if this is omitted. The resulting list can have no more than 999 items (or possibly less, if memory is low).

[82] *The TI-82 can only create lists of up to 99 items.*

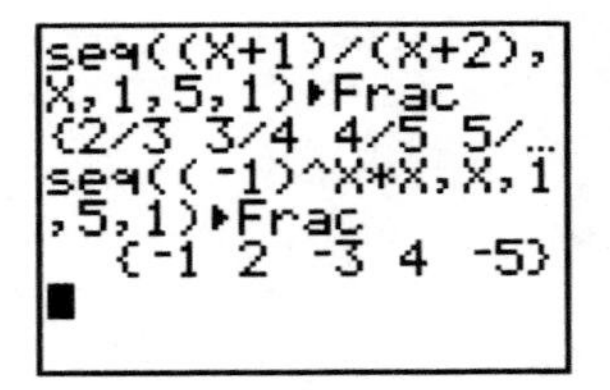

Here are the results for parts (a) and (b); in both of these cases, "`,1)`" could have been omitted from the end of the command. Note that the first output is longer than can be displayed on a single line; the fact that it ends with ellipsis marks (...) means more terms are listed but are not visible. These additional terms can be seen by pressing [▶].

Note that *variable* can be any letter—it may be most convenient to use `X` (as was done here), since it can be typed with [X,T,Θ,*n*].

Section 11.1 Technology Note (page 822) — Sequence mode

The first screen accompanying this Technology Note shows the `seq` command with the variable `n`, which is the character produced by [X,T,Θ,*n*] when the TI-83 is in Sequence mode—see section 9 of the introduction (page 4). Note that it is not necessary to use Sequence mode to use the `seq` command.

Place the TI-83 in Sequence mode, as shown above on the right. In this mode, the [Y=] key accesses the screen shown below on the right, which allows the definition of up to three sequences: `u(n)`, `v(n)`, and `w(n)`. The letters `u`, `v`, and `w` are produced by [2nd] followed by [7], [8], and [9], respectively, and n is produced by [X,T,Θ,n]. Note that the TI-83 represents terms of the sequence as `u(n)` rather than a_n; otherwise, this is very similar to the notation used in the text. Also note that in this screen, the line `u(nMin)=` is left blank; this is not needed unless one is defining a sequence recursively (see the next example).

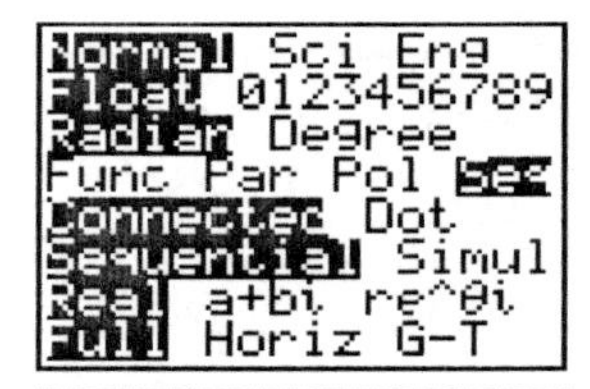


```
Plot1 Plot2 Plot3
nMin=1
u(n)=(-1)^n/2^n
u(nMin)=
v(n)=
v(nMin)=
w(n)=
```

[82] *On the TI-82, you may define up to two sequences, called* `Un` *and* `Vn`.

Once a sequence formula has been entered, it can be referenced from the home screen, as the screen on the right illustrates.

```
seq(u,n,1,5)▸Fra
c
{-1/2 1/4 -1/8 ...
```

Scatter diagrams of sequences (like those shown in the text) can be produced fairly easily once a formula has been entered. For example, here are the appropriate window settings for producing a plot of the first 10 terms of a sequence. (The window variables not visible off the bottom of the screen are the usual `Ymin`, `Ymax`, and `Yscl`.) Note also that the graph style for the sequence being plotted should be set to `DOTTED` (see page 8).

```
WINDOW
nMin=1
nMax=10
PlotStart=1
PlotStep=1
Xmin=0
Xmax=10
↓Xscl=1
```

Section 11.1 Example 2 (page 822) Using a Recursion Formula

In Sequence mode, a recursion formula can be entered as shown on the screen on the right. The initial term of the sequence $a_1 = u(1)$ is entered on the `u(nMin)=` line.

```
Plot1 Plot2 Plot3
nMin=1
u(n)=2u(n-1)+1
u(nMin)=4
v(n)=
```

The TI-83 can also compute terms of such recursive sequences using the `Ans` storage variable. The screen on the right illustrates this approach: After entering the number 4, followed by the formula `2Ans+1`, pressing [ENTER] repeatedly computes successive terms of the sequence. Note that this approach does not lend itself to recursion formulas involving more than previous term (see the next paragraph), nor is it as useful for summations (see the next example).

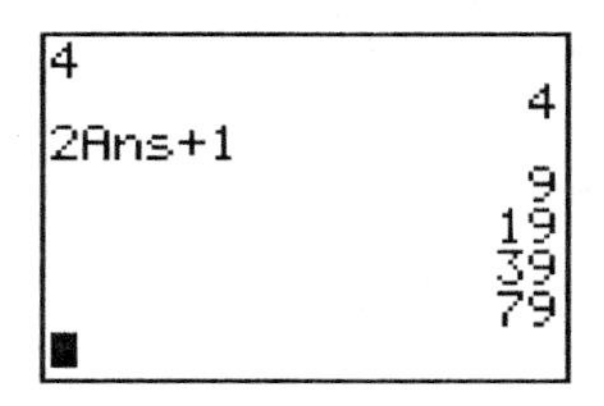

The Fibonacci sequence is described in the "For Discussion" section following this example in the text. It is typically defined recursively by $f_0 = f_1 = 1$, and $f_n = f_{n-1} + f_{n-2}$ for $n \geq 2$. The `Ans` method given above does not work to generate terms of this sequence, but in Sequence mode, the settings shown on the right in the [Y=] screen would allow easy computation of these terms. Note that the `u(nMin)=` line contains a list of the initial values.

```
Plot1 Plot2 Plot3
nMin=0
u(n)=u(n-1)+u(n
-2)
u(nMin)={1,1}
```

```
seq(u(n),n,0,6,1
)
{1 1 2 3 5 8 13}
```

Section 11.1 Example 4 (page 825) Using Summation Notation

The `sum` command found by pressing [2nd][STAT][◄][5]. `sum` can be applied to any list—either one of the built-in lists L1, ..., L6, or a list created with the `seq(` command, or a user-defined list.

Section 11.4 Technology Note (page 851) Computing Factorials

Section 11.4 Example 1 (page 852) Evaluating Binomial Coefficients

Section 11.6 Example 4 (page 865) Using the Permutations Formula

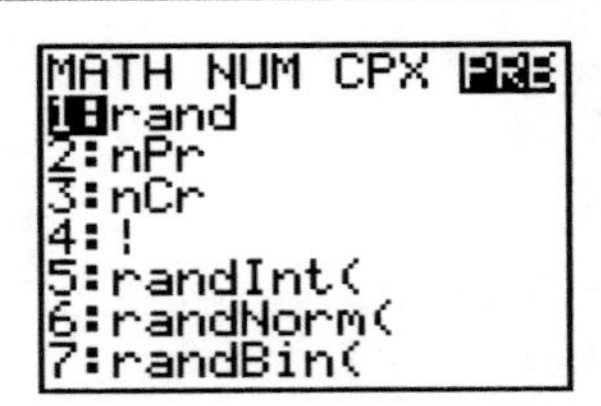

The factorial operator `!`, the combinations function `nCr`, and the permutations function `nPr` are found in the MATH:PRB menu ([MATH][◄]), shown here.

Section 11.7 Example 6 (page 878) Using a Binomial Experiment to Find Probabilities

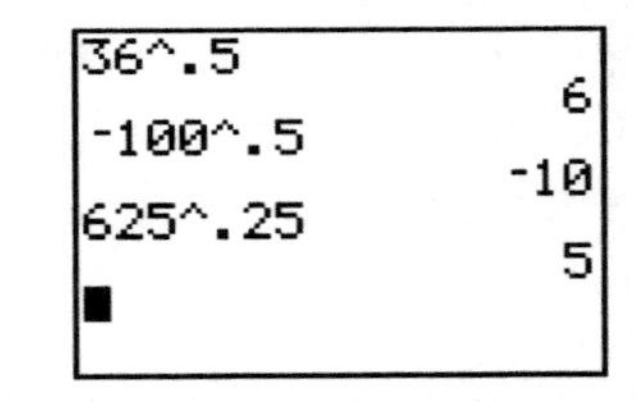

Statistical distribution functions are found by pressing [2nd][VARS]; `binompdf` is option 0 in this menu. As the text explains, the format for this function is

`binompdf(`*# of trials*`,`*prob. of success*`,`*# of successes*`)`

[82] *The TI-82 does not have these statistical distribution functions.*

Section R.4 Example 4 (page 916) Using the Definition of $a^{1/n}$

In evaluating these fractional exponents with a calculator, some time can be saved by entering fractions as decimals. The screen on the right performs the computations for (a), (b) and (c) with fewer keystrokes than entering, for example, `36^(1/2)`. (Of course, $36^{1/2} = \sqrt{36}$, which requires even fewer keystrokes.)

```
36^.5
                6
-100^.5
              -10
625^.25
                5
```

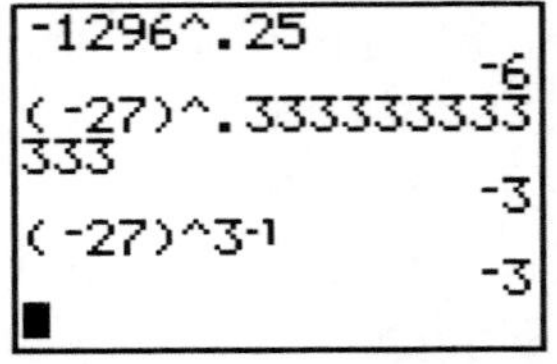

Note for (d), the parentheses around -1296 cannot be omitted: The text observes that this is not a real number, but the calculator will display a real result if the parentheses are left out. (See also the next example.) For (e), the decimal equivalent of $1/3$ is $0.\overline{3}$, which can be entered as a sufficiently long string of 3s (at least 12). The last line shows a better way to do this; the TI-83's order of operations is such that 3^{-1} is evaluated before the other exponentiation. This method would also work for the other computations; e.g., (a) could be entered [3][6][^][2][x^{-1}].

Section R.4 Example 5 (page 917) Using the Definition of $a^{m/n}$

For (f), attempting to evaluate $(-4)^{5/2}$ will produce either a `NONREAL ANS` error, or a complex number result, depending on the `Real a+bi re^θi` mode setting, described on page 4.

[82] *On the TI-82, the only possible result is a* `NONREAL ANS` *error, as the TI-82 does not support complex results.*

Appendix B Example 3 (page 934) Using the SSE Program

See section 13 of the introduction (page 8) for information about installing and running programs on the TI-83, and see page 11 for a description of how to enter data into the TI-83's lists.

Note that the program shown in Figure 2 can be significantly shortened to a single line:

```
PROGRAM:SSE
sum((L2-Y1(L1))2)
```

Appendix C Example 1 (page 937) Finding the Distance between Two Points in Space

The TI-83 does not have built-in support for vectors, but programs might be found to perform tasks like this. See section 13 of the introduction (page 8) for information about installing programs in the TI-83.

Alternatively, the coordinates of the points P and Q can be stored in *lists*. These are sets of numbers enclosed in braces { and }, typed with [2nd][{] and [2nd][}], and named L1, L2, . . . , L6 on the TI-83, typed as [2nd][1], [2nd][2], . . . , [2nd][6]. The screen on the right shows how to compute $d(P, Q)$ using the sum command ([2nd] [STAT][◄][5]).

```
{2,-4,3}→L1:{4,7
,-3}→L2
            {4 7 -3}
(L1-L2)²
         {4 121 36}
√(sum(Ans
        12.68857754
```

Appendix C Example 4 (page 938) Performing Vector Operations

Vector operations might be performed with a program (see page 8), or the components of each vector can be stored in lists, which can then be used to do the desired operations. Shown are the commands to store **v** in L1 and **w** in L2, and then to compute (a) $\mathbf{v} + \mathbf{w}$ and (e) $\mathbf{v} \cdot \mathbf{w}$.

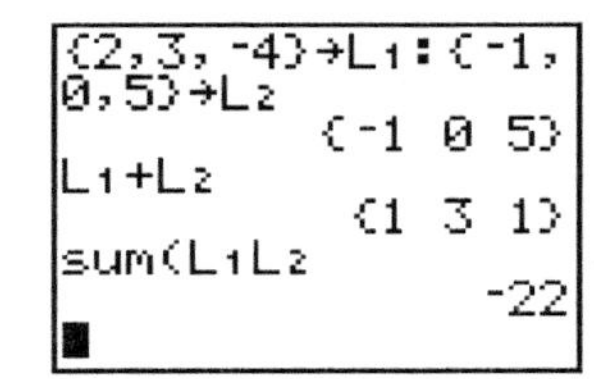

Appendix: Simulating complex numbers with a TI-82

The TI-82 does not support complex numbers; however, it can be made (using matrices) to add, subtract, multiply, and divide complex numbers in rectangular format. Here are the details:

Enter two matrices in the TI-82 by reproducing the screen on the right. The "[" character is [2nd][×], and "]" is [2nd][−]. Enter "[A]" as [MATRX][1], and "[B]" as [MATRX][2] (do not use [2nd][×] and [2nd][−] for these brackets). The screen is shown just before pressing [ENTER].

```
[[1,0][0,1]]→[A]
             [[1 0]
              [0 1]]
[[0,1][-1,0]]→[B
]■
```

The matrix [A] stands for "1", and [B] stands for "i." To enter $2-3i$, for example, type 2[A]-3[B] (**not** just 2-3[B]). Addition and subtraction are simply performed; the screen on the right shows the computation of the addition problem $(3+5i)+(6-2i)$. To obtain the answer—$9+3i$—simply read the first row of the resulting matrix; the first number is the real part, and the second is the imaginary part.

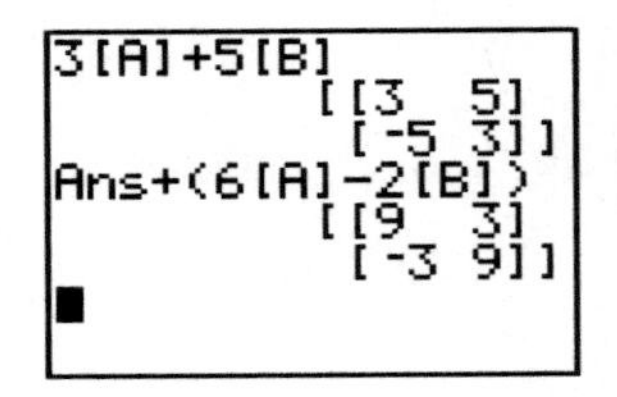

Multiplication is no more complicated than addition and subtraction. Shown below are the calculator entries and outputs for some sample multiplication problems.

```
(2[A]-3[B])(3[A]
+4[B])
        [[18 -1]
         [1  18]]
■
```

$(2-3i)(3+4i)$
$= 18 - i$

```
(5[A]-4[B])(7[A]
-2[B])
        [[27 -38]
         [38 27 ]]
■
```

$(5-4i)(7-2i)$
$= 27 - 38i$

```
(6[A]+5[B])(6[A]
-5[B])
        [[61 0 ]
         [0  61]]
■
```

$(6+5i)(6-5i)$
$= 61$

```
(4[A]+3[B])²
        [[7   24]
         [-24 7 ]]
■
```

$(4+3i)^2$
$= 7 + 24i$

For division problems, do not use the [÷] key. Instead, *multiply* by the inverse ([x^{-1}]) of the denominator.

```
(3[A]+2[B])(5[A]
-[B])⁻¹▸Frac
     [[1/2  1/2]
      [-1/2 1/2]]
■
```

$\frac{3+2i}{5-i} = \frac{1}{2} + \frac{1}{2}i$

```
(4[A]+2[B])(3[A]
-[B])⁻¹
        [[1  1]
         [-1 1]]
■
```

$\frac{4+2i}{3-i} = 1 + i$

```
3[A]*[B]⁻¹
        [[0 -3]
         [3 0 ]]
■
```

$\frac{3}{i} = -3i$

```
(2[A]+[B])(([A]-
2[B])^3)⁻¹▸Frac
 [[-4/25  -3/25]
  [3/25   -4/25]]
■
```

$\frac{2+i}{(1-2i)^3} =$
$-\frac{4}{25} - \frac{3}{25}i$

Chapter 2 The Texas Instruments TI-86 Graphing Calculator

Introduction

The information in this section is essentially a summary of material that can be found in the TI-86 manual. Consult that manual for more details.

While the TI-85 and TI-86 differ in some details, in most cases the instructions given in this chapter can be applied (perhaps with slight alteration) to a TI-85. The icon [85] is used to identify significant differences between the two, but some differences (e.g., a slight difference in keystrokes between the two calculators) are not noted. TI-85 users should watch for these comments.

1 Power

To power up the calculator, simply press the [ON] key. This should bring up the "home screen"—a flashing block cursor, and possibly the results of any previous computations that might have been done.

If the home screen does not appear, one may need to adjust the contrast (see the next section).

To turn the calculator off, press [2nd] [ON] (note that the "second function" of [ON]—written in yellow type above the key—is "OFF"). The calculator will automatically shut off if no keys are pressed for several minutes.

2 Adjusting screen contrast

If the screen is too dark (all black), decrease the contrast by pressing [2nd] then pressing and holding [▾]. If the screen is too light, increase the contrast by pressing [2nd] and then press and hold [▴].

As one adjusts the contrast, the numbers 1 through 9 will appear in the upper right corner of the screen. If the contrast setting reaches 8 or 9, or if the screen never becomes dark enough to see, the batteries should be replaced.

3 Replacing batteries

To replace the four AAA batteries, first turn the calculator off ([2nd] [ON]), then remove the back cover, remove and replace each battery, replace the back cover, then turn the calculator on again. (After replacing batteries, one may need to adjust the contrast down as described above.)

4 Basic operations

Simple computations are entered in essentially the same way they would be written. For example, to compute $2 + 17 \times 5$, press [2][+][1][7][×][5][ENTER] (the [ENTER] key tells the calculator to act on what has been typed). Standard order of operations (including parentheses) is followed.

```
2+17*5
                87
▮
```

The result of the most recently entered expression is stored in Ans, which is typed by pressing [2nd][(-)] (the word "ANS" appears in yellow above this key). For example, [5][+][2nd][(-)][ENTER] will add 5 to the result of the previous computation.

```
2+17*5
                87
5+Ans
                92
▮
```

After pressing [ENTER], the TI-86 automatically produces Ans if the first key pressed is one which requires a number before it; the most common of these are [+], [−], [×], [÷], [^], [x^2], and [STO▸]. For example, [+][5][ENTER] would accomplish the same thing as the keystrokes above (that is, it adds 5 to the previous result).

```
2+17*5
                87
5+Ans
                92
Ans+5
                97
▮
```

Pressing [ENTER] by itself evaluates the previously typed expression again. This can be especially useful in conjunction with Ans. The screen on the right shows the result of pressing [ENTER] a second time.

```
2+17*5
                87
5+Ans
                92
Ans+5
                97
               102
▮
```

Several expressions can be evaluated together by separating them with colons ([2nd][.]). When [ENTER] is pressed, the result of the *last* computation is displayed. The screen shown illustrates the computation $2(5 + 1)^2$.

```
3+2
                 5
Ans+1:Ans²:2 Ans
                72
▮
```

5 Cursors

When typing, the appearance of the cursor indicates the behavior of the next keypress. When the standard cursor (a flashing solid block, ■) is visible, the next keypress will produce its standard action—that is, the command or character printed on the key itself.

If [2nd][DEL] is pressed, the TI-86 is placed in INSERT mode and the standard cursor will appear as a flashing underscore. If the arrow keys ([▲], [▼], [▶], [◀]) are used to move the cursor around within the expression, and the TI-86 is placed in INSERT mode, subsequent characters and commands will be inserted in the line at the cursor's position. When the cursor appears as a block, the TI-86 is in DELETE (or OVERWRITE) mode, and subsequent keypresses will replace the character(s) at the cursor's position. (When the cursor is at the end of the expression, this is irrelevant.)

The TI-86 will return to DELETE mode when any arrow key is pressed. It can also be returned to DELETE mode by pressing [2nd][DEL] a second time.

Pressing [2nd] causes an arrow to appear in the cursor: ⬆ (or an underscored arrow). The next keypress will produce its "second function"—the command or character printed in yellow above the key. (The cursor will then return to "standard.") If [2nd] is pressed by mistake, pressing it a second time will return the cursor to standard.

Pressing [ALPHA] places the letter "A" in the cursor: ▣ (or an underscored "A"). The next keypress will produce the letter or other character printed in blue above that key (if any), and the cursor will then return to standard. Pressing [2nd][ALPHA] puts the calculator in lowercase ALPHA mode, changing the cursor to ▣ and producing the lowercase version of a letter. Pressing [ALPHA] twice (or [2nd][ALPHA][ALPHA]) "locks" the TI-86 in ALPHA (or lowercase ALPHA) mode, so that all of the following keypresses will produce characters until [ALPHA] is pressed again.

6 Accessing previous entries ("deep recall")

By repeatedly pressing [2nd][ENTER] ("ENTRY"), previously typed expressions can be retrieved for editing and re-evaluation. Pressing [2nd][ENTER] once recalls the most recent entry; pressing [2nd][ENTER] again brings up the second most recent, etc. The number of previous entries thus displayed varies with the length of each expression (the TI-86 allocates 128 bytes to store previous expressions).

[85] *The TI-85 allows access only to the last entry typed.*

7 Menus

Keys such as [TABLE], [GRAPH] and [2nd][7] (MATRX) bring up a menu line at the bottom of the screen with a variety of options. These options can be selected by pressing one of the function keys ([F1], [F2], . . . , [F5]). If the menu ends with a small triangle ("▸"), it means that more options are available in this menu, which can viewed by pressing [MORE]. Shown is the menu produced by pressing [2nd][×] (MATH).

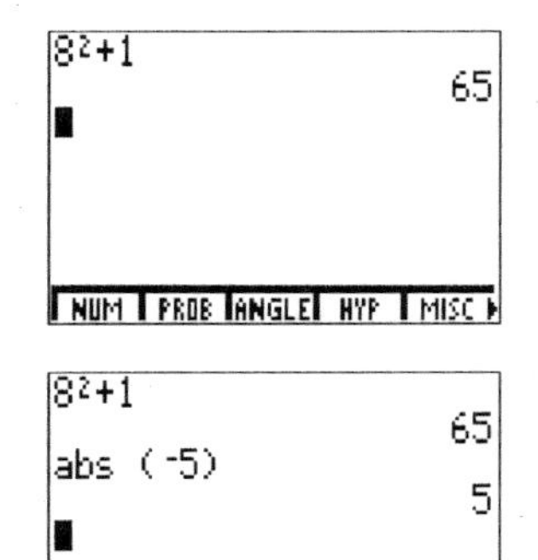

This screen shows the result of pressing [F1] (the "NUM" option, which lists a variety of numerical functions). Note that the MATH menu still appears (with NUM highlighted) and the bottom line now lists the functions available in this sub-menu—including, for example, the absolute value function (abs), which is accessed by pressing [F5]. The command line `abs (-5)` was typed by pressing [F5][(][(-)][5][)][ENTER].

This manual will use (e.g.) MATH:NUM to indicate commands accessed through menus like this. Sometimes the keypresses will be included as well; for this example, it would be [2nd][×][F1][F5].

The various commands in these menus are too numerous to be listed here. They will be mentioned as needed in the examples.

One last comment is worthwhile, however. Some functions that may be used frequently are buried several levels deep in the menus, and may take many keystrokes to access. Worse, the location of the function might be forgotten (is it MATH:NUM or MATH:MISC?), necessitating a search through the menus. It is useful to remember three things:

- Any command can be typed one letter at a time, in either upper- or lowercase; e.g., [ALPHA][ALPHA] [LOG][SIN][6][(-)] will type the letters "`ABS `", which has the same effect as [2nd][×][F1][F5].

- Any command can be found in the CATALOG menu ([2nd][CUSTOM][F1]). Since the commands appear in alphabetical order, it may take some time to locate the desired function. Pressing any letter key brings up commands starting with that letter (it is not necessary to press [ALPHA] first); e.g., pressing [LOG] brings up the list on the right, while pressing [,] shows commands starting with "P."

- Frequently used commands can be placed in the CUSTOM menu, and will then be available simply by pressing [CUSTOM]. To do this, scroll through the CATALOG to find the desired function, then press [F3] (CUSTM) followed by one of [F1]–[F5] to place that command in the CUSTOM menu. In the screen shown, [F1] was pressed, so that pressing [CUSTOM][F1] will type "Solver(." The commands in the MATH:ANGLE menu ([2nd][×][F3]), used frequently for problems in this text, could be made more accessible by placing them in this menu.

8 Variables

The uppercase letters A through Z, as well as some (but not all) lowercase letters, and also sequences of letters (like "High" or "count") can be used as variables (or "memory") to store numerical values. To store a value, type the number (or an expression) followed by [STO▸], then a letter or letters (note that the TI-86 automatically goes into ALPHA mode when [STO▸] is pressed), then [ENTER]. That variable name can then be used in the same way as a number, as demonstrated at right.

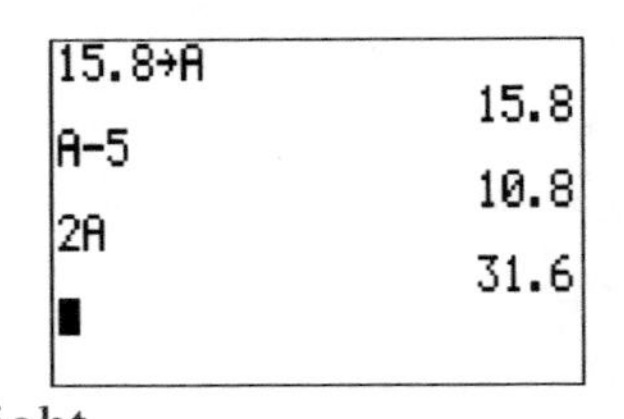

Note: The TI-86 interprets 2A as "2 times A"—the "*" symbol is not required (this is consistent with how we interpret mathematical notation). As for order of operations, this kind of multiplication is treated the same as "*" multiplication.

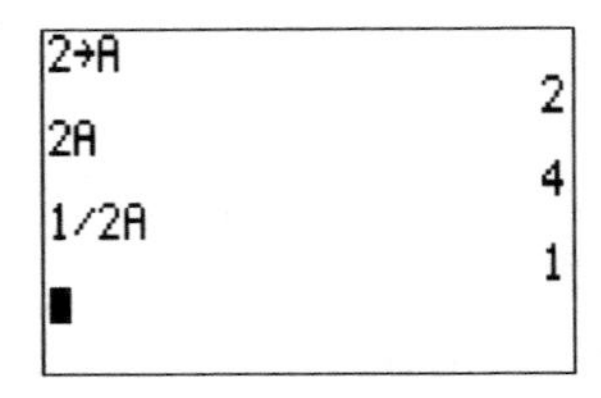

[85] *This latter comment is* **not** *true of the TI-85; on the TI-85, implied multiplication (such as* 2A*) is done before other multiplication and division, and even before some other operations, like the square root function* √. *Therefore, for example, the expression* 1/2A *is evaluated as* 1/4 *on the TI-85 (assuming that* A *is 2).*

9 Setting the modes

By pressing [2nd][MORE] (MODE), one can change many aspects of how the calculator behaves. For most of the examples in this manual, the "default" settings should be used; that is, the MODE screen should be as shown on the right. Each of the options is described below; consult the TI-86 manual for more details. Changes in the settings are made using the arrows keys and [ENTER].

```
12345
                12345
12345
             1.2345E4
12345
             12.345E3
▮
```

The Normal Sci Eng setting specifies how numbers should be displayed. The screen on the right shows the number 12345 displayed in Normal mode (which displays numbers in the range ±999, 999, 999, 999 with no exponents), Sci mode (which displays all numbers in scientific notation), and Eng mode (which uses only exponents that are multiples of 3). Note: "E" is short for "times 10 to the power," so 1.2345E4 $= 1.2345 \times 10^4 = 1.2345 \times 10000 = 12345$.

The Float 012345678901 setting specifies how many places after the decimal should be displayed (the 0 and 1 at the end mean 10 and 11 decimal places). The default, Float, means that the TI-86 should display all non-zero digits (up to a maximum of 12).

Radian Degree indicates whether angle measurements should be assumed to be in radians or degrees. (A right angle measures $\frac{\pi}{2}$ radians, which is equivalent to 90°.) This text does not refer to angle measurement.

RectC PolarC specifies whether complex numbers should be displayed in rectangular or polar format. These two formats are essentially the same as the two used by the textbook. **Note:** The text prefers the term "trigonometric format" rather than "polar format." More information about complex numbers can be found beginning on page 68 (Example 1 from Section 3.1) of this manual.

Func Pol Param DifEq specifies whether formulas to be graphed are functions (y as a function of x), polar equations (r as a function of θ), parametric equations (x and y as functions of t), or differential equations ($Q'(t)$ as a function of Q and t). The text accompanying this manual uses the first and third of these modes.

The RectV CylV SphereV setting indicates the default display format for vectors (not used in this text).

The other two mode settings deal with issues that are beyond the scope of the textbook, and are not discussed here.

A group of settings related to the graph screen are found by pressing [GRAPH][MORE] [F3] (GRAPH:FORMT). The default settings are shown in the screen on the right, and are generally the best choices for most examples in this book (although the last setting could go either way).

RectGC PolarGC specifies whether graph coordinates should be displayed in rectangular (x, y) or polar (r, θ) format. Note that this choice is independent of the Func Pol Param DifEq mode setting.

The CoordOn CoordOff setting determines whether or not graph coordinates should be displayed.

When plotting a graph, the DrawLine DrawDot setting tells the TI-86 whether or not to connect the individually plotted points. SeqG SimulG specifies whether individual expressions should be graphed one at a time (sequentially), or all at once (simultaneously).

GridOff GridOn specifies whether or not to display a grid of dots on the graph screen, while AxesOn AxesOff and LabelOff LabelOn do the same thing for the axes and labels (y and x) on the axes.

10 Setting the graph window

Pressing [GRAPH][F2] brings up the WINDOW settings. The exact contents of the WINDOW menu vary depending on whether the calculator is in function, polar, parametric, or DifEq mode; below are four examples showing this menu in each of these modes.

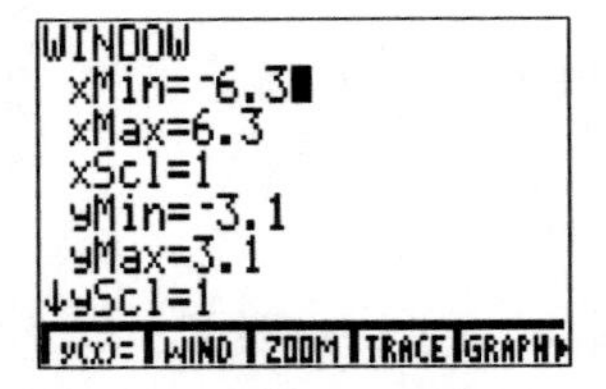

Function mode

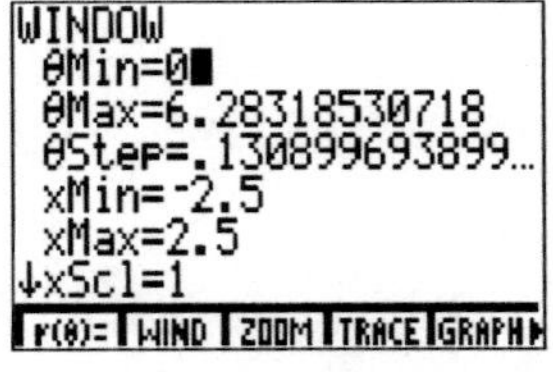

Polar mode

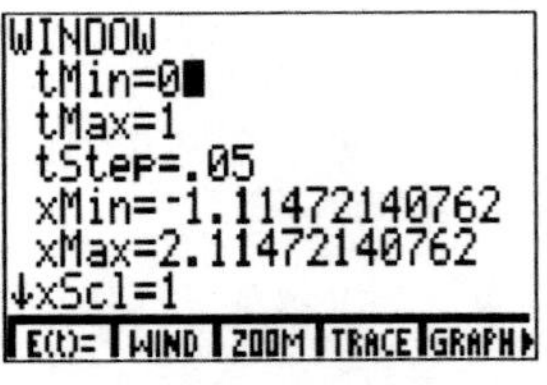

Parametric mode

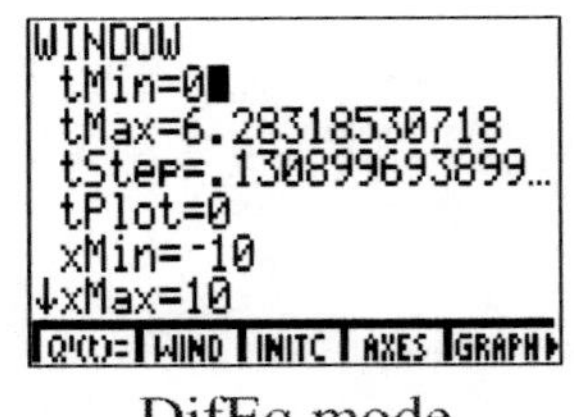

DifEq mode

[85] *On the TI-85, the* WINDOW *settings are called the* RANGE *settings.*

All these menus include the values `xMin`, `xMax`, `xScl`, `yMin`, `yMax`, and `yScl`. When [GRAPH][F5] (GRAPH) is pressed, the TI-86 will show a portion of the Cartesian (x-y) plane determined by these values. In function mode, this menu also includes `xRes`, the behavior of which is described in section 12 of this manual (page 56). The other settings in this screen allow specification of the smallest, largest, and step values of θ (for polar mode) or t (for parametric mode), or initial conditions for the differential equation.

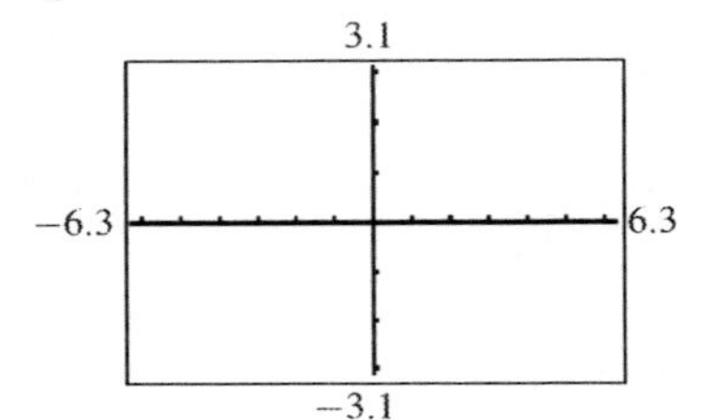

With settings as in the "Function mode" screen shown above, the TI-86 would display the screen at right: x values from -6.3 to 6.3 (that is, from `xMin` to `xMax`), and y values between -3.1 to 3.1 (`yMin` to `yMax`). Since `xScl` $=$ `yScl` $= 1$, the TI-86 places tick marks on both axes every 1 unit; thus the x-axis ticks are at -6, $-5, \ldots, 5$, and 6, and the y-axis ticks fall on the integers from -3 to 3. This window is called the "decimal" window, and is most quickly set by pressing [GRAPH][F3] (ZOOM) [MORE][F4] (ZDECM).

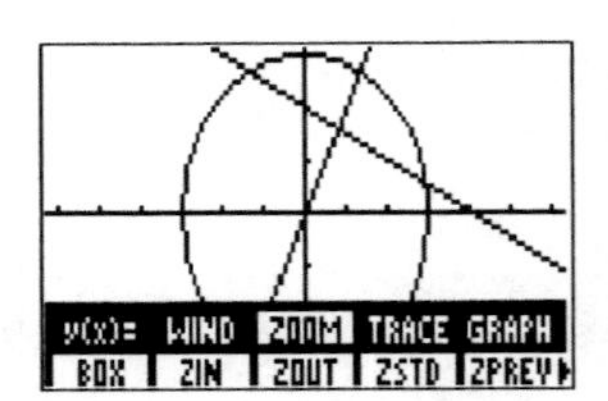

Note: If the graph screen has a menu on the bottom (like that shown on the right), possibly obscuring some important part of the graph, it can be removed by pressing [CLEAR]. The menu can be restored later by pressing [EXIT].

Below are four more sets of window settings, and the graph screens they produce. Note that the first graph on the left has tick marks every 10 units on both axes. The second window is called the "standard" viewing window, and is most quickly set by pressing [GRAPH][F3] (ZOOM) [F4] (ZSTD). The setting ySc1 = 0 in the final graph means that no tick marks are placed on the y-axis.

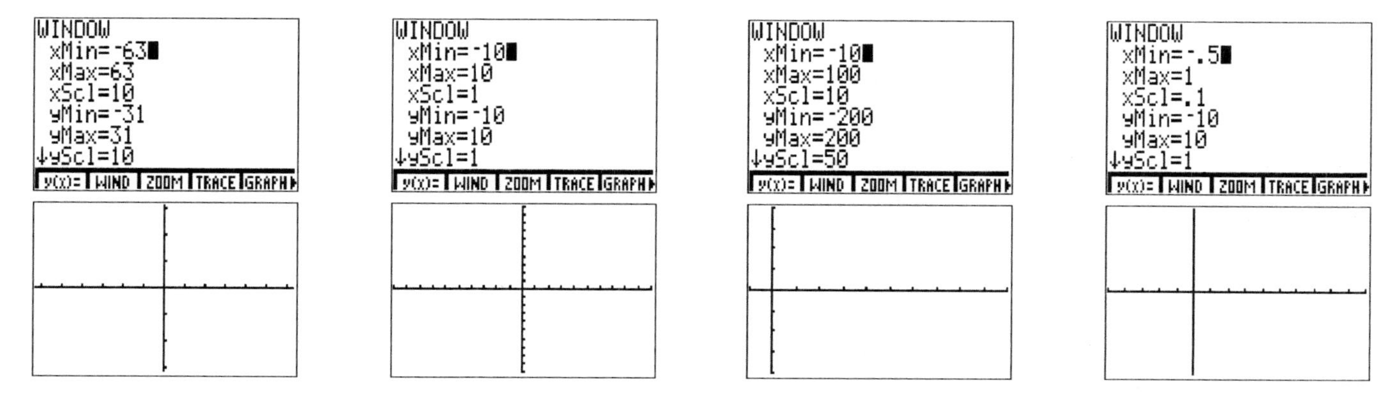

11 The graph screen

The TI-86 screen is made up of an array of rectangular dots (pixels) with 63 rows and 127 columns. All the pixels in the leftmost column have x-coordinate xMin, while those in the rightmost column have x-coordinate xMax. The x-coordinate changes steadily across the screen from left to right, which means that the coordinate for the nth column (counting the leftmost column as column 0) must be xMin $+ n\Delta$x, where Δx = (xMax − xMin)/126. Similarly, the nth row of the screen (counting up from the bottom row, which is row 0) has y-coordinate yMin $+ n\Delta$y, where Δy = (yMax − yMin)/62.

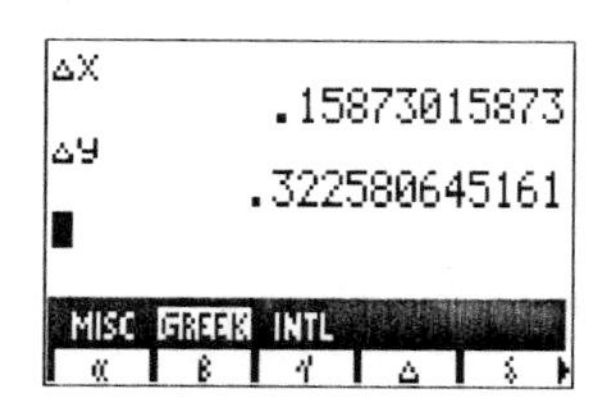

It is not necessary to memorize the formulas for Δx and Δy. Should they be needed, they can be determined by pressing [GRAPH][F5] and then the arrow keys. When pressing [▶] or [◀] successively, the displayed x-coordinate changes by Δx; meanwhile, when pressing [▲] or [▼], the y-coordinate changes by Δy. Alternatively, the values can be found by typing "Δx" and "Δy" on the home screen; this is most easily done by pressing [2nd][0][F2][F4] to access the CHAR:GREEK menu and type the "Δ" character, then typing lowercase x or y. This produces results like those shown on the right; the CHAR:GREEK menu remains on the bottom of the screen.

In the decimal window xMin = −6.3, xMax = 6.3, yMin = −3.1, yMax = 3.1, note that Δx = 0.1 and Δy = 0.1. Thus, the individual pixels on the screen represent x-coordinates −6.3, −6.2, −6.1, . . . , 6.1, 6.2, 6.3 and y-coordinates −3.1, −3, −2.9, . . . , 2.9, 3, 3.1. This is where the decimal window gets its name.

It happens that the pixels on the TI-86 screen are about 1.2 times taller than they are wide, so if Δy/Δx is approximately 1.2 (the exact value is 1.19565. . .), the window will be a "square" window (meaning that the scales on the x- and y-axes are equal). For example, the decimal window (with Δy/Δx = 1) is not square, so that one unit on the x-axis is not the same length as one unit on the y-axis. (Specifically, one y-axis unit is about 20% longer than one x-axis unit.)

Any window can be made square be pressing [GRAPH][F3] (ZOOM) [MORE][F2] (ZSQR). To see the effect of a square window, observe the two pairs of graphs below. In each pair, the first graph is on the standard window, and the second is on a square window (after choosing ZOOM:ZSQR). The first pair shows the lines $y = 2x - 3$ and $y = 3 - \frac{1}{2}x$; note that on the square window, these lines look perpendicular (as they should). The second pair shows a circle centered at the origin with a radius of 8. On the standard window, this looks like an oval since the screen is wider than it is tall. (The reason for the gaps in the circle will be addressed in the next section.)

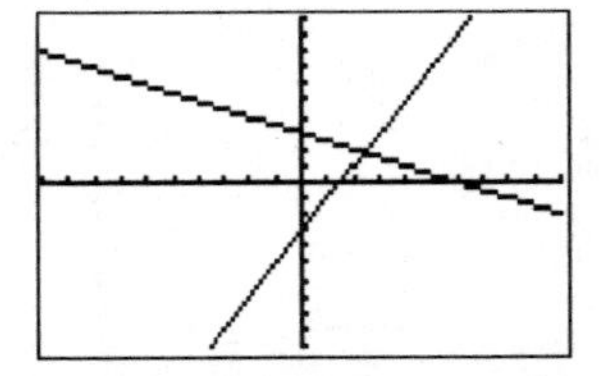
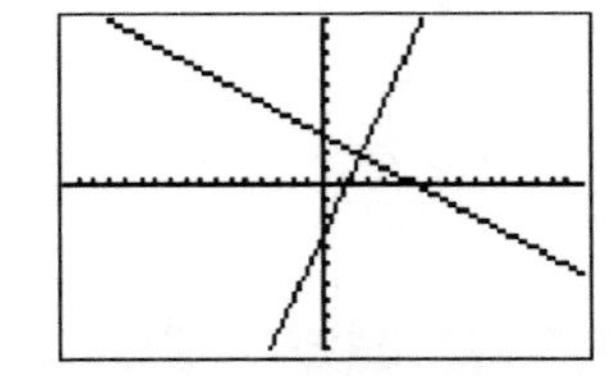
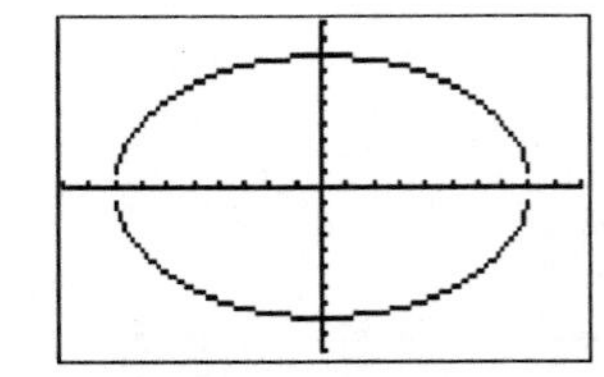
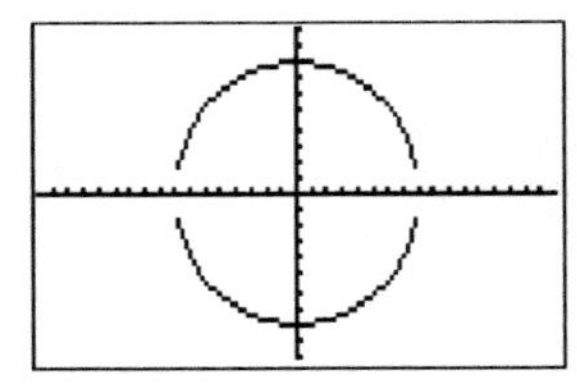

12 Graphing a function

This introductory section only addresses creating graphs in function mode. Procedures for creating parametric and polar graphs are very similar, and are described in this manual in the material related to the examples from the text.

To see the graph of $y = 2x - 3$, begin by entering the formula into the calculator. This is done by pressing [GRAPH][F1] to access the "y equals" screen of the calculator. Enter the formula as y1 (or any other yn); note that the letter x can be typed by pressing [F1] or [x-VAR] (as well as [2nd][ALPHA][+]). If another y variable has a formula, position the cursor on that line and press either [F4] (DELf—to delete the function) or [F5] (SELCT). The latter has the effect of toggling the "highlighting" for the equals sign "=" for that line (an "unhighlighted" equals sign tells the TI-86 not to graph that formula). In the screen on the right, only y1 will be graphed.

The next step is to choose a viewing window. See the previous section for more details on this. This example uses the standard window ([GRAPH][F3][F4]).

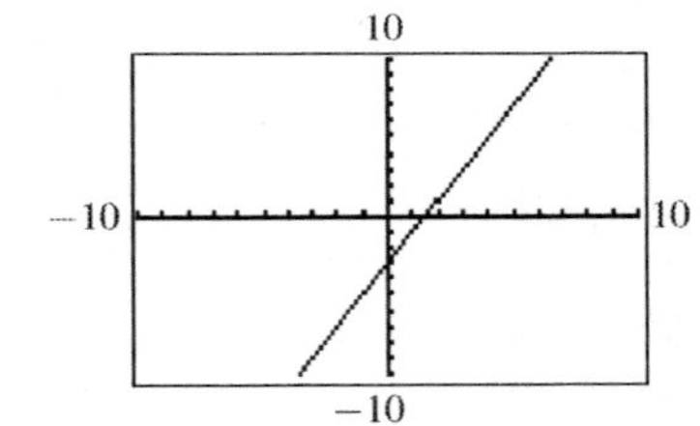

If the graph has not been displayed, press [GRAPH][F5], and the line should be drawn. In order to produce this graph, the TI-86 considers 127 values of x, ranging from xMin to xMax in steps of Δx (assuming that xRes = 1; see below for other possibilities). For each value of x, it computes the corresponding value of y, then plots that point (x, y) and (if the calculator is in Connected [DrawLine] mode) draws a line between this point and the previous one.

If xRes is set to 2, the TI-86 will only compute y for every other x value; that is, it uses a step size of 2Δx. Similarly, if xRes is 3, the step size will be 3Δx, and so on. Setting xRes higher causes graphs to appear faster (since fewer points are plotted), but for some functions, the graph may look "choppy" if xRes is too large, since detail is sacrificed for speed.

Note: If the line does not appear, or the TI-86 reports an error, double-check all the previous steps. Also, check the mode settings (discussed in section 9, page 52).

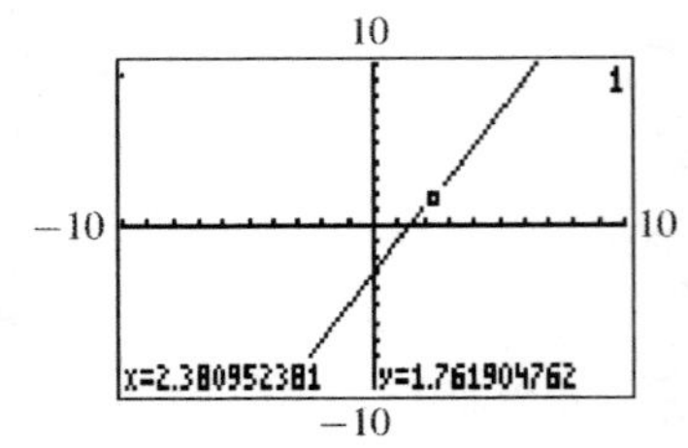

Once the graph is visible, the window can be changed using [F2] (WINDOW) or [F3] (ZOOM). Pressing [F4] (TRACE) brings up the "trace cursor," and displays the x- and y-coordinates for various points on the line as the [◄] and [►] keys are pressed. Tracing beyond the left or right columns causes the TI-86 to adjust the values of xMin and xMax and redraw the graph.

To graph the function

$$y = \frac{1}{x-3},$$

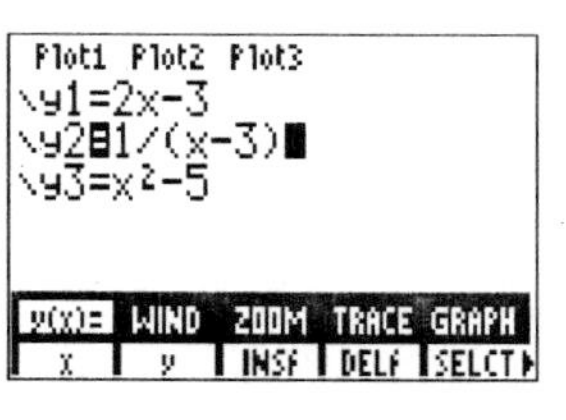

enter that formula into the "y equals" screen (note the use of parentheses). As before, this example uses the standard viewing window.

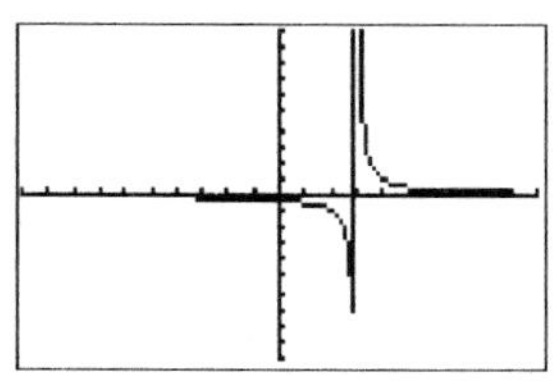

For this function, the TI-86 produces the graph shown on the right. This illustrates one of the pitfalls of the connect-the-dots method used by the calculator: The nearly-vertical line segment drawn at $x = 3$ *should not be there*, but it is drawn because the calculator connects the points

$$x = 2.85714,\ y = -6.99999 \ \text{and}\ x = 3.01587,\ y = 62.99999.$$

Calculator users must learn to recognize these flaws in calculator-produced graphs.

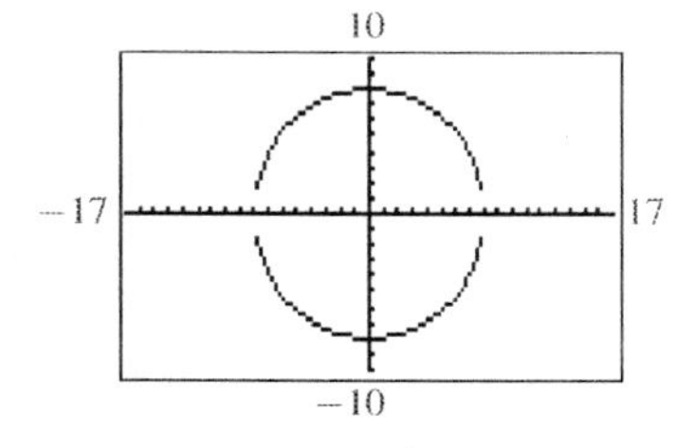

The graph of a circle centered at the origin with radius 8 (shown on the square window ZOOM:ZSTD - ZOOM:ZSQR) shows another problem that arises from connecting the dots. When $x = -8.093841$, y is undefined, so no point is plotted (that is, there is no point on this circle that has x-coordinate less than -8, or greater than 8). The next point plotted on the upper half of the circle is $x = -7.824046$ and $y = 1.668619$; since no point had been plotted for the previous x-coordinate, this is not connected to anything, so there appears to be a gap between the circle and the x-axis. The calculator is not "smart" enough to know that the graph should extend from -8 to 8.

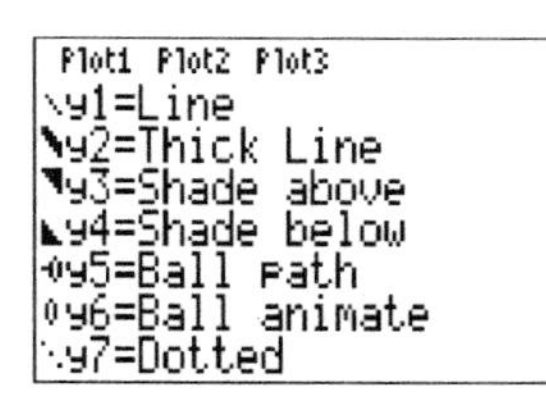

One additional feature of graphing with the TI-86 is that each function can have a "style" assigned to its graph. The symbol to the left of y1, y2, etc. indicates this style, which can be changed by choosing GRAPH:y(x)=:STYLE to cycle through the options. These options are shown on the right (with brief descriptive names); more information can be found in the examples, and complete details are in the TI-86 manual. [85] *The TI-85 does not include graph-style features.*

13 Adding programs to the TI-86

The TI-86's capabilities can be extended by downloading or entering programs into the calculator's memory. Instructions for writing a program are beyond the scope of this manual, but programs written by others and downloaded from the Internet (or obtained as printouts) can be transferred to the calculator in one of three ways:

1. If one TI-86 already has a program, it can be transferred to another using the calculator-to-calculator link cable. To do this, first make sure the cable is firmly inserted in both calculators. On the sending calculator, press [2nd][x-VAR] (LINK), then [F1]:[F2] (SEND:PRGM), and then select (by using the [▲] and [▼] keys and [F2]) the program(s) to be transferred. *Before* pressing [F1] (SEND) on the sending calculator, prepare the receiving calculator by pressing [2nd][x-VAR][F2], and *then* press [F1] on the sending calculator.

2. If a computer with the TI-Graph Link is available, and the program file is on that computer (e.g., after having been downloaded from the Internet), the program can be transferred to the calculator using the TI Connect (or TI Graph Link) software. This transfer is done in a manner similar to the calculator-to-calculator transfer described above; specific instructions can be found in the documentation that

accompanies the software. (They are not given here because of slight differences between platforms and software versions.)

3. View a listing of the program and type it in manually. (**Note:** Even if the TI-Graph Link cable is not available, the software can be used to view program listings on a computer.) While this is the most tedious method, studying programs written by others can be a good way to learn programming. To enter a program, start by choosing [PRGM][F2] (EDIT), then type a name for the new program (up to eight letters, like "`QuadForm`" or "`Midpoint`")—note that the TI-86 is automatically put into ALPHA mode. Then type each command in the program, and press [2nd][EXIT] (QUIT) to return to the home screen when finished.

To run the program, make sure there is nothing on the current line of the home screen, then press [PRGM][F1], select the program using one of the keys [F1]–[F5] and [MORE] (a sample screen is shown; only the first four to six characters of each program name are shown), and press [ENTER]. If the program was entered manually (option 3 above), errors may be reported; in that case, choose `GOTO`, correct the mistake and try again.

Programs can be found at many places on the Internet, including:

- `http://www.bluffton.edu/~nesterd`—the Web site of the author of this manual;
- `http://tifaq.calc.org`—a "Frequently Asked Questions" page maintained by Ray Kremer; and
- `http://www.ticalc.org`.

Examples

Here are the details for using the TI-86 for several of the examples from the textbook. Also given are the keystrokes necessary to produce some of the commands shown in the text's examples. In some cases, some suggestions are made for using the calculator more efficiently.

Throughout this section, it is assumed that the textbook is available for reference. The problems from the text are not restated here, and there are frequent references to the calculator screens shown in the text.

Section 1.1 Technology Note (page 3) — Viewing Windows

Information about setting viewing windows is given in section 10 of the introduction, page 54.

Section 1.1 Example 1 (page 5) — Finding Roots on a Calculator

The three calculator screens shown in Figure 11 illustrate the main methods of computing roots with the TI-86. Aside from fractional exponents, we have the built-in square root function ([2nd][x^2]) and the "x-root" function $^x\sqrt{}$, found in the MATH:MISC menu ([2nd][×][F5][MORE][F4]). The TI-86 does not have a cube-root function like that shown in Figure 11(b), although "3$^x\sqrt{}$ " accomplishes the same thing. It is worth noting that fourth roots can be typed in more efficiently using two square roots, since $\sqrt[4]{a} = a^{1/4} = (a^{1/2})^{1/2} = \sqrt{a^{1/2}} = \sqrt{\sqrt{a}}$.

Also, a quicker way to enter fractional exponents is to use [2nd][EE] (x^{-1}), as the screen on the right shows. The TI-86's order of operations is such that the reciprocal takes place before the exponentiation.

```
23^2-1
                4.79583152331
87^3-1
                4.43104762169
12^4-1
                 1.8612097182
▮
```

Section 1.2 Technology Note (page 18) — Producing Tables

To use the table features of the TI-86, begin by entering the formula ($y = 9x - 5$, in the example shown in Figure 26 of the text) on the GRAPH:y(x)= screen, as one would to create a graph. (The highlighted equals signs determine which formulas will be displayed in the table, just as they do for graphs.)

Next, press [TABLE][F2] to access the TABLE SETUP screen. The table will display y values for given values of x. The `TblStart` value sets the lowest value of x, while `ΔTbl` determines the "step size" for successive values of x. These two values are only used if the `Indpnt` option is set to `Auto`—this means, "automatically generate the values of the independent variable (x)." The effect of setting this option to `Ask` is illustrated at the end of this example.

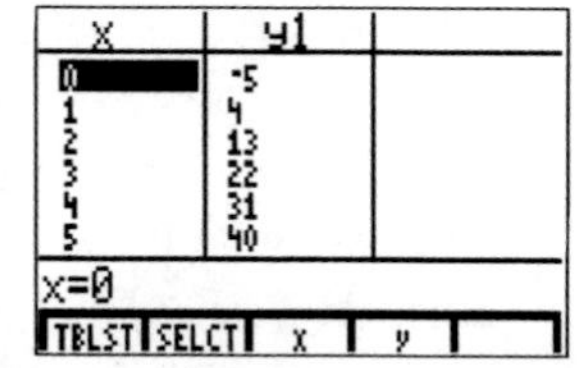
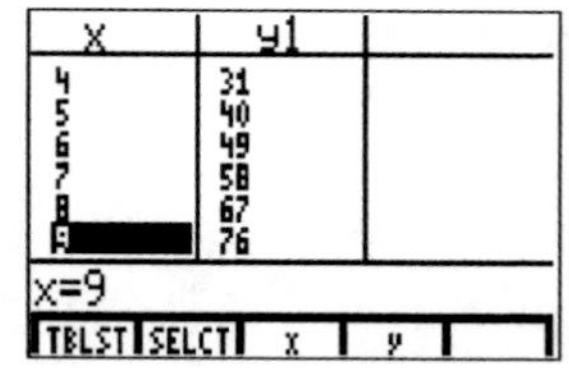

When the TABLE SETUP options are set satisfactorily, press [F1] (TABLE) to produce the table. Shown (above, right) is the table generated based on the settings in the above screen (this is almost identical to Figure 26 of the text). By pressing [▼] repeatedly, the x values are increased, and the y values updated. After pressing [▼] nine times, the table looks like the second screen. Similarly, the x values can be decreased by pressing [▲].

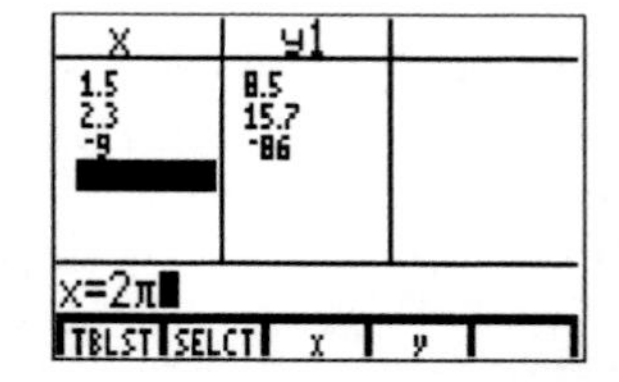

If `Indpnt` is set to `Ask` on the TABLE SETUP screen, the settings of `TblStart` and `ΔTbl` are ignored, and pressing [TABLE][F1] brings up a "blank" table. As values of x are entered, the y values are computed. Up to six x values can be entered; if more are desired, the [DEL] key can be used to make room. Note that one can enter expressions (like 2π or $\sqrt{3}$) in addition to "simple" numbers.

[85] *The TI-85 does not have a built-in table generator, but programs are available that can be used to simulate this feature. See section 13 of this chapter's introduction (page 57) for information about installing and running programs.*

Section 1.4 Example 5 (page 41) Using the Slope Relationship for Perpendicular Lines

See page 55 for information about setting square windows, including an illustration of perpendicular lines on square and non-square windows.

Section 1.4 Example 6 (page 42) Modeling Medicare Costs with a Linear Function

Given a set of data pairs (x, y), the TI-86 can produce a scatter diagram (as well as other types of statistical plots).

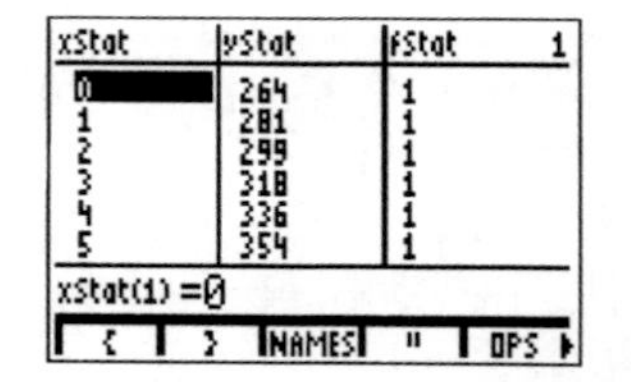

The first step is to enter the data into the TI-86. This is done by pressing [2nd][+] [F2] (STAT:EDIT), which should produce the screen on the right. If `xStat`, `yStat`, and `fStat` are not on the STAT:EDIT screen, and pressing [◄] and [►] does not reveal them, they can be restored to this screen by entering the command `SetLEdit` ("set up the list editor") on the home screen. This command is available by pressing [2nd][−][F5][MORE][MORE][MORE][F3].

In the STAT:EDIT screen, enter the year values (x) into the first column (`xStat`) and the cost (y) into the second column (`yStat`). The third column (`fStat`) should contain six "1"s; this tells the TI-86 that each of these (x, y) pairs occurs only once in the data list. (Actually, the values in `fStat` are ignored when creating scatter diagrams, but they are important for doing other analysis, like finding regression formulas [see the next example].) If any column already contains data, the [DEL] key can be used to delete numbers one at a time, or—to delete the whole column at once—press the [▲] key until the cursor is at the top of the column (on `xStat`, `yStat`, or `fStat`) and press [CLEAR][ENTER]. Make sure that all three columns contain the same number of entries.

To produce the scatter diagram, press [2nd][+][F3] to bring up the STAT:PLOT menu, shown on the right. Select `Plot1` by pressing [F1] (or choose one of the other two plots).

Make the settings shown on the screen on the right. For `Type`, choose [F1] (SCAT) for a scatter diagram.

Next, check that nothing else will be plotted: Press [GRAPH][F1] and make sure that only `Plot1` is highlighted. If `Plot2` or `Plot3` (or an equals sign) is highlighted, use the arrow keys to move the cursor to it, then press [F5] (SELCT).

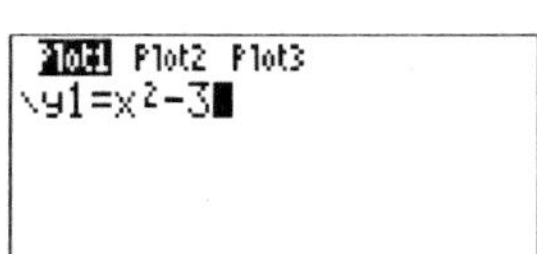

Finally, set up the viewing window (GRAPH:WINDOW) to match the one shown in Figure 58 of the text—or press [GRAPH][F3][MORE][F5] (GRAPH:ZOOM:ZDATA), which automatically adjusts the window to show all the data in the plot. (The resulting window does not quite match the one shown in Figure 58.)

Note: When finished with a statistics plot like this one, it is a good idea to turn it off so that the TI-86 will not attempt to display it the next time [GRAPH] is pushed. This can be done using the SELCT option on the GRAPH:y(x)= screen, or by executing the `PlOff` command, by pressing [2nd][+][F3][F5][ENTER].

[85] *The TI-85 can also produce a scatterplot, but the steps in the process are rather different from those described here. Consult your TI-85 owner's manual for these details.*

Section 1.4 Example 7 (page 43) Finding the Least-Squares Regression Line

Given a set of data pairs (x, y), the TI-86 can find various formulas (including linear, as well as more complex formulas) that approximate the relationship between x and y. These formulas are called "regression formulas."

The first step is to enter the data into the TI-86. See the previous example for a description of this process.

To find the linear regression formula, press [2nd][+][F1] (STAT:CALC). The bottom of the screen now lists the various options for the type of calculation to be done. Pressing [F3] tells the calculator to perform a "`LinR`"—a linear regression. (Don't confuse this with [F4]:`LnR`, a logarithmic regression.)

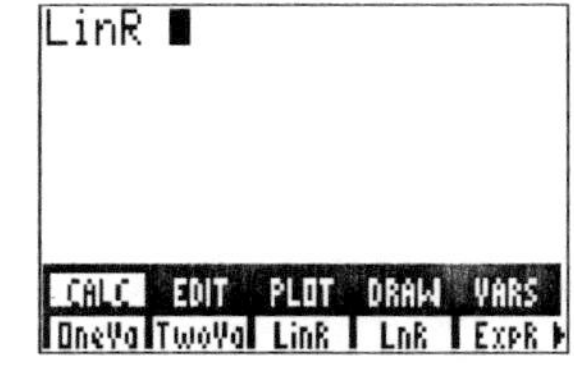

Pressing [ENTER] displays the results of the `LinR` (shown on the right). These numbers agree with those shown in Figure 61(b) of the text (but notice that the TI-86 reports the results using the format `y=a+bx` rather than `y=ax+b`).

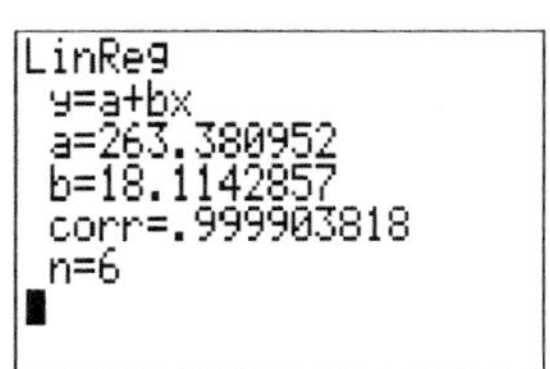

Following this example, the text introduces the idea of the correlation coefficient r. When the TI-86 finds a regression formula, it also computes r and reports it along with the formula (naming it "`corr`" rather than `r`).

[85] *The TI-85 can also perform regression computations, but the steps in the process are rather different from those described here. Consult your TI-85 owner's manual for these details.*

Section 1.5 Example 4 (page 54) Applying the Intersection-of-Graphs Method

We need to solve the equation $f(x) = g(x)$, where $f(x) = 5.91x + 13.7$ and $g(x) = -4.71x + 64.7$. We are looking for an x value that will make the left and right sides of this equation equal to each other, which corresponds to the x-coordinate of the point of intersection of the graphs of $y = f(x)$ and $y = g(x)$.

In order to have the TI-86 locate this intersection, begin by setting up the TI-86 to graph the left side of the equation as `y1`, and the right side as `y2`.

```
Plot1 Plot2 Plot3
\y1=5.91 x+13.7
\y2=-4.71 x+64.7
```

Next, select a viewing window which shows the point of intersection; we use the window shown in Figure 66(a) of the text: [0, 12] × [0, 100]. (The text also shows labels "CDS" and "TAPES" on the graphs. These are produced with the `Text` command; consult your calculator manual for more information.)

The TI-86 can automatically locate the point of intersection using GRAPH:MATH:ISECT ([GRAPH][MORE][F1][MORE][F3]). Use [▲], [▼] and [ENTER] to specify which two functions to use (in this case, the only two being displayed), and then use [◀] or [▶] to specify a guess. After pressing [ENTER], the TI-86 will try to find an intersection of the two graphs. The screens below illustrate these steps.

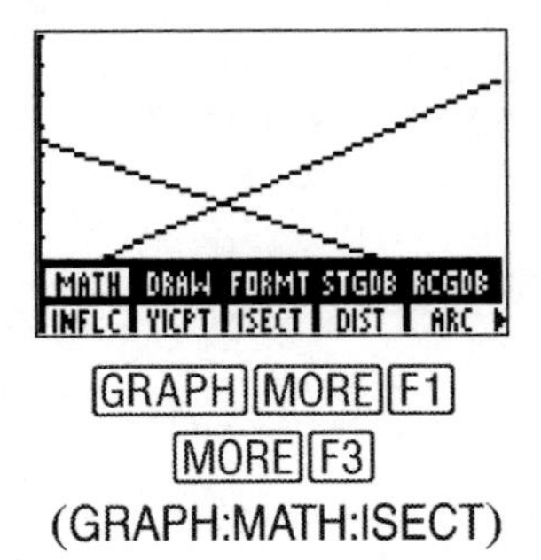

[GRAPH][MORE][F1] [MORE][F3] (GRAPH:MATH:ISECT)

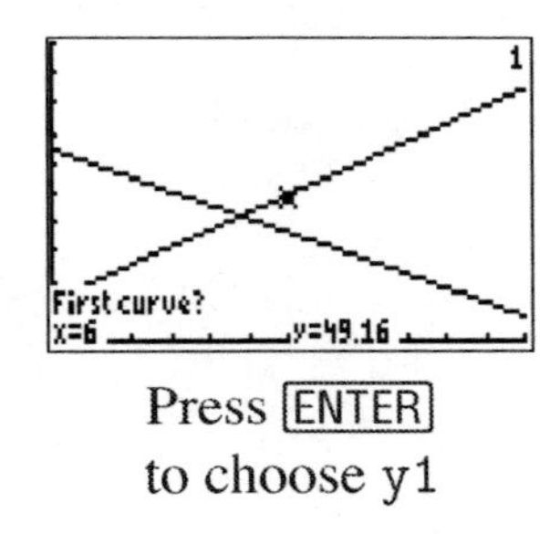

Press [ENTER] to choose `y1`

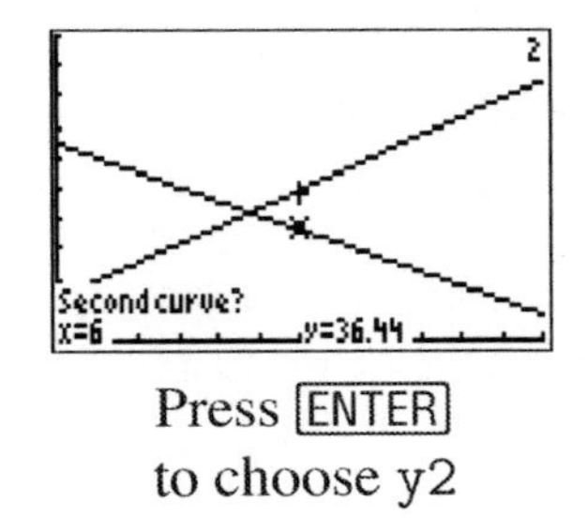

Press [ENTER] to choose `y2`

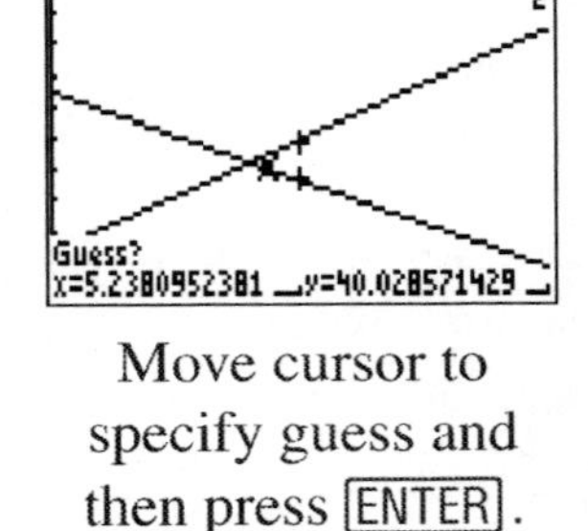

Move cursor to specify guess and then press [ENTER].

The final result of this process is essentially the same as the screen shown in Figure 66(b). The x-coordinate of this point of intersection is calculated to many digits of accuracy.

Also note that following this "intersection" procedure, the calculator variable `x` contains the x-coordinate of this intersection. This might be useful for performing computations with the solution; in the screen shown here it is used to confirm that f and g give identical output values at this x value.

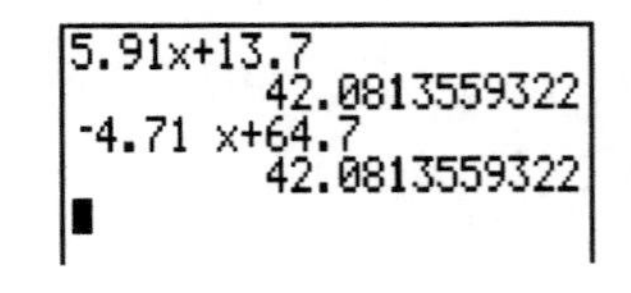

```
5.91x+13.7
              42.0813559322
-4.71 x+64.7
              42.0813559322
```

Note: An approximation for the point of intersection can be found simply by moving the TRACE cursor as near the intersection as possible. The amount of error can be minimized by "zooming in" on the graph. This is the only method available for graphing calculators such as the TI-81.

See the next example for another graphical approach to solving equations, as well as two other (non-graphical) approaches available on the TI-86.

Section 1.5 Example 5 (page 55) Using the *x*-Intercept Method

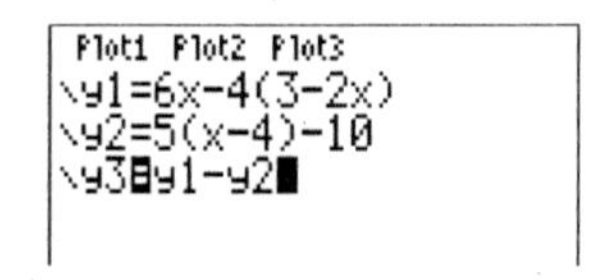

The text suggests graphing `y1=6x-4(3-2x)-(5(x-4)-10)`, noting the importance of placing parentheses around the subtracted expression. An equivalent approach is illustrated on the right: Define `y1` and `y2` using the expressions on the left and right sides of the equation, and then set `y3=y1-y2`. This method avoids the need for so many (potentially confusing) parentheses. The simplest way to type "y" is to press [F2]. Note that `y1` and `y2` have been "de-selected" so that they will not be graphed (see section 12 of the introduction, page 56).

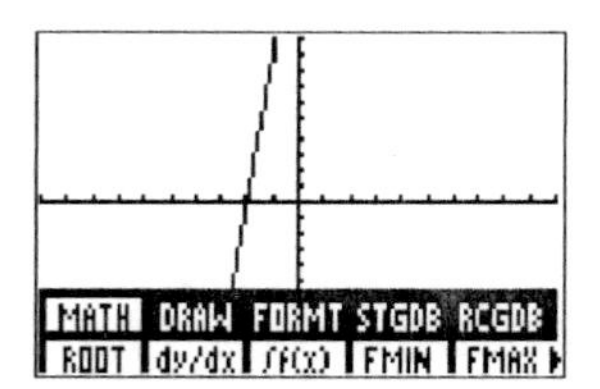

Next, select a viewing window which shows the x-intercept; we use the standard window, as shown in Figure 67(a) of the text. The TI-86 can automatically locate this point with the GRAPH:MATH:ROOT ([GRAPH][MORE][F1][F1]) feature ("root" is a synonym for "x-intercept"). The process is illustrated below: The TI-86 prompts for a left bound (less than the root), a right bound (greater than the root), and a guess (any number between the left and right bounds), and will then display the x-intercept ("root"). (Provided there is only one root between the bounds, and the function is "well-behaved"—meaning it has some nice properties like continuity—the calculator will find it.)

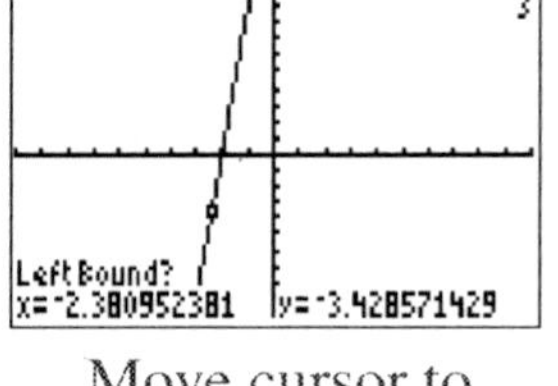

Move cursor to the left of the root, press [ENTER]

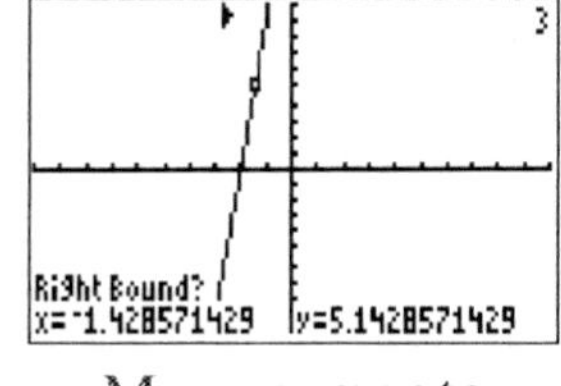

Move cursor to the right of the root, press [ENTER]

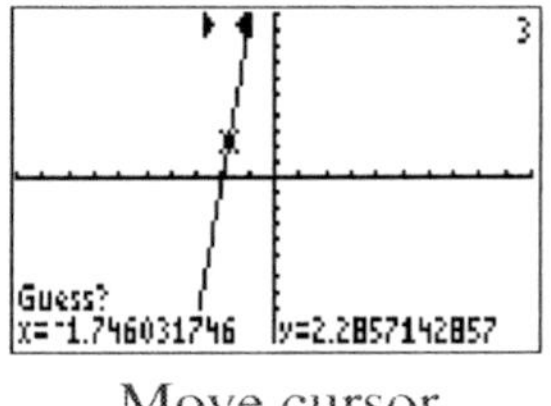

Move cursor close to the root, press [ENTER]

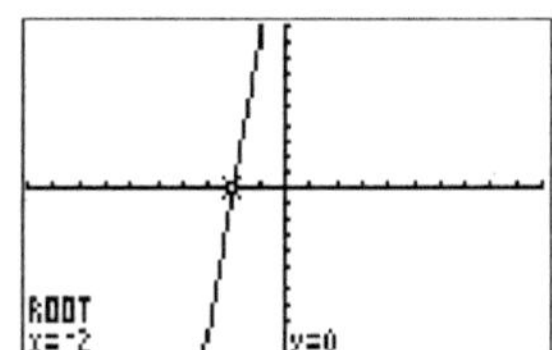

The TI-86 finds the root.

As with the intersection method, after the TI-86 locates the x intercept, the calculator variable `x` contains the x-coordinate.

[85] *The TI-85* GRAPH:MATH:ROOT *feature is* [GRAPH][MORE][F1][F3]. *Unlike the TI-86, the TI-85 does not require the user to identify left and right bounds for the root.*

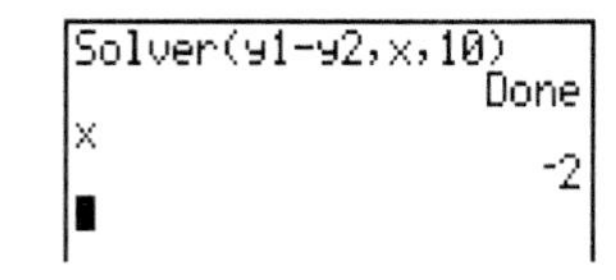

The TI-86 also offers some non-graphical approaches to solving this equation (or confirming a solution): As illustrated on the right, the TI-86's `Solver` function attempts to find a value of `x` that makes the given expression equal to 0, given a guess (10, in this case). The solution is stored in the variable `x`, but as the screen shows, this solution is not automatically displayed. The entry shown makes use of the fact that `y1` and `y2` have been defined as the left and right sides of this equation; if that had not been the case, the same results could have been attained by entering (e.g.) `Solver(6x-4(3-2x)-(5(x-4)-10),x,10)`. Full details on how to use this function (found in the CATALOG) can be found in the TI-86 manual.

Since it is difficult to access the `Solver` command, it might be a good idea to place `Solver` in the CUSTOM menu (see page 52), or to use the TI-86's built-in "interactive solver," found by pressing [2nd][GRAPH]. This prompts for the equation to be solved (use [ALPHA][STO▸] to type the equals sign), then allows the user to enter a guess for the solution (or a range or numbers between which a solution should be sought). To solve the equation, place the cursor on the line beginning with x= and press [F5].

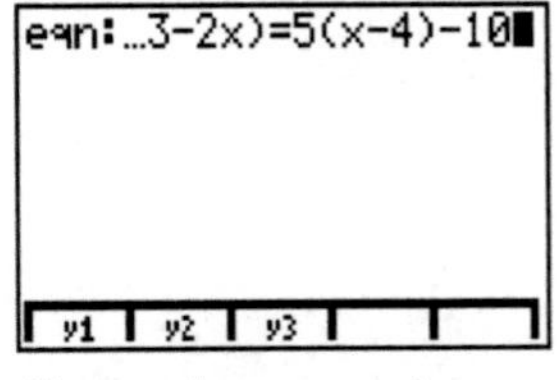

Enter the equation to be solved

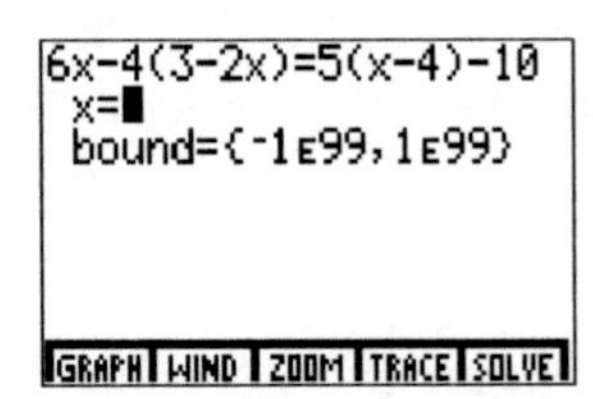

Specify a guess (optional), then press [F5]

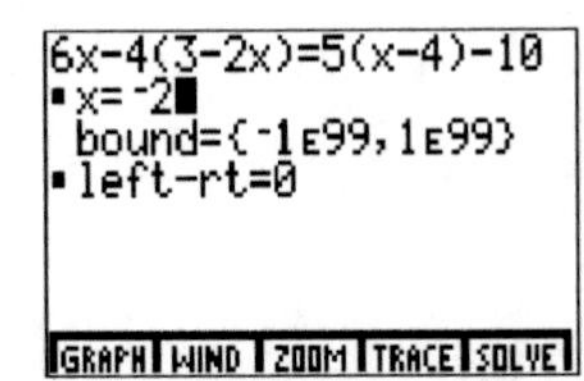

After a brief pause, the solution is found

The solver can also be used with equations containing more than one variable; simply provide values for all but one variable, then place the cursor on the line containing the variable for which a value is needed and press [F5].

Section 1.5 Example 8 (page 59) Using the Intersection-of-Graphs Method

In Figure 71, the text illustrates using a graph to support the solution $[-3, \infty)$ to the inequality $3x - 2(2x + 4) \le 2x + 1$, making the observation that solutions to this inequality correspond to those x-values for which the graph of $y = 3x - 2(2x + 4)$ *intersects or is below* the graph of $y = 2x + 1$. This connection between "<" and "below" (or ">" and "above") is an important one, and students should strive to understand it. However, it can sometimes be confusing, especially when one is just learning it, and the following graphical approach may be useful.

To solve (or confirm the solution of) an inequality like this, enter the formula y1=3x-2(2x+4)≤2x+1, where the "≤" symbol is found in the TEST menu ([2nd] [2]). When one of these symbols is included in an expression, the TI-86 responds with 1 if the statement is true, and 0 otherwise. Therefore, y1 will equal 1 for the values of x which satisfy the inequality, and 0 for all other values of x. With this understanding, one can observe the graph produced, and confirm that the solution is $x \ge -3$. (Care must be taken to determine whether or not -3 should be included in the solution set; the graph does not make that clear. This same observation is made in the Technology Note next to this example in the text.)

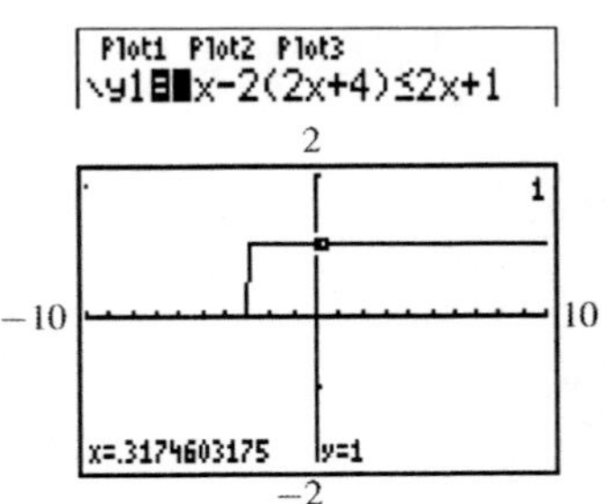

Section 1.5 Technology Note (page 60) Typing Function Variables

For the technique of defining y3=y1-y2, see the comments about Example 5 from Section 1.5 (page 63 of this manual).

Section 1.5 Example 11 (page 61) Solving a Three-Part Inequality

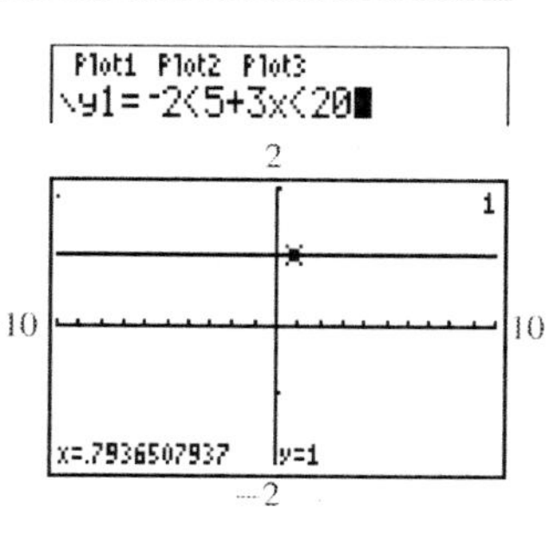

As was mentioned in the discussion of Example 8 from Section 1.5 (page 64 of this manual), the understanding that "<" and ">" go with "below" and "above" (respectively) is very important. The technique mentioned in that discussion can be applied to three-part inequalities, but some modification is needed. With single inequalities, one simply defines y1 by typing in the inequality. For double inequalities, this approach does not work. The results are shown on the right; they would lead one to think that any value of x (or at least any x between -10 and 10) satisfies the inequality.

Instead, one must split the double inequality into two single inequalities, defining either

`y1=(-2<5+3x)(5+3x<20),` or

`y1=-2<5+3x and 5+3x<20`

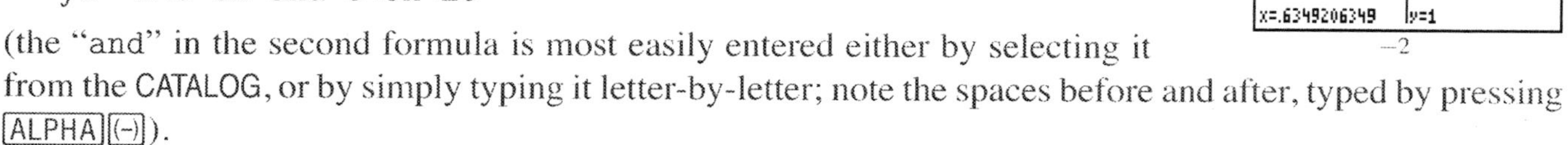

(the "and" in the second formula is most easily entered either by selecting it from the CATALOG, or by simply typing it letter-by-letter; note the spaces before and after, typed by pressing [ALPHA][(-)]).

Section 2.1 Technology Note (page 96) Rational Exponents

See page 59 of this manual for more information about roots and rational exponents.

Section 2.1 Technology Note (page 96) Absolute Values

The absolute value function is typed using [2nd][×][F1][F5]. To graph $y = |x|$, for example, define `y1=abs x`.

Section 2.1 Technology Note (page 98) Function and Parametric Modes

For information about selecting function and parametric modes, see section 9 of the introduction, page 52.

Section 2.2 Technology Note (page 103) Graphing Groups of Similar Functions
Section 2.3 Technology Note (page 113)
Section 2.3 Technology Note (page 114)
Section 2.3 Technology Note (page 115)
Section 2.3 Technology Note (page 117)

The screen shown on page 103 of the text illustrates one approach to graphing groups of similar functions: Set `y2=y1+3`, `y3=y1-2`, and `y4=y1+5`. This allows an entire "family" to be graphed by simply changing `y1`.

An alternative is to define `y1=x²+{0,3,-2,5}`. (The curly braces { and } are found in the LIST menu, [2nd][−]). When a list (like `{0,3,-2,5}`) appears in a formula, it tells the TI-86 to graph this formula several times, using each value in the list; therefore, this one definition will graph the four functions $y = x^2$, $y = x^2 + 3$, $y = x^2 - 2$, and $y = x^2 + 5$. Different families can be produced simply by changing `x²` to another function.

This list approach translates nicely to other types of transformations. For horizontal shifts, use, e.g., `y1=(x+{0,-3,-5,4})²`. For vertical stretches and shrinks, use `y1={1,2,3,4}x²`; the Technology Note on page 115 of the text illustrates a similar approach. For horizontal stretches and shrinks, use the approach shown in that Technology Note, or something like `y2=y1({2,0.5}x)`.

For reflections (page 117 of the text), use the approach shown in the text, or a variation of the one given above. For example, to graph $y = \sqrt{x}$ and $y = \sqrt{-x}$, for example, define `y1=√({1,-1}X)`.

Note that the usefulness of using lists to accomplish horizontal transformations (horizontal shifts and reflection across the y axis) is limited. For example, the graph of $y = \sqrt{x^2 - 3x}$—or any expression in which x appears more than once—is not conveniently reflected across the y-axis or horizontally shifted by this method. An adaptation of the method shown in the text would work better: E.g., define `y1=√(x²-3x)` and `y2=y1(x+{2,5,-3})` to see the graph of $f(x) = \sqrt{x^2 - 3x}$, $f(x + 2)$, $f(x + 5)$, and $f(x - 3)$.

[85] *The TI-85 does not support "function notation" like* `y2=y1(x+{2,5,-3})`*, but the same thing can be accomplished by defining* `y2=evalF(y1,x,x+{2,5,-3})`*. More information about the* `evalF` *function can be found on page 67.*

Section 2.5 Example 1 (page 139) Finding Function Values for a Piecewise-Defined Function

Section 2.5 Example 2 (page 139) Graphing a Piecewise-Defined Function

Recall that the inequality symbols $>$, $<$, $\geq$, $\leq$ are found in the TEST menu ([2nd][2]). The use of Dot mode (mentioned in the Technology Note) is not crucial to the graphing process, as long as one remembers that vertical line segments connecting the "pieces" of the graph (in this case, at $x = 0$) are not really part of the graph.

The discussion in the text for Example 2 shows how to enter this piecewise-defined function into the calculator, and also how the TI-86 evaluates such an expression. Note that the formula entered for `y1` is a fairly simple translation of the written definition of the function. For Example 1, e.g., we have

$$f(x) = \begin{cases} x+2 & \text{if } x \leq 0 \\ \frac{1}{2}x^2 & \text{if } x > 0 \end{cases}$$ becomes `y1=(x+2)(x≤0)+(1/2)x²(x>0)`.

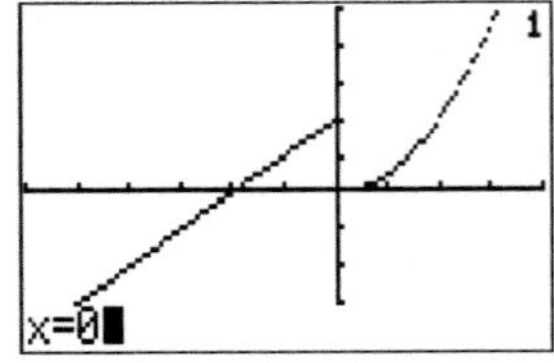

Figure 55 of the text shows that $y = 2$ when $x = 0$—a detail that is not evident from the graph (since the TI-86 does not show open and closed circles as does the graph shown in text). Similarly, the screen in Figure 57(b) shows that $y = 3$ when $x = 2$, using the trace feature, but note that on the window shown, one cannot get to $x = 2$ by pressing [◀] and [▶] to move the trace cursor. Instead, one must use the TI-86's "extended trace" feature, which allows one to trace to any x value between `xMin` and `xMax`. To do this, press [F4] (TRACE), then rather than pressing [◀] or [▶], type a number and press [ENTER]. The screen on the right shows what happens when one brings up the trace cursor and types [0] (just before pressing [ENTER]). This same result can be achieved using GRAPH:EVAL ([GRAPH][MORE][MORE][F1]).

[85] *This latter approach is also available on the TI-85.*

Extension: For more complicated piecewise-defined functions, a similar procedure can be used. Consider

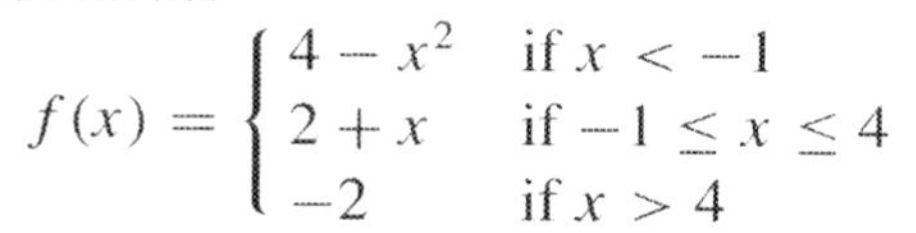

$$f(x) = \begin{cases} 4 - x^2 & \text{if } x < -1 \\ 2 + x & \text{if } -1 \le x \le 4 \\ -2 & \text{if } x > 4 \end{cases}$$

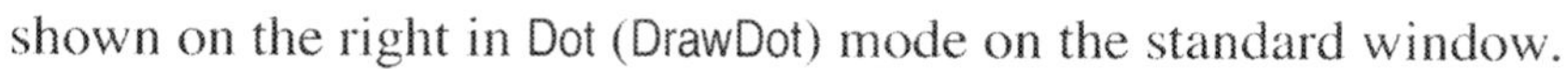

shown on the right in Dot (DrawDot) mode on the standard window.

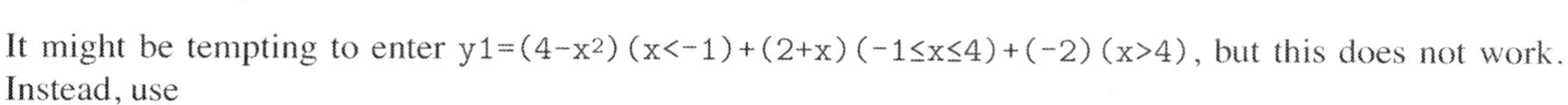

It might be tempting to enter `y1=(4-x²)(x<-1)+(2+x)(-1≤x≤4)+(-2)(x>4)`, but this does not work. Instead, use

```
y1=(4-x²)(x<-1)+(2+x)(-1≤x)(x≤4)+(-2)(x>4)
```

Section 2.5 Example 4 (page 141) Evaluating [[x]]

The `int` function is [2nd][×][F1][F4]. In Figure 59, note the use of lists to find [[x]] for all five values with one entry.

Do not confuse `int` with the `iPart` ("integer part") function, which is slightly different—specifically, `iPart(-6.5)` returns -6 while `int(-6.5)` gives -7.

Section 2.6 Example 1 (page 149) Using the Operations on Functions

Section 2.6 Example 2 (page 150) Using the Operations on Functions

In Example 1, the screens in the text (Figure 64) show how to evaluate function sums, differences, products, and quotients using the TI-86's function notation. The screen on the right shows an alternative way to compute $(f + g)(1)$, etc.; these entries make use of the fact that, when not followed by a number in parentheses), `y1` and `y2` are evaluated using the current value of `x`. The approach shown here has a slight advantage over that shown in the text because fewer changes are required from one entry to the next.

```
1→x:y1+y2
                10
-3→x:y1-y2
                14
5→x:y1*y2
               520
```

The TI-86 cannot give symbolic expressions like those required in Example 2. The entry `y1(x)+y2(x)` simply returns a number, using the current value stored in `x`. (In this case, `x` has the value 5.)

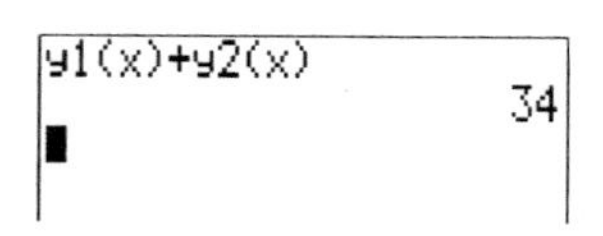

```
y1(x)+y2(x)
                34
```

[85] *The TI-85 does not support function notation; an entry like "`y1(-3)`" is interpreted as (the value of `y1` with the current value of `x`) times -3. However, the approach shown above works, as does the `evalF` ("evaluate formula") function, found in the* CALC *menu ([2nd][÷]). This which requires three arguments: the expression to be evaluated, the variable, and the value to be used in place of that variable. (The `evalF` function is also available on the TI-86, but function notation is generally more convenient.)*

```
evalF(y1+y2,x,1)
                10
evalF(y1-y2,x,-3)
                14
evalF(y1*y2,x,5)
               520
evalF nDer der1 der2 fnInt
```

Section 2.6 Example 5 (page 153) — Evaluating Composite Functions
Section 2.6 Example 6 (page 153) — Finding Composite Functions

The screens shown in Figure 68 illustrate how to evaluate composite functions at specific input values. Note that compositions of three or more functions can be accomplished just as simply: If y1, y2, and y3 are defined as the functions f, g, and h, one can evaluate $(f \circ g \circ h)(3)$ by typing y1(y2(y3(3))).

```
y2(y1(x))
                 987
```

The TI-86 cannot give symbolic expressions like those required in Example 6. The entry y2(y1(x)) simply returns a number, using the current value stored in x. (In this case, x has the value 5.)

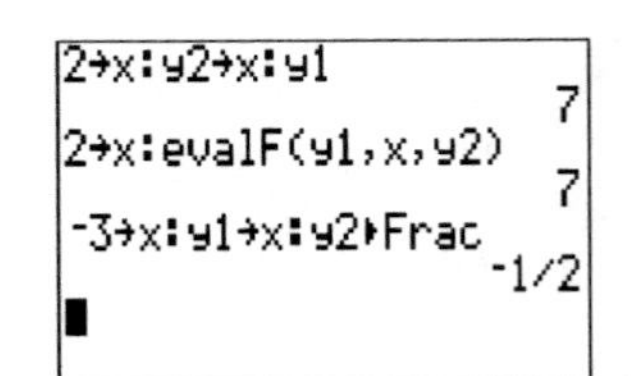

[85] *Because the TI-85 does not understand function notation, it would compute* y1(y2(2)) *as (the current value of* y1*) times (the current value of* y2*) times 2. These composite function values can be found by more roundabout approaches. The screen on the right shows two methods for 5(a), but only the shorter method for 5(b).*

Section 3.1 Technology Note (page 174) — Complex Number Mode

The TI-86 does perform complex computations, but it does not have a separate "complex number mode" like the one illustrated in this Technology Note; complex results are *always* allowed.

Section 3.1 Example 1 (page 175) — Writing $\sqrt{-a}$ as $i\sqrt{a}$
Section 3.1 Example 2 (page 176) — Finding Products and Quotients Involving $\sqrt{-a}$

```
√-16
                 (0,4)
√-80
      (0,8.94427191)
```

The TI-86 does these computations almost as shown in the text, except that it reports the results differently. The screen on the right illustrates the output from the computations for this example. The TI-86 uses the format (a, b) for the complex number $a + bi$.

```
(0,1)→i
                 (0,1)
i*√2
   (0,1.41421356237)
```

While the calculator always reports complex *results* in this format (or in "polar format"—see page 53), it can recognize *input* typed in using the text's format by first defining a variable i as the complex number $(0, 1) = 0 + 1i$. Shown on the right is a computation using this definition. One could also use I (uppercase), which can be typed with one less keystroke, but i will be used in these examples.

```
√-6*√-10
   (-7.74596669242,0)
√-48/√24
   (0,1.41421356237)
```

Note that whenever the TI-86 does a computation involving square roots of negative numbers, it reports the results as a complex number *even if the result is real*. Thus (-7.74596669242,0) represents the real number $-7.74596669242 \approx -2\sqrt{15}$, and (0,1.41421356237) is $i\sqrt{2}$.

Section 3.1 Example 3 (page 176) Adding and Subtracting Complex Numbers

Section 3.1 Example 4 (page 177) Multiplying Complex Numbers

Section 3.1 Example 5 (page 178) Simplifying Powers of *i*

Section 3.1 Example 6 (page 179) Dividing Complex Numbers

These computations are more easily done if the variable `i` is defined as `(0,1)`, as was described in the previous example.

Section 3.2 Example 3 (page 185) Using the Vertex Formula

The text identifies the vertex of the graph of $f(x) = -0.65x^2 + \sqrt{2}\,x + 4$ as a maximum at $(\sqrt{2}/1.3, 4 + 1/1.3) \approx (1.09, 4.77)$, and shows calculator screens that support those values. The TI-86 can automatically locate the vertex of the parabola (to a reasonable degree of accuracy) using the FMIN and FMAX commands in the GRAPH:MATH menu ([GRAPH][MORE][F1]).

Enter the function in `y1`. Graph in a window that shows the extreme value (such as the window shown in the text: $[-2, 4] \times [-2, 5]$). Since the coefficient of x^2 is negative, this is a parabola that opens down, and the extreme point is a maximum value, so we will use FMAX. The process is illustrated in the screens below: The TI-86 prompts for a left bound (less than the min/max), a right bound (greater than the min/max), and a guess (any number between the left and right bounds). When [ENTER] is pressed, the TI-86 will try to find a "peak" (or a "valley") on the graph, which is reported on the last screen.

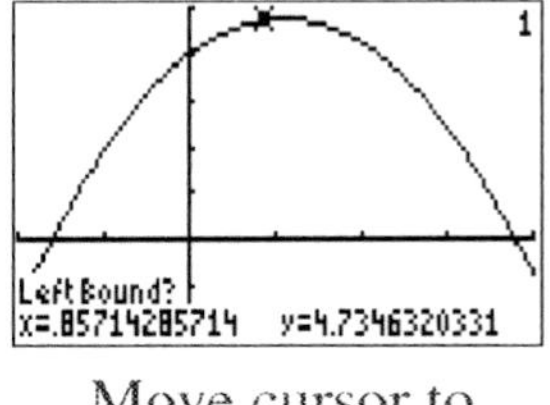

Move cursor to the left of the maximum, press [ENTER]

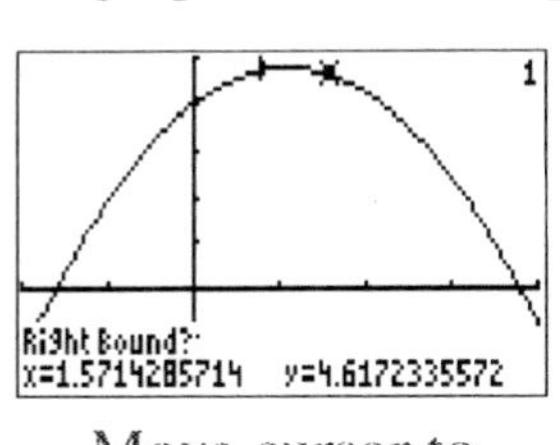

Move cursor to the right of the maximum, press [ENTER]

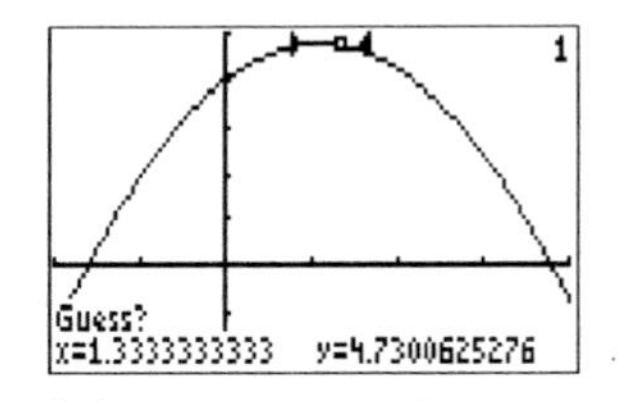

Move cursor close to the maximum (optional), press [ENTER]

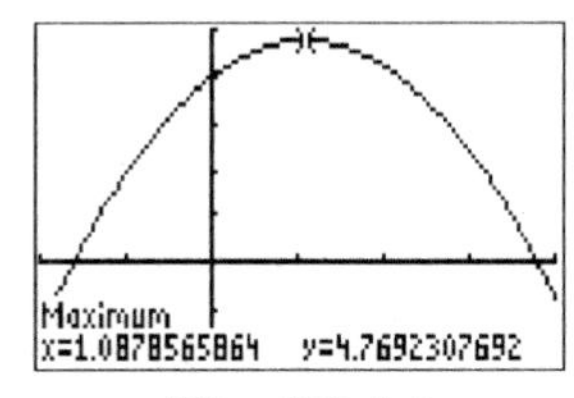

The TI-86 finds the maximum

[85] *On the TI-85,* FMIN *and* FMAX *are found by pressing* [GRAPH][MORE][F1][MORE]. *The TI-85 does not prompt for left and right bounds. Also, for this function, be sure to type either* `x√2` *or* `√2*x`, **not** `√2x`*; see the comments in section 8 of the introduction (page 52).*

After going through the process of locating a maximum or minimum, the calculator variables x and y contain the coordinates of the point. Depending on the window and the initial guess (and other factors), the x value may be off a bit from the exact answer (although for quadratic functions, the TI-86 is extremely accurate). For this function, the values of x and y agree (to as many decimal places as are displayed) with the exact coordinates of the maximum, but for more complicated functions, exact answers should not be expected.

```
x
                1.08785658644
√2/1.3
                1.08785658644
```

```
y
                4.76923076923
y1(x)
                4.76923076923
```

A limitation of the technology is that calculations are programmed to stop within a certain degree of accuracy.* (In other words, the TI-86 looks for the vertex until it decides that it is "close enough"; it will not always find exact answers.) It is important for the user to recognize this limitation for two reasons: First, do not report all digits displayed by the calculator, as they are not all reliable. Second, if the calculator reports a result of (say) 1.49999956, it is reasonable to guess that the exact answer might be 1.5.

*The degree of accuracy depends on the value of `tol` set on the TOLER screen ([2nd][3][F4]). The smaller the value of `tol`, the more accurate the results of this and similar computations.
[85] *On the TI-85, the* TOLER *settings are accessed using* [2nd][CLEAR].

Section 3.2 Example 5 (page 186) Identifying Extreme Points and Extreme Values

Figure 15 of the text illustrates the use of the `fMin` operation ([2nd][÷][MORE][F1]), which is similar to the `fMax` ([2nd][÷][MORE][F2]) operation. These two commands perform the same service as the GRAPH:MATH:FMIN and FMAX features described in the previous example (except that the calculator variables x and y are unaffected by these functions). The format is

`fMin(`*function*`,`*variable*`,`*low x*`,`*high x*`)`

Function can either be one of the function variables (e.g., `y1`) or an expression (like `4x2-18x+3`). *Variable* is usually `x`, but can be any other variable used in *function*. The other two parameters specify the smallest and largest values of `x` (or whatever variable is used) between which a minimum function value is sought. (This is especially needed when the function has many high or low points.)

As was noted with the graphical minimum/maximum locating procedures, be aware that the values returned by these procedures are typically approximations of the exact answers. Note in the screen on the right that the answer reported can vary depending on the specified low and high values of `x`. In both cases, the answer only approximates the exact answer (which, using calculus, can be shown to be $\sqrt{5/3}$).

```
fMin(x^3-5x-12,x,0,3)
          1.29099444734
fMin(x^3-5x-12,x,0,4)
          1.29099522253
█
```

Section 3.2 Example 6 (page 187) Modeling Hospital Spending with a Quadratic Function

See page 60 for instructions on creating scatter diagrams on the TI-86. The TI-86's `P2Reg` (polynomial-degree-2 regression) feature—found by pressing [MORE] in the STAT:CALC menu—can be used to find a quadratic function to approximate a set of data. The procedures for doing this are similar to those for a linear regression, described on page page 61 of this manual, except that the coefficients are reported in a list called `PRegC`. (Note, though, that the quadratic regression formula is not the same as the approximating function given in the text.)

Section 3.3 Technology Note (page 198) Calculator Programs

See section 13 of the introduction (page 57) for information about installing programs in the TI-86. (Also note the information in the following example.)

Section 3.7 Example 2 (page 242) Finding All Zeros of a Polynomial Function

```
POLY
order=4
```

The TI-86 has a built-in polynomial solver which can find all zeros (real and complex) of a polynomial, accessed through [2nd][PRGM] (POLY). This first prompts the user for the "order" (degree) of the polynomial, meaning the highest power of x; for this example, this is 4.

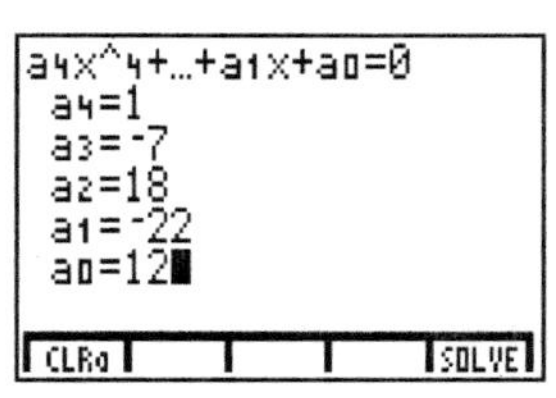

Pressing [ENTER] then brings up the screen on the right, requesting the coefficients of the equation. Note that the top line of the screen contains a reminder that the expression must be equal to 0. The menu at the bottom indicates that [F1] will clear the coefficients, while [F5] solves the equation.

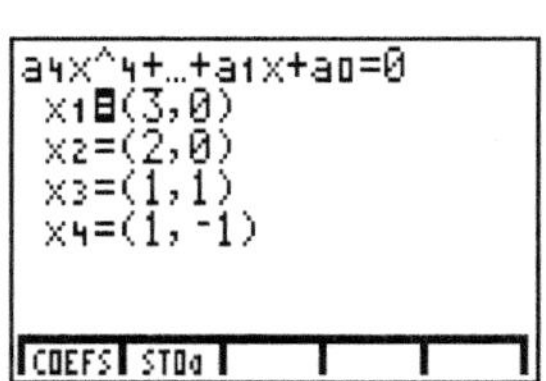

Pressing [F5] reports the solutions; note that since some of the zeros are complex, **all** solutions are reported in the TI-86's complex number format. [F1] allows the user to change the coefficient values (that is, it goes back to the previous screen), and [F2] provides a way to store the coefficients in a variable (as a list).

Section 3.8 Example 7 (page 255) Examining Polynomial Models for Debit Card Use

The procedures for creating scatter diagrams and performing regressions are described on pages 60 and 61 of this manual. Note that the first screen shown in Figure 88 includes the value "R^2", which is interpreted in a manner similar to the correlation described earlier (that is, values of R^2 close to 1 indicate a good "fit"). The TI-86 does not report this value.

Section 4.1 Example 2 (page 273) Graphing a Rational Function

Note that this function is entered as `y1=2/(x+1)`, **not** `y1=2/x+1`.

The issue of incorrectly drawn asymptotes is addressed in section 12 of this chapter's introduction (page 56). This asymptote can also be eliminated by setting the window to `xMin` $= -5$ and `xMax` $= 3$ (or any choice of `xMin` and `xMax` which has -1 halfway between them). Since this function is not defined at -1, it cannot plot a point there, and as a result, it does not attempt to connect the dots across the "break" in the graph.

Section 4.2 Example 8 (page 285) Graphing a Rational Function Defined by an Expression That Is Not in Lowest Terms

Figure 23 shows the graph of `y1=(x²-4)/(x-2)` on a variation of the decimal window, on which the "hole" in the graph can be seen. There are many possible windows on which the hole is visible; any window for which $x = 2$ is halfway between `xMin` and `xMax` would work. Likewise, there are many windows for which the hole would not be visible.

It is important to realize that some holes cannot be made visible. For example, take the graph of the function `y1=(x^4-4)/(x²-2)`—which looks like the function $y = x^2 + 2$, except at $x = \pm\sqrt{2}$. It is difficult (if not impossible) to find a window showing the holes at $x = \pm\sqrt{2}$.

Section 4.4 Example 5 (page 309) Modeling the Period of Planetary Orbits

Section 4.4 Example 6 (page 309) Modeling the Length of a Bird's Wing

The procedures for creating scatter diagrams are covered on page 60 of this manual. The "power regression" illustrated in Example 4 (Figures 47&48) is performed in a manner similar to linear regression (see page 61); on the TI-86, it is found in the the STAT:CALC menu ([2nd][+][F1][MORE][F1]).

Here is the output for the power regression performed on the data in Example 3. Note that this gives further confirmation that the formula $f(x) = x^{1.5}$ does a good job of modeling the relationship between average distance from the sun x and period of revolution y.

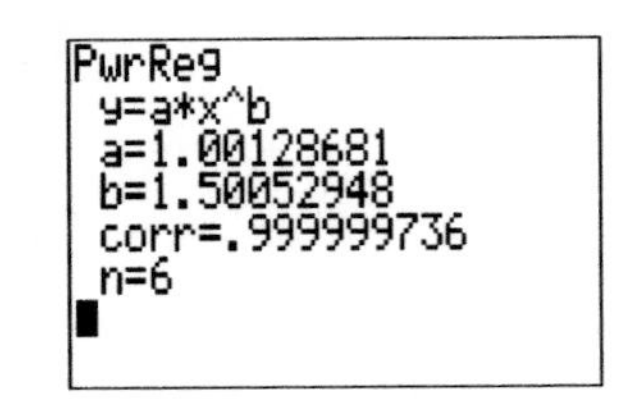

Section 4.4 Example 9 (page 313) Graphing a Circle

The text suggests graphing `y1=√(4-x²)` and `y2=-y1`. Here are two alternative ways to graph this circle; these approaches might be useful in other situations where one wishes to graph to or more complicated (but similar) functions:

- After typing the formula in `y1`, move the cursor to `y2`, press [(-)], then press [2nd][STO▸] (RCL) [F2][ALPHA] [1][ENTER]. This will "recall" the formula of `y1`, placing that formula in `y2`. This takes a few additional keystrokes, but can be a useful approach in cases where the second function to be graphed is similar to the first, but cannot easily be written in terms of `y1`.
- Enter the single formula `y1={-1,1}√(4-x²)`. (The curly braces { and } are found in the LIST menu, [2nd][-]). See page 65 for more information about this approach.

Section 5.1 Example 6 (page 345) Finding the Inverse of a Function with a Restricted Domain

The graph in Figure 8 shows the functions $f(x) = \sqrt{x+5}$ and $f^{-1}(x) = x^2 - 5,\ x \geq 0$. To produce a similar graph on the TI-86, enter the second function as a piecewise-defined function with only one "piece": Enter `y2=(x²-5)(x≥0)`, where ≥ is [2nd][2][F5]. (See page 66 for more about piecewise-defined functions.)

An even more accurate graph can be created by entering `y2=(x²-5)/(x≥0)`. (Note the division symbol in the middle.) This function is undefined (because of division by 0) whenever $x < 0$.

Section 5.2 Example 3 (page 354) Comparing the Graphs of $f(x)=2^x$ and $g(x)=(1/2)^x$

The labels $(-2, 4)$ and $(2, 4)$ on the calculator screen shown were produced with the `Text(` command. Consult your TI-86 manual for more information.

Section 5.2 Example 5 (page 355) Using Graphs to Evaluate Exponential Expressions

The two calculator screens shown in Figure 20 can be produced using the "extended trace" features of the TI-86, mentioned previously on page 66 of this manual. After putting the TI-86 in trace mode, simply type a number or expression (like √6 or -√2). Pressing [ENTER] causes the trace cursor to jump to that x-coordinate. This same result can be achieved using the GRAPH:MATH:EVAL feature.

[85] *This latter approach is also available on the TI-85.*

Section 5.3 Technology Note (page 365) Logarithms of Nonpositive Numbers

This Technology Note states that (under some circumstances) a calculator will give an error message when asked to compute (e.g.) `log 0` or `log -1`. It is true that `log 0` gives a `DOMAIN` error, but the TI-86 will compute logarithms of negative numbers, returning complex results. Shown on the right are the common logarithms of -1, -2, and -3. In general, if $x > 0$, then $\log(-x) = \log(x) + \log(-1)$, as the product rule for logarithms would suggest.

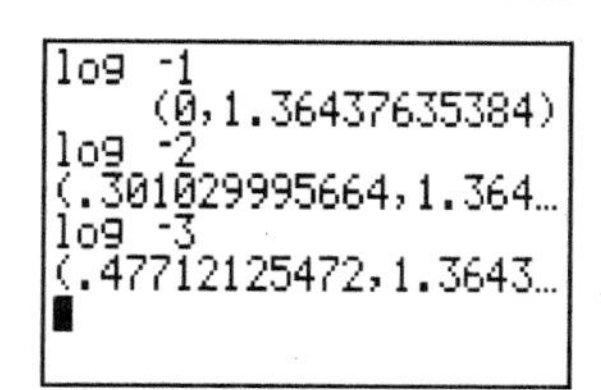

Section 5.3 Example 3 (page 365) Finding pH and [H_3O^+]

For (a), the text shows `-log(2.5*10^(-4))`, but this could also be entered as shown on the first line of the screen on the right, since "E" (produced with [EE]) and "*10^" are nearly equivalent. The two are not completely interchangeable, however; in particular, in part (b), "10^" **cannot** be replaced with "E", because "E" is only valid when followed by an *integer*. That is, `E-7` produces the same result as `10^-7`, but the last line shown on the screen produces a syntax error.

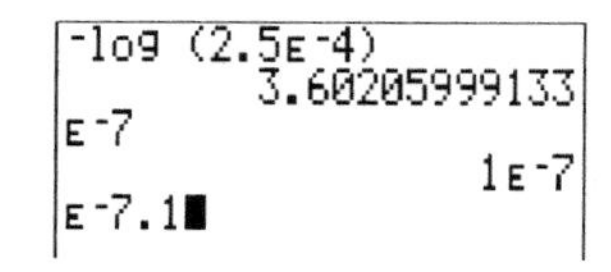

(Incidentally, "10^" is [2nd][LOG], but [1][0][^] produces the same results.)

Section 5.4 Technology Note (page 375) Asymptotes in Logarithmic Graphs

The calculator screen behavior referred to here is more evident when accompanied by a horizontal shift. Shown here (on the decimal window) is the graph of $y = \ln(x + 2)$; note that the graph seems to have an endpoint at about $(-1.9, -2.3)$. In reality, the graph has a vertical asymptote at $x = -2$, and that portion of the graph "goes down to $-\infty$"—that is, as $x \to -2$ from the right, $y \to -\infty$.

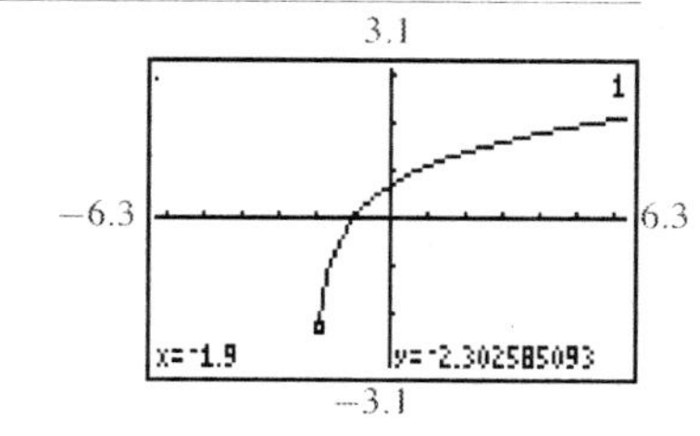

Section 5.6 Technology Note (page 398) Financial Calculations

The TI-86 does not have these finance menus. It may be possible to find a program to do this (see section 13 of the introduction, page 57), or the TI-86's Solver might be used to mimic this TI-83 feature.

If you have access to a TI-Graph Link (computer/calculator transfer cable), you may wish to install the finance package. While this package was originally produced by TI, it does not appear to be available at their web site, but it can be found at `www.ticalc.org`, as well as other places on the Internet (do a search for "TI-86 finance1"). Installation instructions are also available for download.

This package makes many financial functions available on your TI-86, including "TVM" ("time value of money") functions. With this package installed, the TI-86's TVM features are nearly equivalent to those found in the TI-83; see page 23 of this manual for information about how they work.

[85] *The finance package will not work on the TI-85.*

Section 5.6 Example 6 (page 399) Using Amortization to Finance an Automobile

The TI-86's Solver feature (see page 63) can be used to solve problems like this, and to explore some useful generalizations. The screens below illustrate the process. Note that it is a good idea to enter a "guess" for R before pressing [F5] (SOLVE).

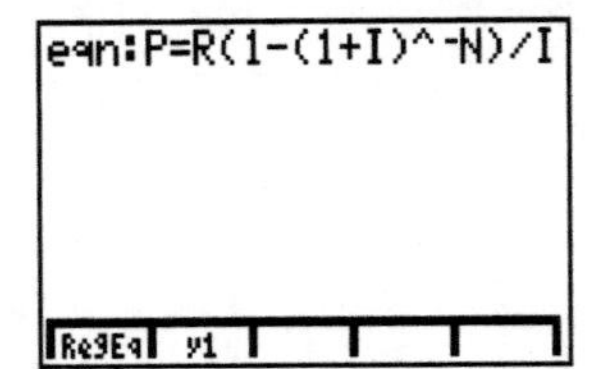

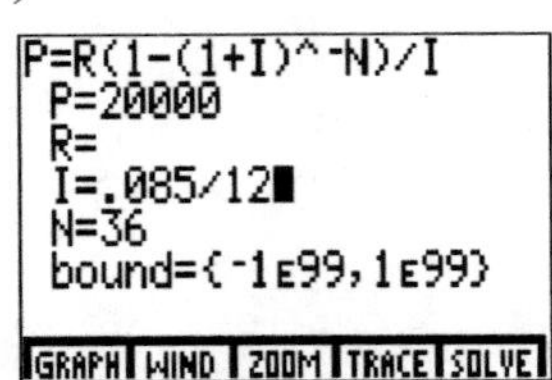

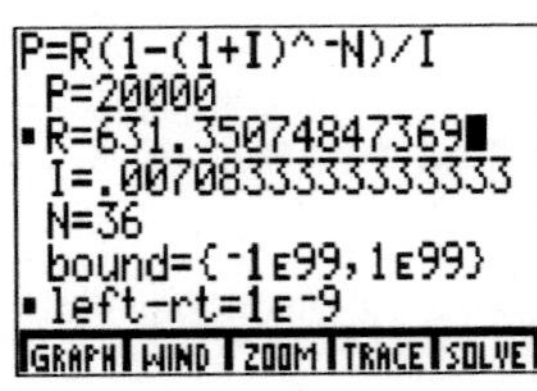

By going back and changing values of the variables, one can explore (for example) how long it would take to pay off the loan if one was making $700 monthly payments, or how big the payments should be if one makes two payments each month instead of one.

Section 5.6 Example 8 (page 400) Modeling Atmospheric CO_2 Concentrations

Section 5.6 Example 9 (page 401) Modeling Interest Rates

The procedures for creating scatter diagrams and performing regressions are described on pages 60 and 61 of this manual. Note that the screens in Figures 55(a) and 57(b) include the values `r²` and `r`; the TI-86 gives only the value of `r` (calling it `corr`).

Section 6.1 Example 3 (page 420) Graphing a Circle

For part (b), the text suggests graphing `y1=4+√(36-(x+3)²)` and `y2=4-√(36-(x+3)²)`. Here are three options to speed up entering these formulas (see also page 72):

- After typing the formula in `y1`, move the cursor to `y2` and press [2nd][STO▸] (RCL), then type `y1` [ENTER] (note that [F2] will produce the lowercase `y`). This will "recall" the formula of `y1`, placing that formula in `y2`. Now edit this formula, changing the first "+" to a "−."
- After typing the formula in `y1`, enter `y2=8-y1`. This produces the desired results, since $8 - \texttt{y1} = 8 - \left(4 + \sqrt{36 - (x+3)^2}\right) = 8 - 4 - \sqrt{36 - (x+3)^2} = 4 - \sqrt{36 - (x+3)^2}$.
- Enter the single formula `y1=4+{-1,1}√(36-(x+3)²)`. (The curly braces { and } are [F1] and [F2] in the LIST menu, [2nd][−]). When a list (like `{-1,1}`) appears in a formula, it tells the TI-86 to graph this formula several times, using each value in the list.

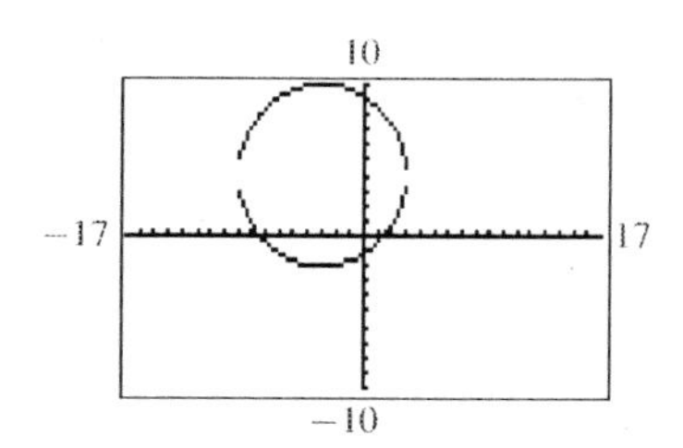

The window chosen in Figure 6 of the text is a square window (see section 11 of the introduction of this chapter), so that the graph looks like a circle. (On a non-square window, the graph would look like an ellipse—that is, a distorted circle.) Note, however, that on the TI-86, this window is not square. Other square windows would also produce a "true" circle, but some will leave gaps similar to those circles shown in sections 11 and 12 of the introduction (pages 55–56). Shown is the same circle on the square window produced by "squaring up" the standard window—GRAPH:ZOOM:ZSTD - ZOOM:ZSQR. The Technology Note next to this example shows a similar graph for the circle in part (a).

Section 6.4 Technology Note (page 454) — Parametric Mode

See the next example, as well as section 9 of the introduction (page 52), for information about selecting Parametric mode. The process of creating a graph in this mode is described in the next example.

Section 6.4 Example 1 (page 454) — Graphing a Plane Curve Defined Parametrically

Place the TI-86 in Parametric mode, as the screen on the right shows. In this mode, [GRAPH][F1] allows entry of pairs of parametric equations (x and y as functions of t). No graph is produced unless both functions in the pair are entered and selected (that is, both equals signs are highlighted).

```
Normal Sci Eng
Float 012345678901
Radian Degree
RectC PolarC
Func Pol Param DifEq
Dec Bin Oct Hex
RectV CylV SphereV
dxDer1 dxNDer
```

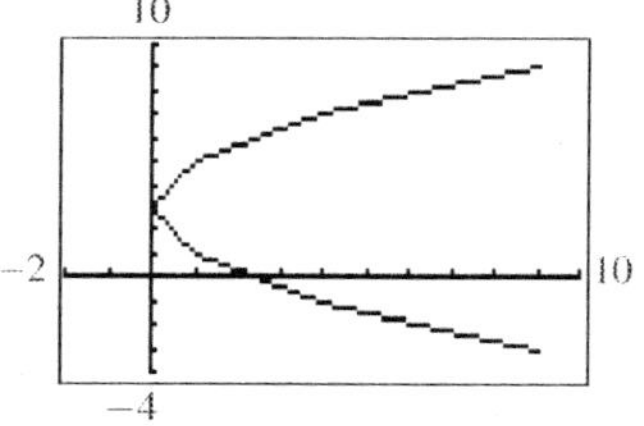

The text shows the window (GRAPH:RANGE) settings with `tMin=-3`, `tMax=3`, and `tStep=0.05`. The first two of these are specified in the example, but the value of `tStep` does not need to be 0.05, although that choice works well for this graph. Too large a choice of `tStep` produces a less-smooth graph, like the one shown on the right (drawn with `tStep=1`); note the angularity of the parabola near its vertex. Setting `tStep` too small, on the other hand, produces a smooth graph, but it is drawn very slowly. Sometimes it may be necessary to try different values of `tStep` to choose a good one.

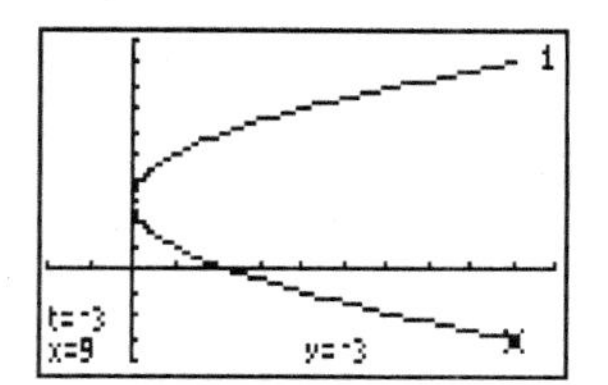

One can trace on a parametric graph, just as on a function-mode graph. The screen shown here is what appears when GRAPH:TRACE is first selected: The trace cursor begins at the (x, y) coordinate corresponding to `tMin`, and pressing [▶] increases the value of t (and likewise, [◀] decreases t). This can be somewhat disorienting, since for this graph, pressing [▶] moves the cursor to the *left*.

Section 7.1 Example 1 (page 468) — Solving a System by Substitution
Section 7.1 Example 2 (page 468) — Solving a System by Elimination

See page 62 for a description of the TI-86's intersection-locating procedure.

Section 7.1 Example 6 (page 472) Solving a Nonlinear System by Elimination

See the discussion related to Example 9 from Section 4.4 (page 72 of this manual) for tips on entering formulas like these.

In Figure 7, the text shows the intersections as found by the procedure built in to the calculator (described on page 62 of this manual, called "ISECT" on the TI-86). However, that is somewhat misleading; for these equations, the TI-86 can only find these intersections if the "guesses" supplied by the user are the exact x coordinates of the intersections (that is, -2 and 2). This is because the two circle equations are only valid for $-2 \le x \le 2$, while the hyperbola equations are only valid for $x \le -2$ and $x \ge 2$. Since only ± 2 fall in both of these domains, any guess other than these two values results in a `BAD GUESS` error.

Section 7.2 Example 1 (page 481) Solving a System of Three Equations in Three Variables

To solve systems of linear equations on the TI-86, use the built-in SIMULT feature, which can solve systems with up to 30 unknowns. Pressing [2nd][TABLE] ([85] *or* [2nd][STAT]) causes the TI-86 to prompt for the number of equations, after which it prompts for the coefficients, one equation at a time. The screens below show the first and third equations being entered; the arrow keys (or [F1] and [F2]) allow the user to move from one coefficient to another. When all the constants have been entered, pressing [F5] solves for the three unknowns (which the TI-86 calls x1, x2, and x3, rather than x, y, and z).

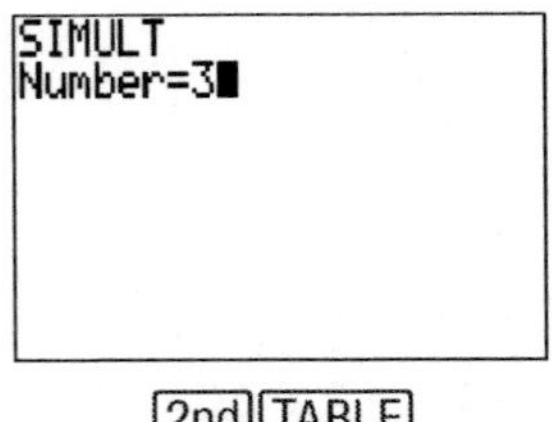

[2nd][TABLE] (or [2nd][STAT])

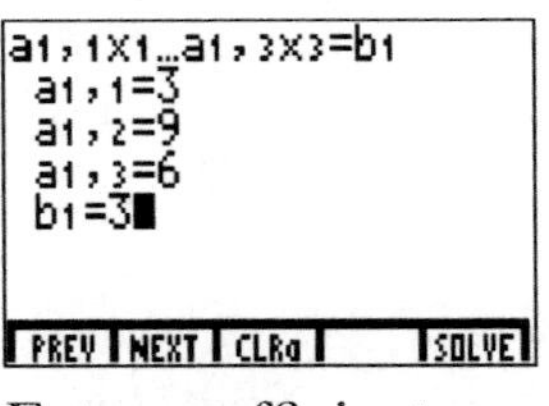

Enter coefficients. . .

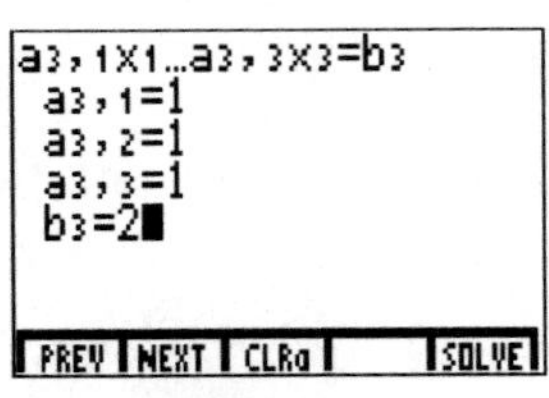

Press [F5] when done.

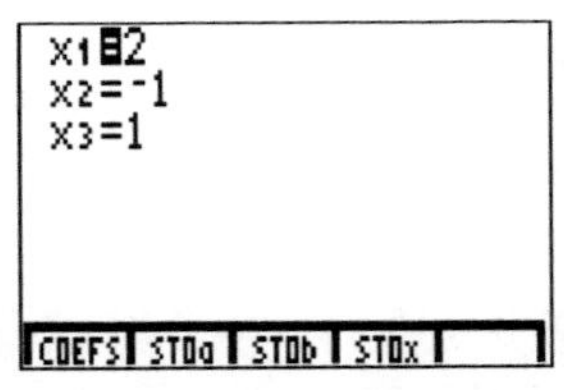

Here are the results.

Section 7.2 Example 5 (page 484) Using a System to Fit a Parabola to Three Data Points

The text mentions that quadratic regression (`P2Reg` on the TI-86) can be used to find the quadratic function fitting the three given data points. Shown on the right is the output of the TI-86's `P2Reg` procedure.

```
QuadraticReg
 y=ax²+bx+c
 n=3
 PRegC=
      {1.25 -.25 -.5}
```

It is worth noting that the `P3Reg` and `P4Reg` procedures could likewise be used to find functions to exactly fit sets of four or five data points (much more easily than solving the related systems of equations). Furthermore, the `LinR` (linear regression) procedure can be used with a pair of points to find the equation of the line through those points.

Section 7.3 Technology Note (page 488) — Entering Matrices

The matrices shown in the calculator screens in the text have the names "[A]" and "[B]." On the TI-82/83/84, all matrices have names of the form [*letter*]; on the TI-86, matrices can have any variable name. (In fact, variable names *cannot* include the bracket characters on the TI-86.)

There are two ways to define or edit a matrix on the TI-86. The first method it to store a matrix to a variable using a home screen command (as is illustrated on the right); use [2nd][(] and [2nd][)] to type the square brackets [and].

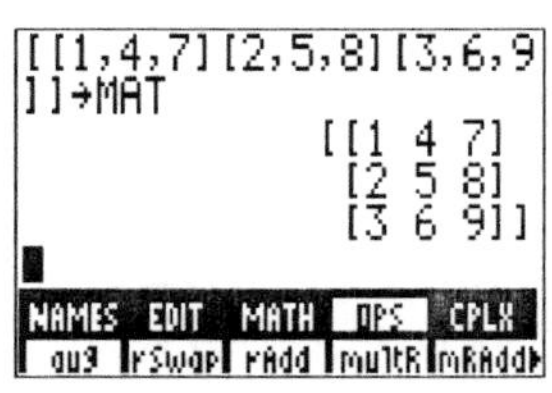

The other method is to select the MATRX:EDIT feature ([2nd][7][F2]), then choose the name of a matrix (or type a name for a new matrix), specify the number of rows and columns, and type in the entries. The screen on the right shows the process of editing MAT.

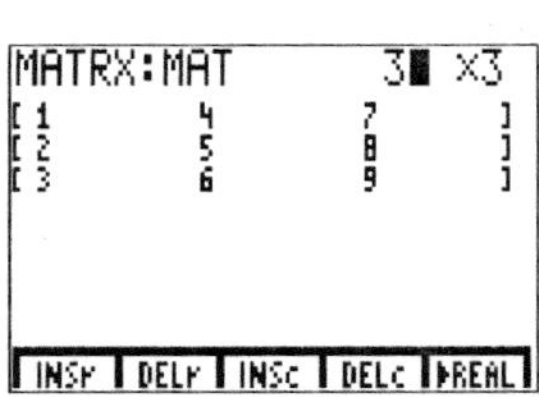

These two methods are equivalent to those illustrated in Figure 13 of the text.

Section 7.3 Technology Note (page 489) — Matrix Row Transformations

Section 7.3 Example 1 (page 489) — Using Row Transformations

The TI-86 has four matrix row operations, located in the MATRX:OPS menu ([2nd][7][F4][MORE]). The screen on the right shows this menu.

The formats for the matrix row-operation commands are:

- rSwap(*matrix*,*A*,*B*) produces a new matrix that has row *A* and row *B* swapped.
- rAdd(*matrix*,*A*,*B*) produces a new matrix with row *A* added to row *B*.
- multR(*number*,*matrix*,*A*) produces a new matrix with row *A* multiplied by *number*.
- mRAdd(*number*,*matrix*,*A*,*B*) produces a new matrix with row *A* multiplied by *number* and added to row *B* (row *A* is unchanged).

Shown below are examples of each of these operations on the matrix MAT$=\begin{bmatrix}1&4&7\\2&5&8\\3&6&9\end{bmatrix}$.

rSwap(MAT,1,2)
[[2 5 8]
[1 4 7]
[3 6 9]]

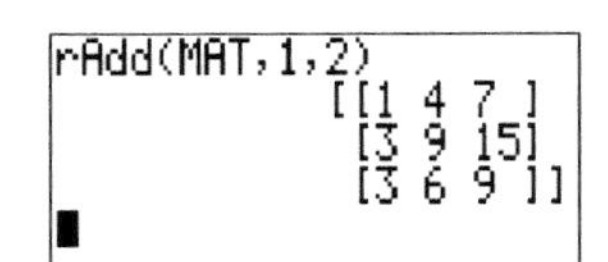

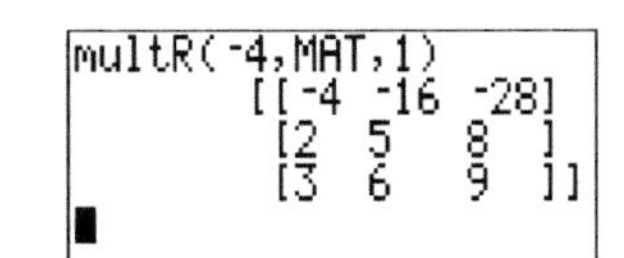

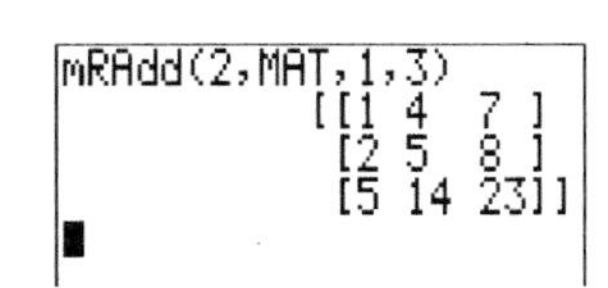

Keep in mind that these row operations leave the matrix MAT untouched. To perform a sequence of row operations, each result must either be stored in a matrix, or use the result variable Ans as the matrix in successive steps.

For example, with MAT equal to the augmented matrix given in the text just before this Technology Note and Example, the following screens illustrate the initial steps in the solving the system. Note the use of Ans in the last three screens.

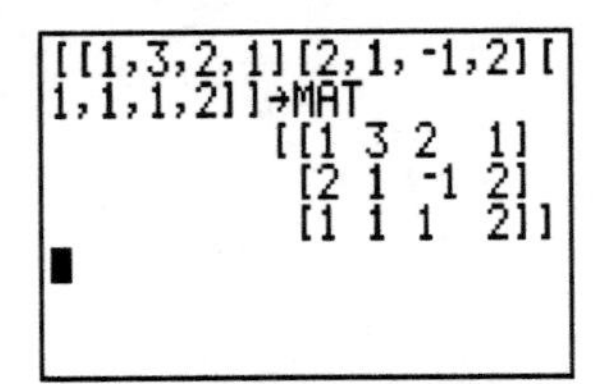

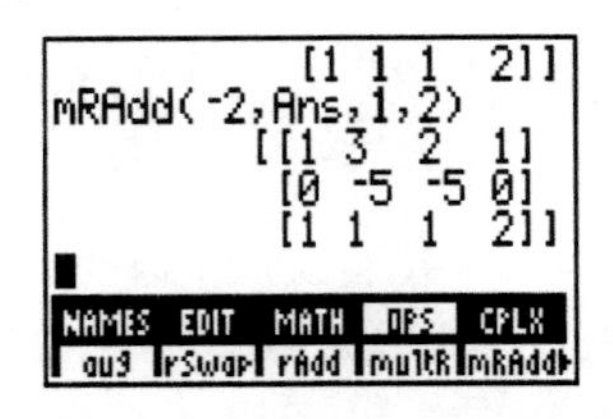

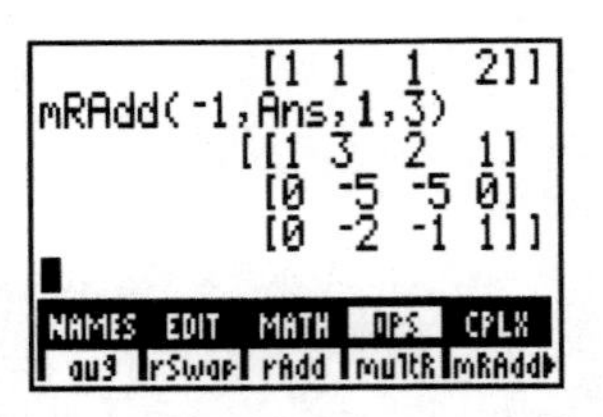

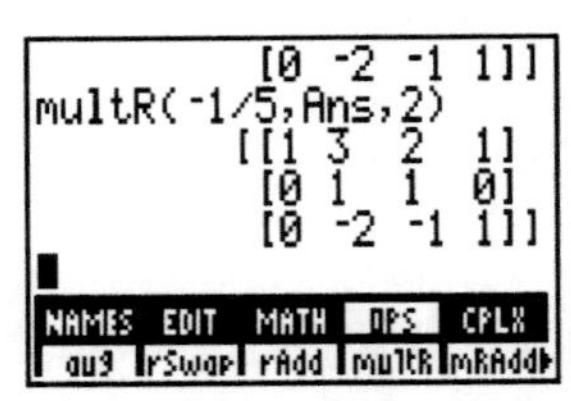

Section 7.3 Example 2 (page 490) Solving a System by the Row Echelon Method

Section 7.3 Example 5 (page 493) Solving a System by the Reduced Row Echelon Method

The ref and rref commands are found in the MATRX:OPS menu ([2nd][7][F4], then [F4] for ref or [F5] for rref). On the TI-86, the parentheses are not needed, so the necessary command is ref MAT or rref MAT. These will do all the necessary row operations at once, making these individual steps seem tedious. However, doing the whole process step-by-step can be helpful in understanding how it works.

Section 7.4 Example 1 (page 501) Classifying Matrices by Dimension

If a matrix has been entered in the TI-86, the dimensions can be found using the dim command ([2nd][7][F4][F1]), which returns a list containing two numbers: { row count, column count }. Of course, counting rows and columns is arguably much simpler than using the dim command.

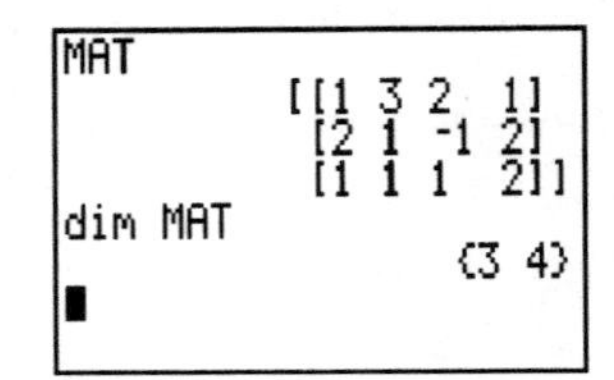

Section 7.4 Example 2 (page 501) Determining Equality of Matrices

If the variables mat1 and mat2 contains two matrices, one can test for equality by entering the command

```
mat1==mat2
```

on the home screen. The double-equals sign can be found in the TEST menu ([2nd][2]), or can be typed by pressing [ALPHA][STO▸][ALPHA][STO▸]. This will give 1 if the matrices are equal, and 0 otherwise.

This test will produce a DIM MISMATCH error if the two matrices do not have the same dimensions—but of course, they are obviously not equal in that case!

When testing equality, note that the TI-86 is quite picky about when it determines that two numbers (or matrices) are equal. Observe the situation on the right, in which it appears that matrices V1 and V2 are equal, but they are found to be unequal by the TI-86. Only by subtracting these matrices do we see that, in fact, they are not equal.

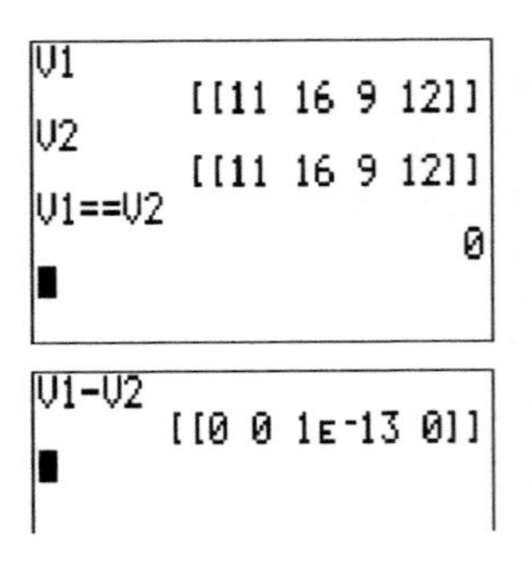

This problem arises because the TI-86 keeps track of the first 14 digits of any number entered into it; it considers two numbers to be exactly equal if those 14 digits match. When the TI-86 computes (e.g.) $\sqrt{7/64}$ and $\sqrt{7}/8$, unavoidable numerical errors (due to rounding and other causes) yield slightly different answers, even though $\sqrt{7/64} = \sqrt{7}/8$. The screen on the right shows this difference to be $-1 \times 10^{-14} = -0.00000000000001$—that is, they differ in the 14th digit. Such a small difference would usually be forgivable, but since the two numbers are not *exactly* the same, the TI-86 reports 0 when asked if they are equal. Similar numerical errors also cause the TI-86 to report (incorrectly) that $(\sqrt{\pi})^2$ is not equal to π.

```
√(7/64)-√7/8
                    -1E-14
(√π)²-π
                    -1E-13
```

Because of these numerical errors, a more reliable way to test for equality is to subtract the two quantities and decide if the difference is sufficiently small. An equivalent approach would be to compute each quantity separately, and visually compare the two results. If the numbers look the same on the screen, it is a fairly safe bet (though not a sure thing) that the quantities are the same.

```
√(7/64)
            .330718913883
√7/8
            .330718913883
```

Section 7.4 Example 3 (page 502) Adding Matrices

Section 7.4 Example 4 (page 504) Subtracting Matrices

Section 7.4 Example 5 (page 505) Multiplying Matrices by Scalars

Section 7.4 Example 8 (page 508) Multiplying Matrices

As the screens in the text indicate, basic arithmetic with matrices is relatively straightforward. To perform these computations on a TI-86, the entries are essentially identical to those shown in the text, except for the names of the matrices.

Section 7.4 Example 9 (page 508) Multiplying Square Matrices

One additional observation about multiplying square matrices: One can raise a square matrix to any non-negative integer power using the [^] key (or [x^2]). Negative or non-integer powers result in a `DOMAIN` error.

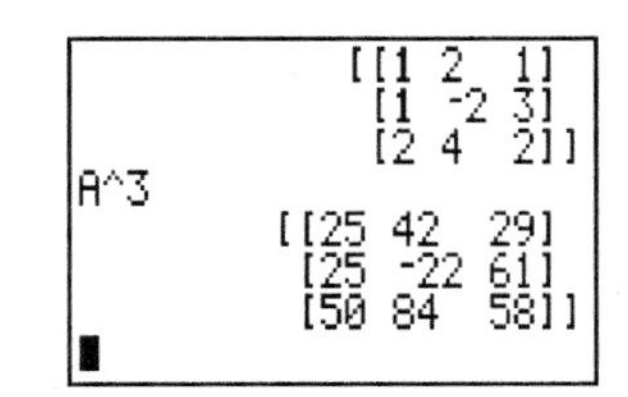

Section 7.5 Example 1 (page 514) Evaluating the Determinant of a 2×2 Matrix

Section 7.5 Example 4 (page 517) Evaluating the Determinant of a 3×3 Matrix

Section 7.5 Example 5 (page 518) Evaluating the Determinant of a 4×4 Matrix

On a TI-86, the parentheses around the matrix name are not needed. The function "`det`" is available in the MATRX:MATH menu ([2nd][7][F3][F1]). Shown on the right is the TI-86 version of the screen in the text.

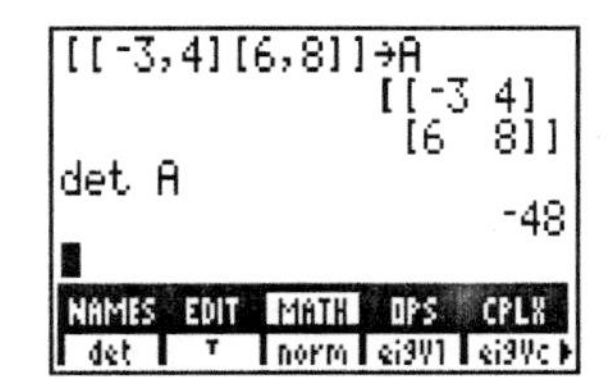

The determinant of a 3 × 3—or larger—matrix is as easy to find with a calculator as that of a 2 × 2 matrix. (At least, it is as easy for the user; the calculator is doing all the work!) Note, however, that trying to find the determinant of a non-square matrix (for example, a 3 × 4 matrix) results in a `DIMENSION` error.

Section 7.6 Example 1 (page 525) — Using the 2×2 Identity Matrix

On the TI-86, the function to produce identity matrices is `ident`, which does not require parentheses, and is located in the MATRX:OPS menu ([2nd][7][F4][F3]).

Section 7.6 Example 3 (page 528) — Finding the Inverse of a 3×3 Matrix
Section 7.6 Example 4 (page 530) — Finding the Inverse of a 2×2 Matrix

An inverse matrix is found using [2nd][EE]—the same method used to find the inverse (reciprocal) of a real number. (Do not try to use [^][(-)][1] in place of [2nd][EE].) For example, to find the inverse of matrix `MAT`, type `MAT` (or choose that name from the MATRX:NAMES menu) then press [2nd][EE][ENTER]. An error will occur if the matrix is not square, or if it is a singular matrix, as in part (b) of this example.

Section 7.7 Example 1 (page 538) — Graphing a Linear Inequality

```
Plot1 Plot2 Plot3
\y1=-(1/4)x+1
```

With the TI-86, there are two ways to shade above or below a function. The simpler way is to use the "shade above" graph style (see page 57). The screen on the right shows the "shade above" symbol next to `y1`, which produces the graph shown in the text (Figure 52). Note that the TI-86 is not capable of showing the detail that the line is "dashed."

The other way to shade is the `Shade(` command, accessed in the GRAPH:DRAW menu ([GRAPH][MORE][F2][F1]). The format is

`Shade(`*lower*`,`*upper*`,`*min x*`,`*max x*`,`*pattern*`,`*resolution*`)`

Here *lower* and *upper* are the functions between which the TI-86 will draw the shading (above *lower* and below *upper*).

The last four options can be omitted. *min x* and *max x* specify the starting and ending x values for the shading. If omitted, the TI-86 uses `xMin` and `xMax`.

The last two options specify how the shading should look. *pattern* determines the direction of the shading: 1 (vertical—the default), 2 (horizontal), 3 (negative-slope 45°—that is, upper left to lower right), or 4 (positive-slope 45°—that is, lower left to upper right). *resolution* is a positive integer (1,2,3,. . .) which specifies how dense the shading should be (1 = shade every column of pixels, 2 = shade every other column, 3 = shade every third column, etc.). If omitted, the TI-86 shades every column; i.e., it uses *resolution* = 1.

[85] *The TI-85 does not support graph styles, and the* `Shade` *command has fewer options. Specifically, the TI-85 only does "solid" shading; the* pattern *and* resolution *options are not available.*

To produce the graph shown in Figure 52 of the text, the appropriate command (typed on the home screen) would be something like

`Shade(-(1/4)x+1,10,-2,6,1,2).`

The use of 10 for the *upper* function simply tells the TI-86 to shade up as high as necessary; this could be replaced by any number greater than 4 (the value of yMax for the viewing window shown). Also, if the function y1 had previously been defined as -(1/4)x+1, this command could be shortened to Shade(y1,10,-2,6,1,2).

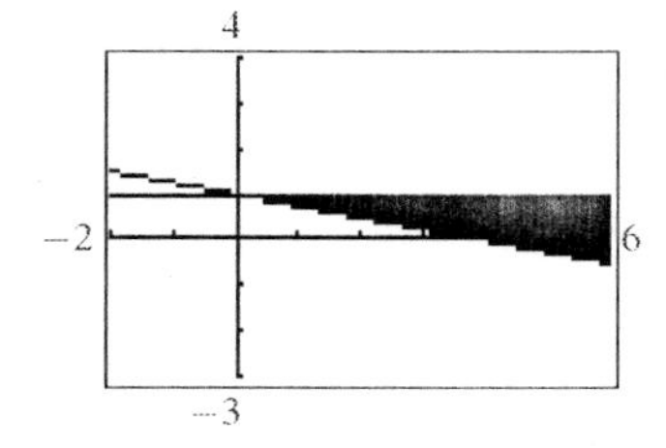

One more useful piece of information: Suppose one makes a mistake in typing the Shade(command (e.g., switching *upper* and *lower*, or using the wrong value of *resolution*), resulting in the wrong shading. The screen on the right, for example, arose from typing Shade(-(1/4)x+1,1,-2,6,1,1). In order to achieve the desired results, the mistake must be erased using the CLDRW (clear drawing) command, found in the GRAPH:DRAW menu ([GRAPH][MORE][F2][MORE][MORE][F1]). Then—perhaps using deep recall (see page 51)—correct the mistake in the Shade(command and try again.

Section 7.7 Example 2 (page 539) Graphing a System of Two Inequalities

The easiest way to produce (essentially) the same graph as that shown in the text is to use the "shade above" graph style (see page 57). The screen on the right (above) shows the "shade above" symbol next to y1 and y2, with the results shown on the second screen on the right (the same as that shown in Figure 52 of the text). When more than one function is graphed with shading, the TI-86 rotates through the four shading patterns (see page 80); that is, it graphs the first with vertical shading, the second with horizontal, and so on. All shading is done with a resolution of 2 (every other pixel).

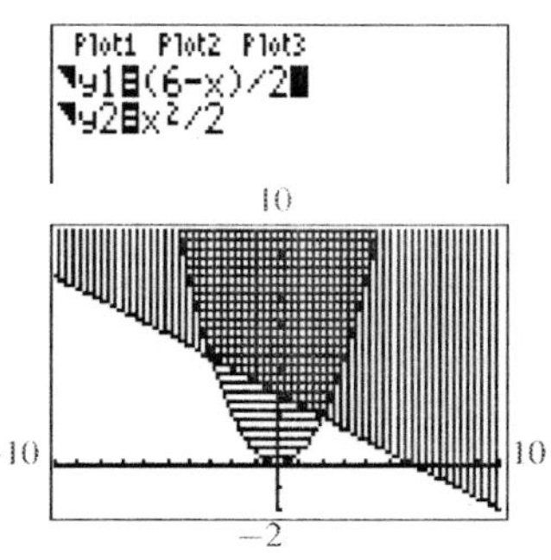

The Shade command (described in the previous example) can be used to produce this from the home screen. If y1=(6-x)/2 and y2=x²/2, the commands at right produce the graph shown above and in the text.

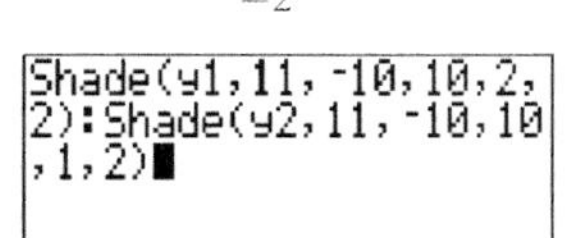

A nicer picture can be created, with a little more work, by making the observation that if y is greater than both $x^2/2$ and $(6-x)/2$, then for any x, we wish to shade those points for which y is greater than the *larger* of these two expressions. The TI-86 provides a convenient way to find the larger of two numbers with the max(function, located in the LIST:OPS menu ([2nd][-][F5][F5]). Then max(y1,y2) will return the larger of y1 and y2, and the "y equals" screen entries shown on the right (above) will produce the graph shown below it. (Note the graph style settings: y1 and y2 display as solid curves, while y3 has the "shade above" style.) The home-screen command Shade(max(y1,y2),11,-10,10,1,2) would produce similar results.

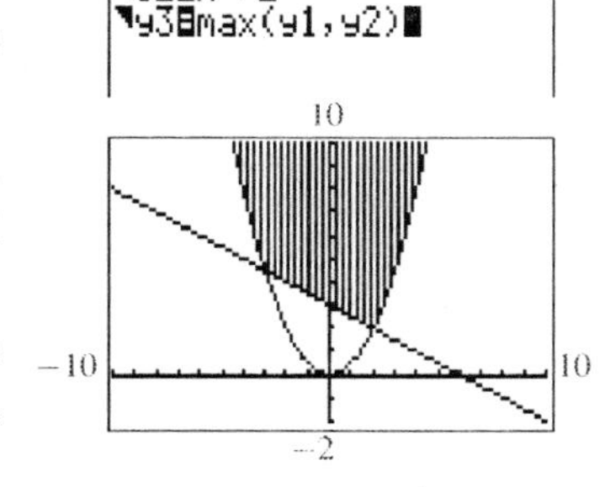

Extension: The table below shows how (using graph styles or home-screen commands) to shade regions that arise from variations on the inequalities in this example, assuming that y1=(6-x)/2 and y2=x²/2. (The results of these commands are not shown here.)

For the system...	or equivalently...	the command would be...
$x < 6 - 2y$ $x^2 < 2y$	$y < (6 - x)/2$ $y > x^2/2$	shade below y1 and above y2, or enter `Shade(y2,y1,-10,10,1,2)`
$x > 6 - 2y$ $x^2 > 2y$	$y > (6 - x)/2$ $y < x^2/2$	shade above y1 and below y2, or enter `Shade(y1,y2,-10,10,1,2)`
$x < 6 - 2y$ $x^2 > 2y$	$y < (6 - x)/2$ $y < x^2/2$	shade below y1 and below y2, or enter `Shade(-10,min(y1,y2),-10,10,1,2)`

[85] *Because the TI-85 does not have graph styles, one must use the* `Shade` *commands given above, combined with* `min` *or* `max`*, to produce the appropriate graphs.*

Section 8.1 Technology Note (page 567) — Radian and Degree modes

See section 9 of the introduction (page 52) for information about selecting Degree and Radian modes.

Section 8.1 Example 2 (page 568) — Calculating with Degrees and Minutes

The degrees and minutes symbols, and the ▸DMS operator (which causes an angle to be displayed in degrees, minutes, and seconds, rather than as a decimal), are all found in the MATH:ANGLE menu ([2nd][×][F3]), shown on the screen below.

Entering angles on a TI-86, however, is different than on a TI-83 (which was used to create the screens shown in the text). The degree and seconds symbols (° and ") are not used; the minutes symbol is used for all three positions, as in the screen on the right. (The results of the computations are displayed with the "usual" symbols.) The screen shown illustrates how to use a TI-86 to perform angle calculations like those in this example.

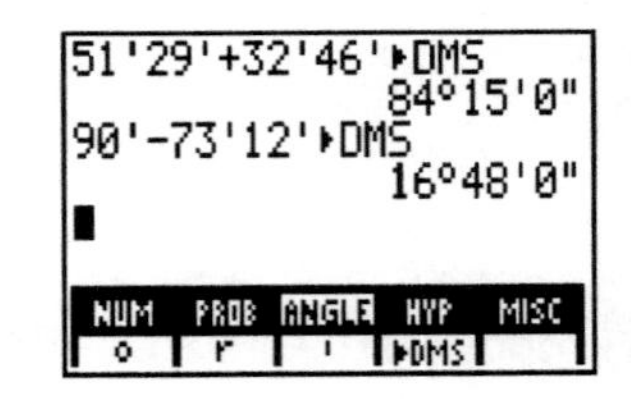

See section 9 of the introduction (page 52) for information about selecting Degree mode. Note that the results are displayed correctly regardless of the mode, *provided* the angle measurement includes the "minutes" symbol. The two entries and outputs shown here were both produced with the TI-86 in Radian mode; the first entry is converted to radians, but the second is not.

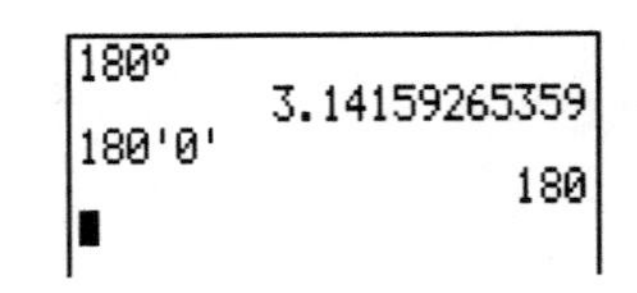

Section 8.1 Example 3 (page 568) — Converting between Decimal Degrees and Degrees, Minutes, Seconds

The screen on the right shows the proper way to enter problems like these for computation on the TI-86. Both the minutes symbol "'" and the ▸DMS operator are found in the MATH:ANGLE menu ([2nd][×][F3]).

74'8'14'
74.1372222222
34.817▸DMS
34°49'1.2"

Section 8.1 Example 5 (page 571) Converting Degrees to Radians

The number π is available as [2nd][^], and the degree symbol is [2nd][×][F3][F1]. With the calculator in Radian mode (see page 52), entering 45° causes the TI-86 to automatically convert to radians.

A useful technique to aid in recognizing when an angle is a multiple of π is to divide the result by π. This approach is illustrated in the screen on the right, showing that 45° is $\pi/4$ radians, and 30° is $\pi/6$ radians. This screen also makes use of the `▸Frac` command from the MATH:MISC menu ([2nd][×][F5][MORE][F1]), which simply means "display the result of this computation as a fraction, if possible." This is a useful enough command that one may wish to put it in the CUSTOM menu (see page 52).

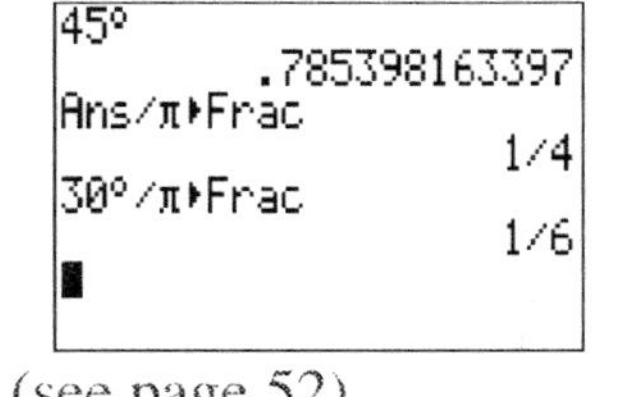

An alternative to using the degree symbol is to store $\pi/180$ in the calculator variable `D` (see page 52). Then typing, for example, `45D` [ENTER] will multiply 45 by $\pi/180$. This approach will work regardless of whether the calculator is in Degree or Radian mode. (A value stored in a variable will remain there until it is replaced by a new value.)

```
π/180→D
                .01745329252
45D
               .785398163397
249.8D
               4.35983247148
```

Section 8.1 Example 6 (page 571) Converting Radians to Degrees

With the TI-86 in Degree mode (see page 52), the radian symbol, produced with [2nd][×][F3][F2], will automatically change a radian angle measurement to degrees.

Alternatively, with the value $180/\pi$ stored in the calculator variable `R` (see page 52), typing `(9π/4)R` [ENTER] will convert from radians to degrees regardless of whether the calculator is in Degree or Radian mode. (The same result can be achieved by *dividing by* the calculator variable `D` as defined in the previous example.)

```
180/π→R
               57.2957795131
(9π/4)R
                         405
4.25R
               243.507062931
```

Section 8.2 Example 3 (page 584) Finding Function Values of an Angle

To enter a restriction (like $x \geq 0$) on the TI-86, enter `y1=(-1/2)x(x≥0)` —as is shown in the screen in the text. The greater-than-or-equal-to symbol is found in TEST menu ([2nd][2]). The TI-86 evaluates an expression like `x≥0` as 1 if it is true, and 0 if it is false; therefore, `y1` has the value 0 for all $x < 0$, and has the value $-0.5x$ for $x \geq 0$.

Section 8.2 Example 4 (page 585) Finding Function Values of Quadrantal Angles

The alternative to putting the calculator in Degree mode is to use the degree symbol ([2nd][×][F3][F1]) following each angle measure.

Since the cotangent, secant, and cosecant functions are the reciprocals of the tangent, cosine, and sine, they can be entered as (e.g.) `1/sin 90`. Note, though, that this will not properly compute cot 90°, since `1/tan 90` produces a domain error. Entering $\cot x$ as $\cos x / \sin x$ will produce the correct result at 90°.

One might guess that the other three trigonometric functions are accessed with [2nd] followed by [SIN], [COS], or [TAN] (which produce, e.g., $\sin^{-1}$). This is **not** what these functions do; in this case, the exponent -1

does not mean "reciprocal," but instead indicates that these are inverse functions (which are discussed in detail in Chapter 9 of the text).

Section 8.2 Technology Note (page 589) — Powers of trigonometric functions

The screen on the right shows the correct way to enter this, two incorrect ways (which compute the sine of 900°), and one method which produces a syntax error.

```
(sin 30)²
                .25
sin(30)²
                  0
sin 30²
                  0
sin ²30
```

Section 8.3 Technology Note (page 598) — Decimal approximations and exact values

As this note suggests, exact values cannot always be found. Sometimes, one can recognize exact values by squaring the result, as the screen on the right illustrates: We can see that $\cos 30^\circ$ is the square root of $0.75 = \frac{3}{4}$ and that $\sin 45^\circ = \sqrt{0.5} = \frac{1}{\sqrt{2}} = \frac{\sqrt{2}}{2}$.

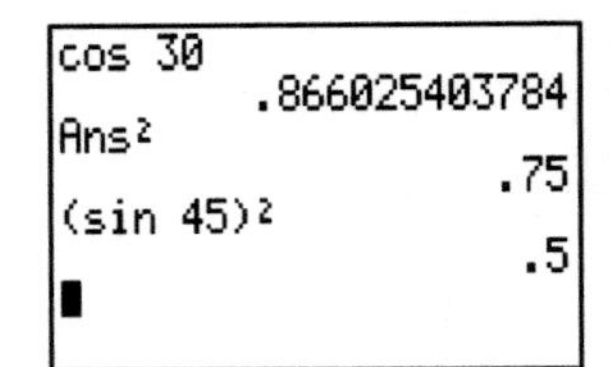

Section 8.3 Example 7 (page 601) — Approximating Trigonometric Function Values with a Calculator

Recall that the TI-86 uses a slightly different format for entering angles in degrees and minutes, as the screen on the right illustrates. This screen also shows a different way of entering the cosecant in part (b). These computations were done in Degree mode.

```
cos 49'12'
      .65342060399
(sin 197.977)⁻¹
     -3.24007122123
```

The screen on the right (with computations done in Radian mode) illustrates a somewhat unexpected behavior: Even if an angle is entered in DMS format, the TI-86 assumes that the angle is in radians. (See the related discussion on page 82.) In order to remedy this, either put the calculator in Degree mode, or use the degree symbol ([2nd][×][F3][F1]) at the very end of the angle measurement, as was done in the second entry.

```
cos 49'12'
      .484082106939
cos 49'12'°
      .65342060399
```

Section 8.3 Technology Note (page 602) — Inverse versus reciprocal functions

The inverse trigonometric functions were mentioned earlier, and are covered in the next example; more details are given in Chapter 9 of the text.

Section 8.3 Example 8 (page 602) — Using Inverse Trigonometric Functions
Section 8.3 Example 10 (page 603) — Finding Angle Measures

The inverse trigonometric functions are accessed with [2nd][SIN], [2nd][COS], and [2nd][TAN]; they *cannot* be entered as [SIN][2nd][EE], etc.

Note that the first output shown in text Figure 55 was produced in Degree mode, and the second was produced in Radian mode. None of the options available in the MATH:ANGLE menu ([2nd][×][F3]) can be used to avoid changing the mode; for example, when in Degree mode, the inverse trigonometric functions will always give an angle measure in degrees.

Section 8.4 Technology Note (page 609) Programs to solve right triangles

See section 13 of the introduction (page 57) for information about installing and running programs on the TI-86. The text notes that one must "consider the various cases"; there are five such cases: two legs, leg and hypotenuse, angle and hypotenuse, angle and adjacent leg, angle and opposite leg.

Section 8.4 Example 7 (page 613) Solving a Problem Involving the Angle of Elevation

The TI-86's GRAPH:MATH:ISECT ([GRAPH][MORE][F1][MORE][F3]) feature will automatically locate the intersection of two graphs. This feature was previously illustrated on page 62, but we repeat the description here: Use [▲], [▼] and [ENTER] to specify which two functions to use (in this case, the only two being displayed), and then use [◀] or [▶] to specify a guess. After pressing [ENTER], the TI-86 will try to find an intersection of the two graphs. The screens below illustrate these steps; the final result is the screen shown as Figure 71 of the text. The guessing step in the fourth screen below is not crucial in this case, since the calculator would locate the intersection even if a very poor guess was given.

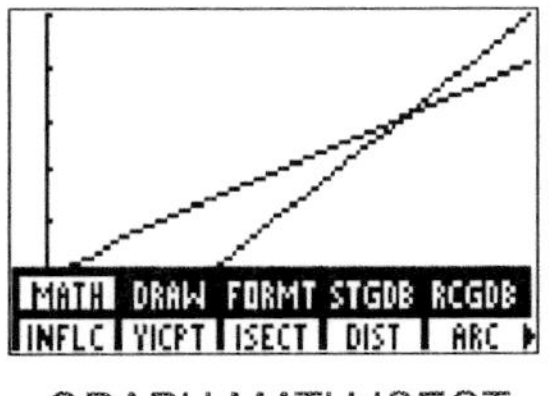

GRAPH:MATH:ISECT

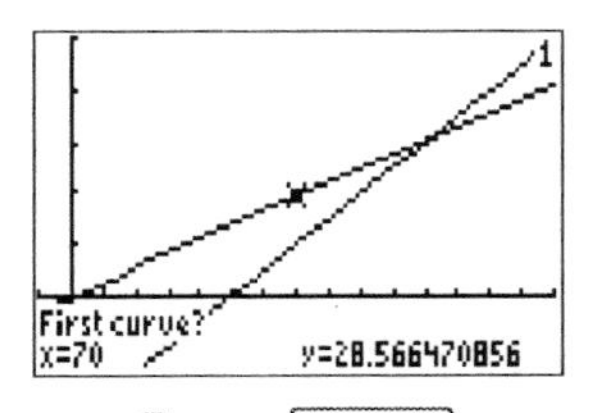

Press [ENTER] to choose y1

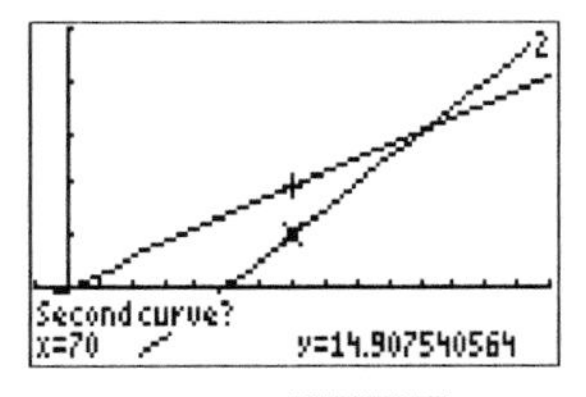

Press [ENTER] to choose y2

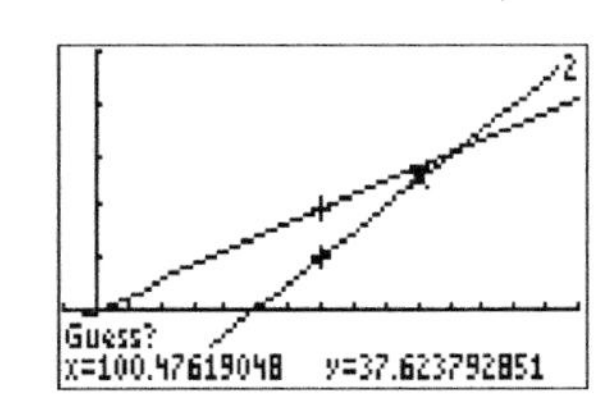

Move cursor to specify guess and then press [ENTER].

Section 8.6 Technology Note (page 630) The trig viewing window

The trig viewing window is set by pressing [GRAPH][F3][MORE][F3] (GRAPH:ZOOM:ZTRIG). On the TI-86, the values of xMin and xMax are $\pm 2.625\pi$ rather than $\pm 2\pi$. These values are chosen so that Δx (see page 52) equals $\pi/24$.

If the TI-86 is in Degree mode, choosing ZTRIG sets xMin and xMax to about ± 472.5, so $\Delta x = 7.5$.

Section 8.6 Example 1 (page 631) Graphing $y = a \sin x$

Section 8.6 Example 2 (page 633) Graphing $y = \sin bx$

The TI-86's graph styles can produce screens like the one shown in Figure 89 or the graph screen accompanying Example 2. See page 57 for information on setting the thickness of a graph. Note that the TI-86 must be in Radian mode in order to produce the correct graph.

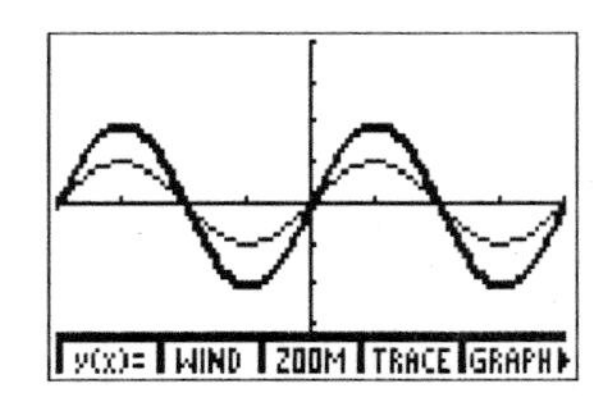

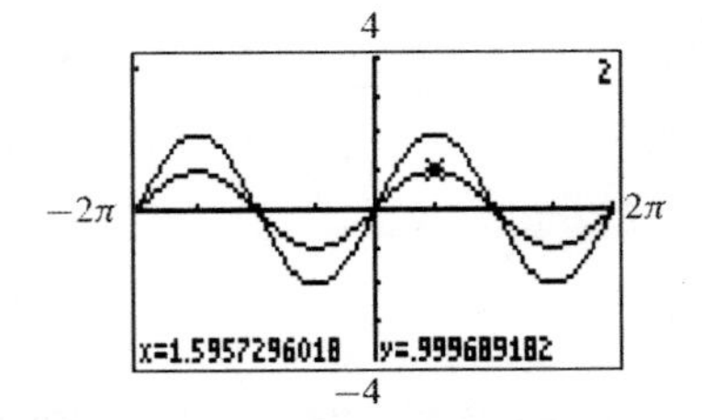

It is possible to distinguish between the two graphs without having them drawn using different styles by using the TRACE ([GRAPH][F4]) feature. On the right, the trace cursor is on graph 2—that is, the graph of $y_2 = \sin x$.
[85] *The TI-85 does not support graph styles, so one must rely on this* TRACE *approach.*

Section 8.6 Example 8 (page 638) Modeling Temperature with a Sine Function

The "sine regression" illustrated at the end of this example is built-in to the TI-86, found in the STAT:CALC menu ([2nd][+][F1][MORE]). Also shown here is the output of this command (rounded to two decimal places, as in the text); note that the calculator takes several seconds to perform this computation. Also note that the reported values are not exactly the same as those given by the TI-83. See page 61 for information about using the TI-86 for regression computations.

[85] *The TI-85 does not perform sine regressions. It might be possible to find a program to do this; see section 13 of this chapter's introduction, on page 57.*

Section 8.7 Technology Note (page 647) Cosecant and secant functions
Section 8.7 Example 1 (page 649) Graphing $y = a \sec bx$
Section 8.7 Example 2 (page 650) Graphing $y = a \csc (x - d)$

See section 9 of the introduction (page 52) for information about "connected" (DrawLine) versus "dot" (DrawDot) mode. To graph the cosecant function, the actual entry on a TI-86 would be

`y1=1/sin x` or `y1=(sin x)⁻¹`.

The function in Example 1 can be entered as `y1=2(cos (x/2))⁻¹` (or as shown in the text). The function in Example 2 can be entered as `y1=3/(2sin (x-π/2))` (or as shown in the text). A reminder: `sin⁻¹` ([2nd] [SIN]) is *not* the cosecant function.

Section 9.1 Example 2 (page 674) Rewriting an Expression in Terms of Sine and Cosine
Section 9.1 Example 3 (page 676) Verifying an Identity (Working with One Side)
Section 9.1 Example 4 (page 677) Verifying an Identity (Working with One Side)
Section 9.1 Example 5 (page 677) Verifying an Identity (Working with One Side)
Section 9.1 Example 6 (page 678) Verifying an Identity (Working with Both Sides)

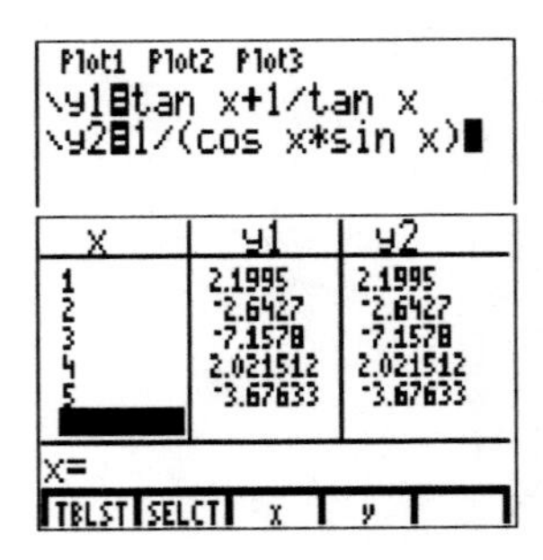

The top screen on the right shows how the expressions for Example 2 are entered on the "y equals" screen, so that they can be graphed and compared, as is illustrated in the text. As an alternative to graphing these two functions, the TI-86's table feature (see page 59 of this manual) can be used: If the y values are the same for a reasonably large sample of x values, one can be fairly sure (though not certain) that the two expressions are equal. To make this approach more reliable, be sure to choose x values that are not, for example, all multiples of π. This method is used in Example 4.

The fact that the two graphs are identical on the calculator screen does provide strong support for the identity, especially when confirmed by tracing, as described in the Technology Note next to Example 3 in the text. The tables shown in Figures 3 and 4 also show that the two expressions produce identical output—at least to the number of digits visible in the table. The exception is Figure 3 when x equals $\pi/2 \approx 1.5708$, for which `y1` shows `ERROR` while `y2` equals 1. In fact, $\cot(\pi/2) + 1 = 1$, but the value of `y1` is reported as `ERROR` since $\tan(\pi/2)$ is undefined.

The footnote on page 676 points out that the graph (or the table) cannot be used to prove the identity. In particular, the graph in Figure 2 only plots points for values of x that are $\Delta x = \pi/24$ units apart, while the table in Figure 3 shows outputs for input values spaced $\pi/8$ units apart. For example, the function `y3=cos x/sin x+cos (48x)` would look identical to `y1` and `y2` if graphed on the trig viewing window, although this function is different from these two functions at any point other than those shown in the graph.

[85] *The TI-85 does not produce tables, but a graph can still be used to support an identity.*

Section 9.2 Example 1 (page 685) Finding Exact Cosine Function Values
Section 9.2 Example 3 (page 687) Finding Exact Sine and Tangent Function Values

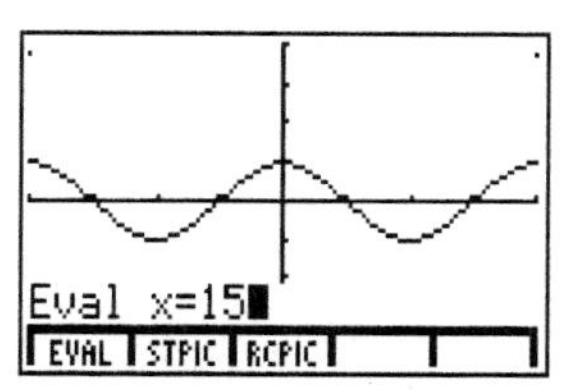

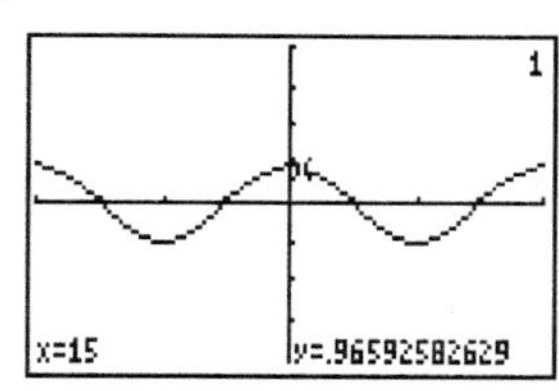

The TI-86 can graphically and numerically support exact value computations such as $\cos 15^\circ = \frac{\sqrt{6}+\sqrt{2}}{4}$. Starting with a graph of `y1=cos x`, the TI-86 makes it possible to trace to any real number value for x between `xMin` and `xMax`. While viewing the graph, press [F4] (TRACE), then type a number and press [ENTER]. This causes the TRACE cursor to jump to that x-coordinate. This same result can be achieved using GRAPH:EVAL ([GRAPH][MORE][MORE][F1]). (This feature was previously discussed on page 66.)

[85] *This latter approach is also available on the TI-85.*

Alternatively, a table of values like those shown here can be used to find the value of $\cos 15^\circ$. Of course, the screen shown in the text in support of Example 1(b) shows the other part of this process: Computing the decimal value of $(\sqrt{6}+\sqrt{2})/2$ and observing that it agrees with those found here.

x	y1
10	.9848078
11	.9816272
12	.9781476
13	.9743701
14	.9702957
15	[illegible]

y1=.96592582628907

TBLST SELCT x y

Note that Example 8 in Section 9.3 (page 699) shows that $\cos 15^\circ = \cos \frac{\pi}{12}$ can also be written as $\frac{\sqrt{2+\sqrt{3}}}{2}$.

Section 9.3 Example 4 (page 695) Deriving a Multiple-Number Identity

The table in the text shows `y1=sin (3x)` and `y2=3sin x-4(sin x)^3`—recall the proper way to enter this second expression—with Δ`Tbl` $= \pi/8$ and `TblStart=` $-3\pi/8$. Note that choosing x values which are multiples of a fraction of π is somewhat risky, since the periods of $\sin 3x$ and $\sin x$ involve fractions of π. Stronger support can be obtained by trying input values that are not multiples of π—say, $x = 1$, or $x = \sqrt{2}$.

Section 9.4 Example 1 (page 706) Finding Inverse Sine Values

Section 9.4 Example 2 (page 707) Finding Inverse Cosine Values

Of course, it is not necessary to graph $y = \sin^{-1} x$ to find these values; one can simply enter, e.g., `sin-1 (1/2)` on the home screen. The first entry of the screen on the right shows what happens when the calculator is in Degree mode; note that the result is not in $[-\pi/2, \pi/2]$. With the calculator in Radian mode, results similar to those in the text are found, and the method employed on page 83 (Example 5 from Section 8.1) confirms that these values are $\pi/6$ and $3\pi/4$.

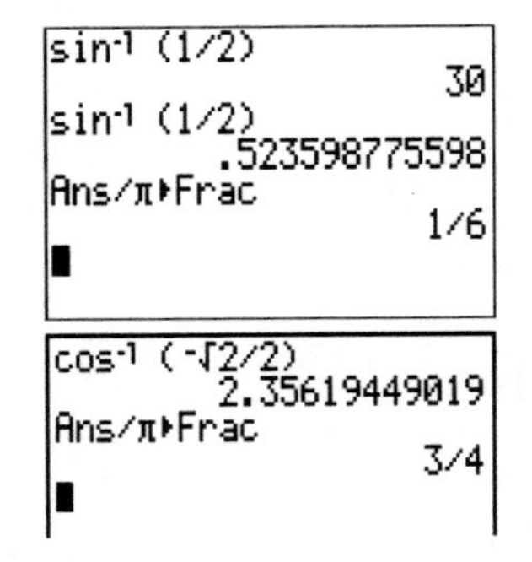

Note that the TI-86 does *not* give an error for the input `sin-1 -2`, but instead gives a complex result: $-\pi/2 + i\ln(2+\sqrt{3})$. The reason that this is technically a correct result is beyond the scope of the textbook; TI-86 owners should recognize that a complex answer is not appropriate for this problem, and so should ignore this result.

```
sin-1 -2
        (-1.57080,1.31696)
```

Section 9.4 Example 4 (page 710) Finding Inverse Function Values with a Calculator

Note that the answer given for (b), 109.499054°, overrepresents the accuracy of that value. A typical rule for doing computations involving decimal values (like −0.3541) is to report only as many digits in the result as were present in the original number—in this case, four. This means the reported answer should be "about 109.5°," and in fact, any angle θ between about 109.496° and 109.501° has a cotangent which rounds to −0.3541. (See also the discussion in the text on page 608.)

Section 9.6 Example 8 (page 727) Describing a Musical Tone from a Graph

Note that the calculator screen shown in Figure 40 illustrates the importance of choosing a "good" viewing window. If we choose the wrong vertical scale (`yMin` and `yMax`), we might not be able to see the graph at all—it might be squashed against the x-axis. If we make the window too wide—that is, if `xMax` minus `xMin` is too large—we might see the "wrong" picture, like the one on the right: We see a periodic function in this view, but not the one we want. There are actually 30 periods in this window, but the TI-86's limited resolution cannot show them all.

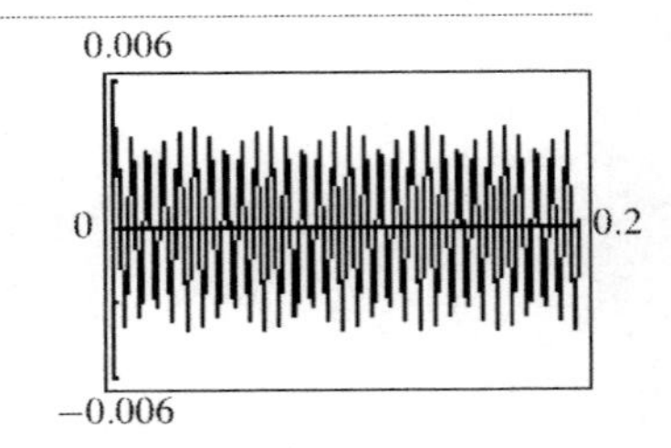

This observation—that a periodic function, viewed at fixed intervals, can appear to be a *different* periodic function—is the same effect that causes wagon wheels to appear to run backwards in old movies.

Note that for the TI-86, this function must be entered with parentheses: `y1=.004sin (300πx)`. For the TI-85, it will be graphed correctly with or without the parentheses (see section 8 of the introduction, page 52).

Section 10.1 Technology Note (page 743) — Programs to solve triangles
Section 10.2 Technology Note (page 756)

See section 13 of the introduction (page 57) for information about installing and running programs on the TI-86. If you wish to type these in on your own, here are two programs that give the same output as those shown in the text. (The first line of the second program ensures that the TI-86 is in Degree mode. The first program will operate correctly regardless of the mode.)

```
PROGRAM:ASA
Input "1st angle:",A
Input "2nd angle:",B
Input "Common side:",D
180-A-B→C
D sin A°/sin C°→E
D sin B°/sin C°→F
Disp "Other angle",C
Disp "Other sides",E,F
```

```
PROGRAM:SSS
Degree
Input "1st side:",D
Input "2nd side:",E
Input "3rd side:",F
cos⁻¹ ((D²+E²-F²)/(2D*E))→C
cos⁻¹ ((D²+F²-E²)/(2D*F))→B
180-B-C→A
Disp "Angles are",A,B,C
```

Section 10.3 Example 1 (page 766) — Finding Magnitude and Direction Angle
Section 10.3 Example 2 (page 767) — Finding Vertical and Horizontal Components

The TI-86 does not have the conversion functions shown in Figures 26 and 28, but can do the desired conversions in a manner that is perhaps even more convenient. The TI-86 recognizes vectors entered in either of two formats:

[*horizontal component*, *vertical component*] or [*magnitude* ∠ *angle*]

(The square brackets are [2nd][(] and [2nd][)], and "∠" is [2nd][,].) Regardless of how the vector is entered, the TI-86 displays it according to the `RectV CylV SphereV` mode setting (see page 52): It displays the vector in component form in RectV mode, and in magnitude/angle form for either of the other two modes.

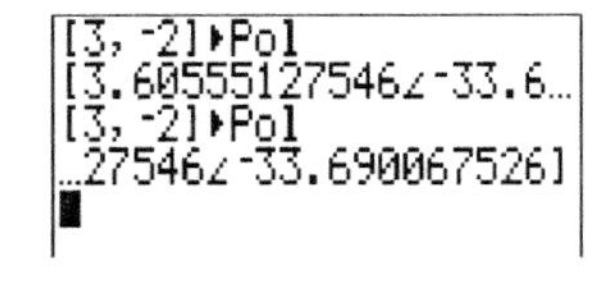

For Example 1, then, either put the TI-86 in CylV or SphereV mode, or (perhaps more conveniently), use the `▶Pol` command ([2nd][8][F4][F3]), as the screen on the right illustrates. This causes a vector to be displayed in magnitude/direction angle format regardless of the mode setting. On the fourth line, we see the results of pressing [▶] to see all the digits of the angle. (With the TI-86 in Degree mode, the angle returned is in degrees.)

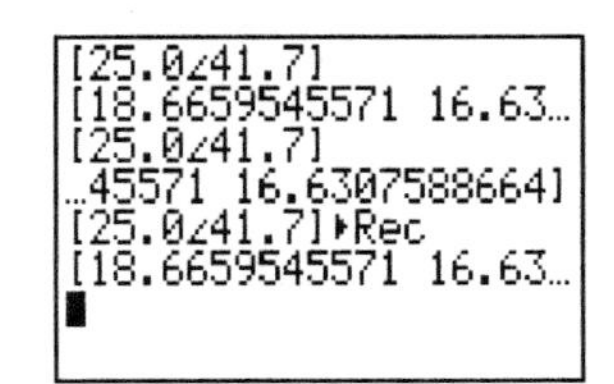

For Example 2, both the horizontal and vertical components can be found at once, as the screen on the right shows. (The vertical component is too long to fit on the screen, but the other digits can be seen by pressing [▶]. Note that the screen shown in the text was created with the calculator set to display only one digit after the decimal.) These computations were done with the TI-86 in Degree mode, so it was not necessary to include the degree symbol on the angle. The third entry shown uses the `▶Rec` command, which forces the vector to be displayed in component form regardless of the mode setting. (It was unnecessary in this case, because the TI-86 with in RectV mode.)

Also note that the conversion for Example 2 could also be done by typing `25cos 41.7` and `25sin 41.7`.

Section 10.3 Example 3 (page 767) Writing Vectors in the Form ⟨*a*, *b*⟩

Note how easily these conversions are done in the TI-86's polar vector format. (The TI-86 was in Degree mode.)

```
[5∠60]
  [2.5 4.33012701892]
[2∠180]
               [-2 0]
[6∠280]
[1.041889066 -5.9088…
```

Section 10.3 Example 5 (page 768) Performing Vector Operations

Section 10.3 Example 6 (page 769) Finding the Dot Product

Figure 35 shows vectors being added, subtracted, and multiplied by constants using lists. This method can also be used with the TI-86, but the built-in support for vectors is a better approach. Using the TI-86's vector notation, the vectors can be stored in calculator variables (see page 52) which can then be used to do the desired operations. Shown are the commands to compute 5(c): 4**u** − 3**v**, and 6(a): ⟨2, 3⟩ · ⟨4, −1⟩ (the `dot` command is [2nd][8][F3][F4]).

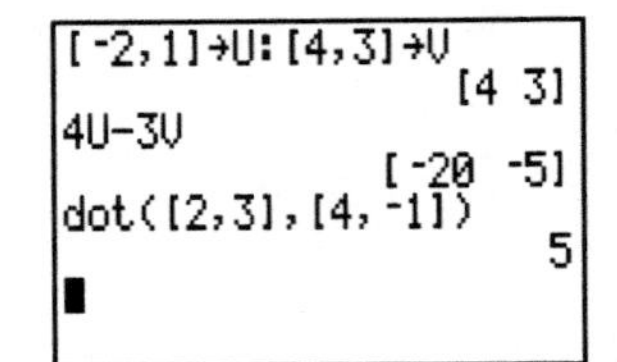

Section 10.4 Example 2 (page 779) Converting from Trigonometric Form to Rectangular Form

```
(2∠300)
    (1,-1.73205080757)
```

Complex numbers can be entered in trigonometric form in a manner similar to the polar format for vectors (see page 89; "∠" is [2nd][,]). The screen on the right illustrates how to quickly convert from the TI-86's trigonometric form to its rectangular form (assuming the calculator is in RectC mode; see page 52). Note that angles are assumed to be radians or degrees according the TI-86's mode; the entry on the right would have produced incorrect results in the TI-86 had been in Radian mode.

The command `▸Rec` (from the CPLX menu, [2nd][9][MORE]) can also be appended to a complex number in order to force the TI-86 to display it in rectangular format, regardless of the mode setting.

Section 10.4 Example 3 (page 780) Converting from Rectangular Form to Trigonometric Form

Section 10.4 Technology Note (page 782) The angle and abs commands

The TI-86's `▸Pol` command, found by pressing [MORE] in the CPLX ([2nd][9]) menu, will cause a complex number to be displayed in polar (trigonometric) format. (This command is also found in the VECTR:OPS menu.) This gives both the magnitude and angle at the same time. One can use the variable `i` = (0, 1) in conjunction with this command. This screen was produced in Degree mode; if Radian mode had been used, the angles would be decimal approximations for $5\pi/6$ and $-\pi/2$, which are harder to recognize.

```
(-√3,1)▸Pol
                 (2∠150)
-√3+i▸Pol
                 (2∠150)
-3i▸Pol
                 (3∠-90)
▸Rec ▸Pol
```

Alternatively, the `abs` and `angle` commands, also found in the TI-86's CPLX menu, will find the modulus and argument separately. On the screen on the right, note that if one makes use of the variable `i` = (0, 1), the `angle` and `abs` functions require parentheses; the third output is incorrect because it first computes the "angle" of the real number $-\sqrt{3}$ (which is π) and then adds this to `i`.

```
angle (-√3,1)
                    150
abs (-√3,1)
                      2
angle -√3+i
     (3.14159265359,1)
conj real imag abs angle
```

[85] *The TI-85 mistakenly computes the "angle" of negative numbers to be 0 instead of π.*

Section 10.4 Example 5 (page 783) Using the Product Theorem

Section 10.4 Example 6 (page 784) Using the Quotient Theorem

If doing such computations on a calculator, the TI-86's polar format can save a lot of typing. Shown on the right are the appropriate input, and the two possible outputs that can result: The first is in rectangular format, and the second is in polar format. Both of these results are ways of expressing the real number −6.

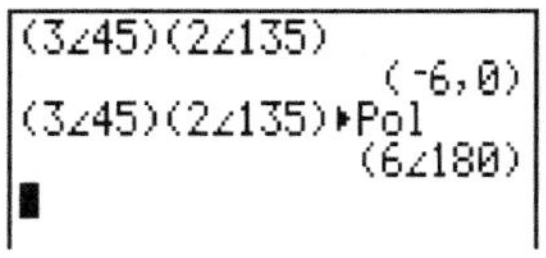

Section 10.5 Example 1 (page 787) Finding a Power of a Complex Number

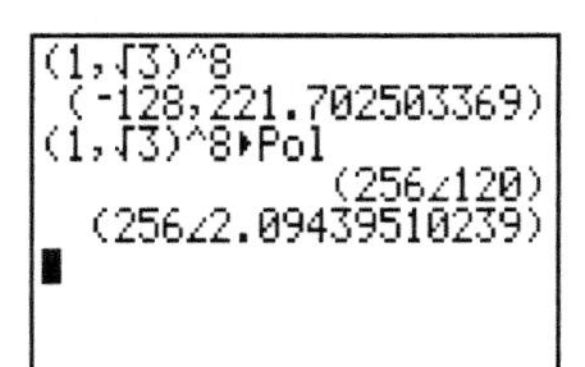

The screen on the right shows several options for computing $(1 + i\sqrt{3})^8$ with the TI-86. The first is fairly straightforward, but the reported result shows the complex part of the answer ($128i\sqrt{3}$) given in decimal form. For an "exact" answer, the TI-86's polar format can be used. The TI-86 was in Degree mode for the first of the two polar-format answers, and in Radian mode for the second.

Section 10.5 Example 4 (page 790) Solving an Equation by Finding Complex Roots

The TI-86's POLY polynomial solver (see page 71) can be used to see the pattern of these roots, if the calculator is in PolarC and Degree modes. The two negative angles shown in the second screen are coterminal with the angles 216° and 288° given in the text.

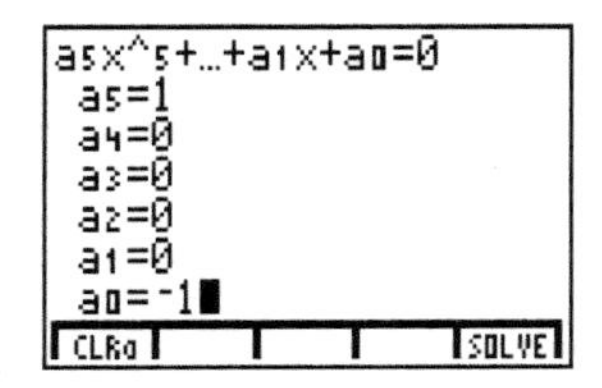

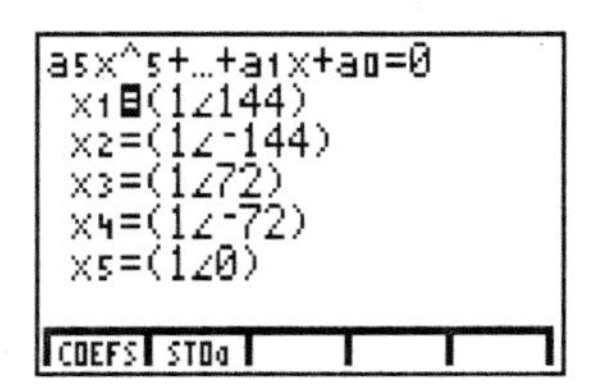

Section 10.6 Example 1 (page 794) Plotting Points with Polar Coordinates

Section 10.6 Example 2 (page 795) Giving Alternative Forms for Coordinates of a Point

Converting between polar and rectangular coordinates can be done using any of the conversion methods for vectors and complex numbers (covered beginning on page 89).

Section 10.6 Example 3 (page 796) Examining Polar and Rectangular Equations of Lines and Circles

Section 10.6 Example 4 (page 796) Graphing a Polar Equation (Cardioid)

Section 10.6 Example 5 (page 797) Graphing a Polar Equation (Rose)

Section 10.6 Example 6 (page 798) Graphing a Polar Equation (Lemniscate)

To produce these polar graphs, the TI-86 should be set to Degree and Polar modes (see the screen on the right). In this mode, [GRAPH][F1] allows entry of polar equations (r as a function of θ). One could also use Radian mode, adjusting the values of θMin, θMax, and θStep accordingly (e.g., use 0, 2π, and $\pi/30$ instead of 0, 360, and 5).

For the cardioid, rose, and lemniscate, the window settings shown in the text show these graphs on "square" windows (see section 11 of the introduction, page 55), so one can see how their proportions compare to

those of a circle. Note, however, that on the TI-86, these window are not square; the ZOOM:ZSQR option can be used to adjust the window dimensions to make it square.

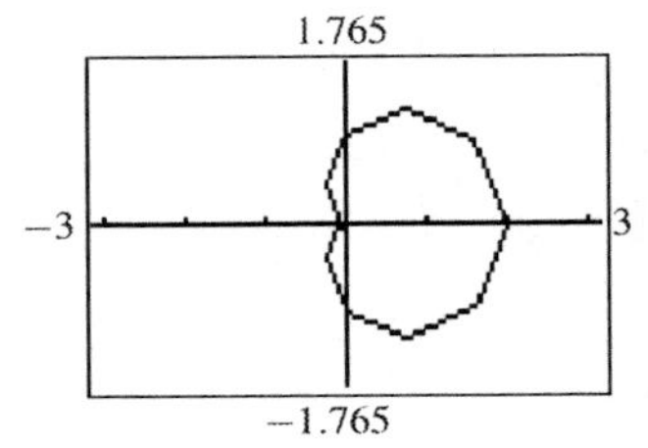

For the cardioid, the value of θStep does not need to be 5, although that choice works well for this graph. Too large a choice of θStep produces a graph with lots of sharp "corners," like the one shown on the right (drawn with θStep=30). Setting θStep too small, on the other hand, produces a smooth graph, but it is drawn very slowly. Sometimes it may be necessary to try different values of θStep to choose a good one.

The lemniscate can be drawn by setting θMin=0 and θMax=180, or θMin=-45 and θMax=45. In fact, with θ ranging from -45 to 225, the graph of r1=√cos (2θ) (alone) will produce the entire lemniscate. (θStep should be about 5.) The rose can be produced by setting θMin=0 and θMax=360, or using any 360°-range of θ values (with θStep about 5).

Section 10.6 Example 7 (page 798) Graphing a Polar Equation (Spiral of Archimedes)

To produce this graph on the viewing window shown in the text, the TI-86 must be in Radian mode. (In Degree mode, it produces the same shape, but magnified by a factor of $180/\pi$ —meaning that the viewing window needs to be larger by that same factor.)

Section 10.7 Example 2 (page 805) Graphing an Ellipse with Parametric Equations

See page 75 for information about using parametric mode. This curve can be graphed in Degree mode with tMin=0 and tMax=360, or in Radian mode with tMax=2π. In order to see the proportions of this ellipse, it might be good to graph it on a square window. This can be done most easily with the ZOOM:ZSQR option ([GRAPH][F3][MORE][F2]). On a TI-86, initially with the window settings shown in the text, this would result in the window $[-6.8, 6.8] \times [-4, 4]$.

Section 10.7 Example 3 (page 805) Graphing a Cycloid

The TI-86 *must* be in Radian mode in order to produce this graph.

Section 10.7 Example 4 (page 806) Creating a Drawing With Parametric Equations

To turn off the display of the coordinate axes, set AxesOff on the GRAPH:FORMT screen ([GRAPH][MORE][F3]). It is a good idea to restore this setting to AxesOn when finished, since for most uses, the absence of the axes can be confusing.

Section 10.7 Example 5 (page 807) — Simulating Motion With Parametric Equations

Section 10.7 Example 7 (page 808) — Analyzing the Path of a Projectile

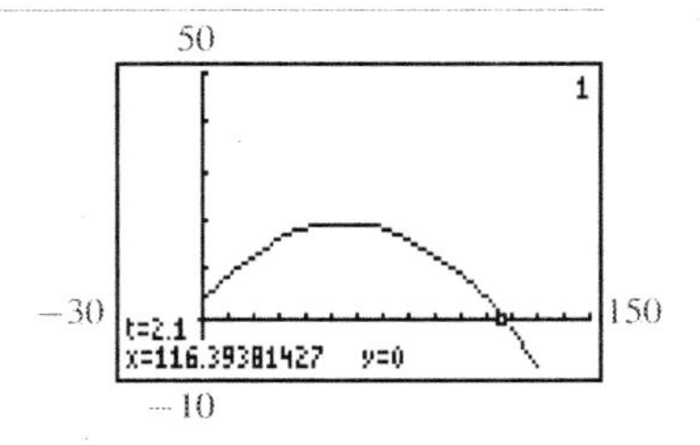

Parametric mode is particularly nice for analyzing motion, because one can picture the motion by watching the calculator create the graph, or by using TRACE ([GRAPH][F4]) and watching the motion of the trace cursor. (When tracing in parametric mode, the [▶] and [◀] keys increase and decrease the value of t, and the trace cursor shows the location (x, y) at time t.) The screen on the right is essentially the same as Figure 82, and illustrates tracing on the projectile path in Example 7. Note that the value of t changes by ±`tStep` each time [▶] or [◀] is pressed, so obviously the choice of `tStep` affects which points can be traced. This graph was produced by setting `tMin=0`, `tMax=3`, and `tStep=0.1`.

Rather than entering three separate pairs of equations for Example 5, the TI-86's list features can be used to graph all three curves with a single pair of equations: Define

`xt1=132cos ({30,50,70})*t` and `yt1=132sin ({30,50,70})*t-16t²`

with the calculator in Degree mode.

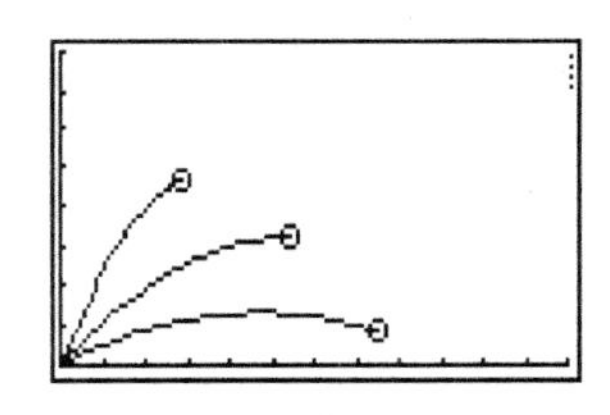

The TI-86's graph styles (see page 57) can be useful, too. The screen on the right shows the three paths for Example 5 being plotted in "ball path" style.

Section 11.1 Example 1 (page 821) — Finding Terms of Sequences

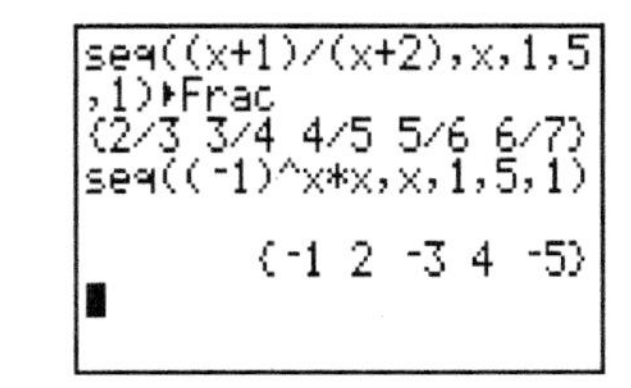

The `seq(` command can be found in the LIST:OPS menu ([2nd][−][F5][MORE][F3]). Given a formula a_n for the nth term in a sequence, the command

`seq(`*formula*`,`*variable*`,`*start*`,`*end*`,`*step*`)`

produces the list $\{a_{start}, a_{start+step}, \ldots, a_{end}\}$. In most uses, the value of *step* is 1, which is the assumed value if this is omitted. The size of the resulting list can have no more than 2800 items (or fewer, as available memory allows). The results for (a) and (b) are shown here; in both of these cases, "`,1)`" could have been omitted from the end of the command.

Note that *variable* can be any letter (or letters). The text uses the character n (not available on the TI-86), but x may be more convenient (since it can be typed with [x-VAR]).

Section 11.1 Technology Note (page 822) — Sequence mode

The TI-86 does not have a sequence mode. However, it can produce scatter diagrams of sequences like the one shown in Figure 3 and elsewhere in the text.

To produce a plot like this on the TI-86, begin by storing values in the list variables `xStat` and `yStat` as shown on the right, so that `xStat` contains values of n, and `yStat` contains the corresponding values of a_n. Here we illustrate using the sequence $a_n = 1/n$; the slightly shorter command `1/xStat→yStat` has the same effect as the second command shown.

```
seq(x,x,1,10,1)→xStat
{1 2 3 4 5 6 7 8 9 1…
seq(1/x,x,1,10,1)→ySt
at
{1 .5 .333333333333 …
```

Once the lists `xStat` and `yStat` have been defined, follow the procedures of Example 6 from Section 1.4 (page 60 of this manual) to produce the scatterplot.

Section 11.1 Example 2 (page 822) Using a Recursion Formula

The TI-86 can generate successive terms of some recursively-defined sequences, using the `Ans` storage variable. The screen on the right illustrates this approach: After entering the number 4, followed by the formula `2*Ans+1`, pressing [ENTER] repeatedly computes successive terms of the sequence. (Note that this approach does not lend itself to summation; see the next example.)

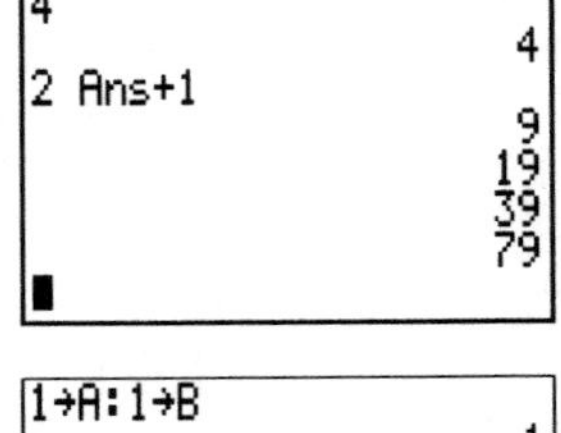

Since the `Ans` variable only stores the result of the *last* computation, it cannot be used to access the second-to-last result. Therefore, this approach cannot be extended to sequences like the Fibonacci sequence, described in the "For Discussion" section following this example in the text. It is typically defined recursively by $f_0 = f_1 = 1$, and $f_n = f_{n-1} + f_{n-2}$ for $n \geq 2$. With some work (see the screen on the right), it is possible to compute terms of such a sequence.

```
1→A:1→B
                    1
A+B→C:B→A:C→B
                    2
                    3
                    5
                    8
```

Section 11.1 Example 4 (page 825) Using Summation Notation

The `sum` command can be found in the LIST:OPS menu ([2nd][-][F5][MORE][F1]). `sum` can be applied to any list—either to a list variable, or directly to a list created with the `seq(` command. Evaluating summations on the TI-86 requires first generating the sequence as a list, then summing the list. The first two entries perform the summation for this example, using the `Ans` storage variable. Alternatively, the `sum` and `seq(` commands can be combined on a single line. The third entry on the right demonstrates this.

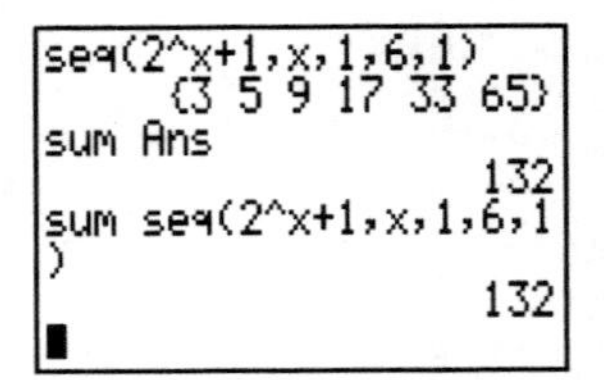

Section 11.4 Technology Note (page 851) Computing Factorials

Section 11.4 Example 1 (page 852) Evaluating Binomial Coefficients

Section 11.6 Example 4 (page 865) Using the Permutations Formula

The factorial operator `!`, the combinations function `nCr`, and the permutations function `nPr` are found in the MATH:PROB menu ([2nd][×][F2]), shown on the right.

Section 11.7 Example 6 (page 878)

Using a Binomial Experiment to Find Probabilities

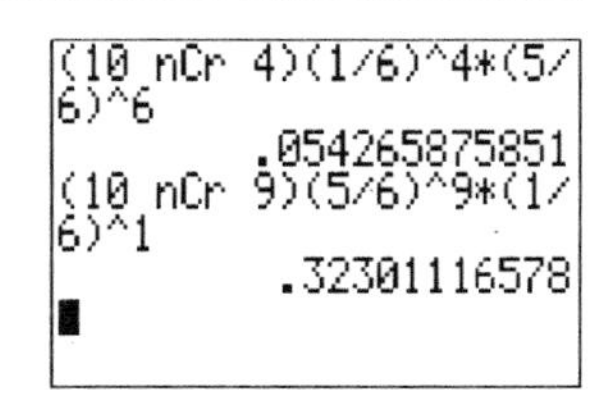

The TI-86 does not have statistical distribution functions like `binompdf`. The computations shown in this example must be done by manually entering the entire binomial probability formula (or by obtaining a program to automate such computations).

If you have access to a TI-Graph Link (computer/calculator transfer cable), you may wish to install the advanced statistics package. While this package was originally produced by TI, it does not appear to be available at their web site, but it can be found at `www.ticalc.org`, as well as other places on the Internet (do a search for "TI-86 infstat1"). Installation instructions are also available for download.

This package makes many statistical and probability functions available on your TI-86, including the `bipdf` function, which is equivalent to the `binompdf` function illustrated in the text.

Section R.4 Example 4 (page 916)

Using the Definition of $a^{1/n}$

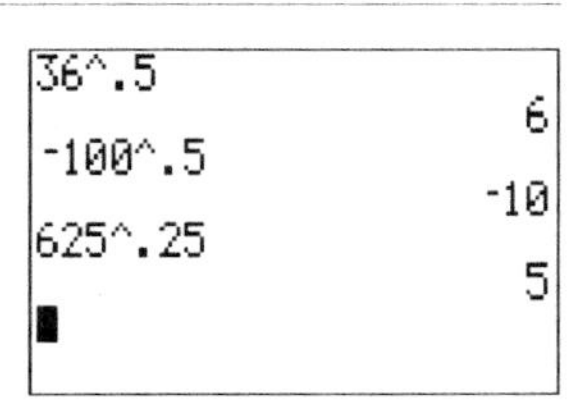

In evaluating these fractional exponents with a calculator, some time can be saved by entering fractions as decimals. The screen on the right performs the computations for (a), (b) and (c) with fewer keystrokes than entering, for example, `36^(1/2)`. (Of course, $36^{1/2} = \sqrt{36}$, which requires even fewer keystrokes.)

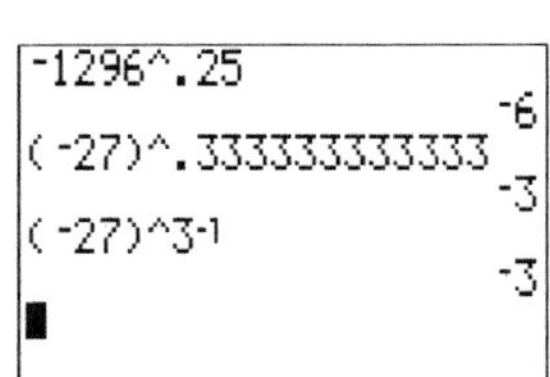

Note for (d), the parentheses around -1296 cannot be omitted: The text observes that this is not a real number, but the calculator will display a real result if the parentheses are left out. (See also the next example.) For (e), the decimal equivalent of $1/3$ is $0.\overline{3}$, which can be entered as a sufficiently long string of 3s (at least 12). The last line shows a better way to do this; the TI-86's order of operations is such that 3^{-1} is evaluated before the other exponentiation. This method would also work for the other computations; e.g., (a) could be entered [3][6][^][2][2nd][EE].

Section R.4 Example 5 (page 917)

Using the Definition of $a^{m/n}$

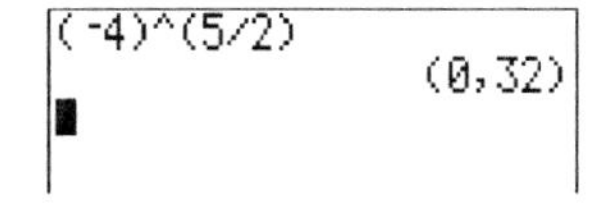

For (f), attempting to evaluate $(-4)^{5/2}$ will produce the result shown on the right. This is the TI-86's way of displaying complex numbers (which are discussed in the text beginning in page 68); the result `(0,32)` would be written as $0 + 32i$ in the format used by the text.

Appendix B Example 3 (page 934) — Using the SSE Program

See section 13 of the introduction (page 57) for information about installing and running programs on the TI-86. and see page 60 for a description of how to enter data into the calculator's lists. (For the TI-85 and TI-86, the data would be stored in `xStat` and `yStat` instead of L_1 and L_2).

Here is the program shown in Figure 2, translated for the TI-85 and TI-86:

```
PROGRAM:SSE
sum (yStat-evalF(y1,x,xStat))²
```

This slightly shorter version works on the TI-86 (but not the TI-85):

```
sum (yStat-y1(xStat))²
```

`yStat` and `xStat` can be accessed on the LIST:NAMES menu ([2nd][−][F3]), and `evalF` is found in the CALC menu ([2nd][×]).

Appendix C Example 1 (page 937) — Finding the Distance between Two Points in Space

The TI-86 has built-in support for vectors, and can do many computations with them. These commands are found in the VECTR menu ([2nd][8]).

For example, coordinates of points can be entered as vectors in the form $[x, y, z]$ (the square brackets are [2nd][(] and [2nd][)]), and then the `norm` function ([2nd][8] [F3][F3]) can be used as shown on the right. (The `norm` function adds the squares of each number in the vector, then takes the square root—in other words, it performs the computations of the distance function.)

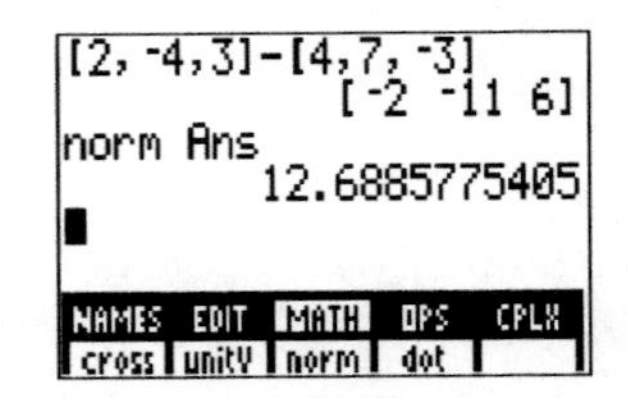

Appendix C Example 4 (page 938) — Performing Vector Operations

Using the TI-86's vector notation, the vectors can be stored in calculator variables (see page 52) which can then be used to do the desired operations. Shown are the commands to compute (a) $\mathbf{v} + \mathbf{w}$ and (e) $\mathbf{v} \cdot \mathbf{w}$ (the `dot` command is [2nd][8][F3][F4]).

```
[2,6,-4]→V:[-1,0,5]→W
                [-1 0 5]
V+W
                [1 6 1]
dot(V,W)
                     -22
```

Introduction

The information in this section is essentially a summary of material that can be found in the TI-89 manual. Consult that manual for more details.

Owners of a TI-92 will find that most of this material applies to that calculator as well.

1 Power

To power up the calculator, simply press the [ON] key. The screen displayed at this point depends on how the TI-89 was last used. It may show the "home screen"—a menu (the toolbar) across the top, the results of previous computations (if any) in the middle (the history area), a line for new entries (which may be blank, or may show the previous entry), and a status line at the bottom. An example of this home screen is shown on the right.

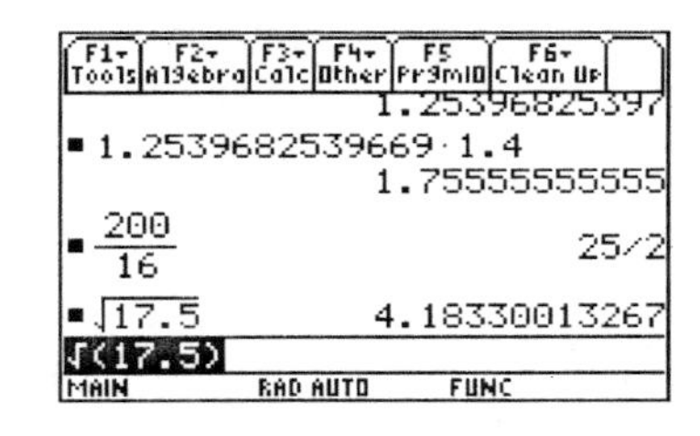

The TI-89 may show some other screen—perhaps a graph, an error message, a menu, or something else. If so, one can return to the home screen by pressing either the [HOME] key (from a graph), or the [ESC] key (from an error message), or [2nd][ESC] (from a menu). Note that the "second function" of [ESC]—written in yellow type above the key—is "QUIT."

If the screen is blank, or is too dark to read, one may need to adjust the contrast (see the next section).

To turn the calculator off, press [2nd][ON] (OFF), in which case the TI-89 will start up at the home screen next time. This will not work if an error message is displayed. Pressing [◆][ON] will also turn the TI-89 off, but when [ON] is next pressed, the screen will show exactly what it showed before. The calculator will automatically shut off if no keys are pressed for several minutes, in which case it will behave as if [◆][ON] had been pressed.

2 Adjusting screen contrast

If the screen is too dark (all black), decrease the contrast by pressing and holding [◆] and [−]. If the screen is too light, increase the contrast by pressing and holding [◆] and [+]. If the screen never becomes dark enough to see, the batteries should be replaced.

3 Replacing batteries

To replace the four AAA batteries, first turn the calculator off ([2nd] [ON]), then remove the back cover, remove and replace each battery, replace the back cover, then turn the calculator on again. (After replacing batteries, one may need to adjust the contrast down as described above.) Note: The status line at the bottom of the home screen should display `BATT` when the batteries are getting low.

4 Basic operations

Simple computations are entered in essentially the same way they would be written. For example, to compute $2 + 17 \times 5$, press [2][+][1][7][×][5][ENTER] (the [ENTER] key tells the calculator to act on what has been typed). Standard order of operations (including parentheses) is followed. Note that the entry and the result (87) are displayed in the last line of the history area, and the entry is also displayed (and highlighted) in the entry line. If any new text is typed, this highlighted text will be deleted.

```
F1 Tools | F2 Algebra | F3 Calc | F4 Other | F5 PrgmIO | F6 Clean Up
■ 1.25396825396 69·1.4        1.75555555555
■ 200/16                               25/2
■ √17.5                       4.18330013267
■ 2 + 17·5                               87
2+17*5
MAIN      RAD AUTO      FUNC      8/30
```

The result of the most recently entered expression is stored in `ans(1)`, which is typed by pressing [2nd][(-)] (the word "ANS" appears in yellow above this key). For example, [5][+][2nd][(-)][ENTER] will add 5 to the result of the previous computation. Note that in the history area, "`ans(1)`" has been replaced by "87."

```
F1 Tools | F2 Algebra | F3 Calc | F4 Other | F5 PrgmIO | F6 Clean Up
■ 200/16                               25/2
■ √17.5                       4.18330013267
■ 2 + 17·5                               87
■ 5 + 87                                 92
5+ans(1)
MAIN      RAD AUTO      FUNC      9/30
```

After pressing [ENTER], the TI-89 automatically produces `ans(1)` if the first key pressed is one which requires a number before it; the most common of these are [+], [−], [×], [÷], [^], and [STO▸]. For example, [+][5][ENTER] would accomplish the same thing as the keystrokes above (that is, it adds 5 to the previous result). Again, note that the history area shows the value of `ans(1)` rather than the text "`ans(1)`."

```
F1 Tools | F2 Algebra | F3 Calc | F4 Other | F5 PrgmIO | F6 Clean Up
■ 200/16                               25/2
■ √17.5                       4.18330013267
■ 2 + 17·5                               87
■ 5 + 87                                 92
■ 92 + 5                                 97
ans(1)+5
MAIN      RAD AUTO      FUNC      10/30
```

Although the previously entered expression disappears from the entry line if anything is typed, that expression can be re-evaluated by simply pressing [ENTER]. This can be especially useful in conjunction with `ans(1)`.

Several expressions can be evaluated together by separating them with colons ([2nd][4]). When [ENTER] is pressed, the result of the *last* computation is displayed. (The other results are lost. An example showing how this can be used is shown later.)

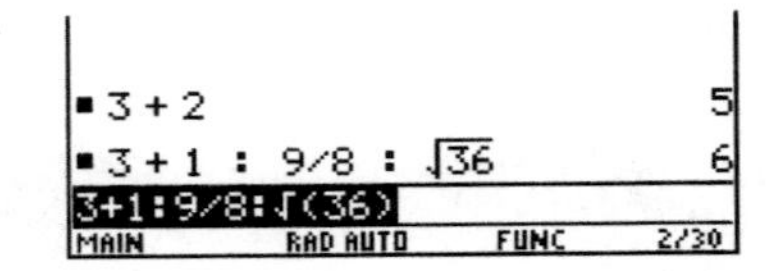

5 Cursors

When typing, the appearance of the cursor and the status line indicates the behavior of the next keypress.

The cursor appears as either a flashing vertical line (the default) or a flashing solid block. The vertical line indicates that subsequent keypresses will be *inserted* at the current cursor location. The block cursor indicates that subsequent keypresses will *overwrite* the character(s) to the right of the cursor. (Of course, if the cursor is located at the right end of the entry line text, these two behaviors are equivalent.) To switch between between these two modes of operation, press [2nd][←] (INS).

By default, most keys produce the character shown on the key itself. The four modifier keys [2nd], [◆], [↑], and [alpha] change this. Pressing any one of these keys causes a corresponding indicator to appear in the status line, and the next keypress will then do something different from its primary function. Pressing [2nd] or [◆] causes the next keypress to produce the results—the character or operation—indicated by (respectively) the yellow or green text above that key. If [2nd] or [◆] is pressed by mistake, pressing it a second time will undo that modifier.

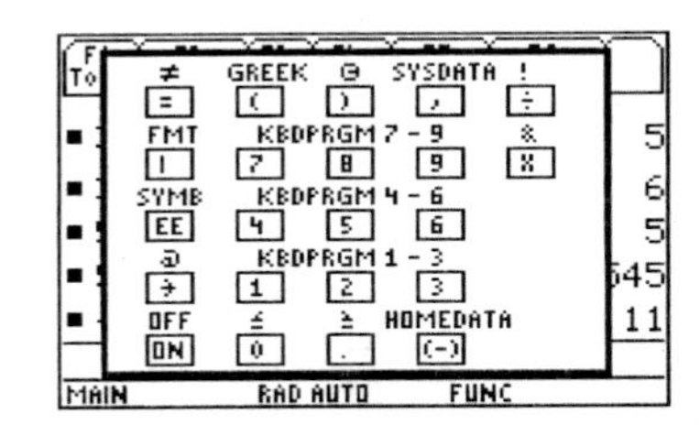

For [2nd], that makes each key's function fairly clear, but many of the keys have no green text above them, leading one to think that the [◆] modifier would accomplish nothing with that key. In fact, nearly every key does something in response to the [◆] key. For an easy way to see the [◆] functions of the lower half of the keypad, press [◆][EE], which produces the display on the right. (Note that the letter associated with [EE] is "K"—think "K" for "keys.") For example, [◆] followed by [=], [)], [÷], [×], [STO▸], [0], or [.] produces the character shown. [◆][(] followed by any letter produces the Greek equivalent of that letter (or as near an equivalent as there is); e.g., [◆][(][Z] produces ζ ("zeta"). [◆][|] allows one to change the number of previous entries saved in the history area. The other [◆] functions are beyond the scope of this manual.

Pressing [alpha] means that the next keypress will produce the (lowercase) letter or other character printed in purple above that key (if any). [alpha] has no effect on (and is not needed for) [X], [Y], [Z], and [T]. Following that letter, subsequent keypresses will produce their primary functions (i.e., not letters). To produce an uppercase letter, press [alpha][↑] followed by a letter key. ([alpha] is not needed for [X], [Y], [Z], and [T].)

The TI-89 can be "locked" into (lowercase) alphabetic mode by pressing [2nd][alpha] (or [alpha][alpha]). From then on, each key produces its letter. This continues until [alpha] is pressed again, which takes the TI-89 out of alphabetic mode.

To lock the TI-89 in uppercase alphabetic mode, press [↑][alpha]. As before, pressing [alpha] again takes the TI-89 out of alphabetic mode.

6 Accessing previous entries ("deep recall")

By repeatedly pressing [2nd][ENTER] (ENTRY), previously typed expressions can be retrieved for editing and re-evaluation. Pressing [2nd][ENTER] once recalls the most recent entry; pressing [2nd][ENTER] again brings up the second most recent, etc.

More conveniently, pressing ⊕ and ⊖ allows one to select previous entries and results from the history area; simply highlight the desired expression and press [ENTER]. For example, pressing ⊕⊕[ENTER] would copy the previous entry to the new-entry line, while ⊕⊕⊕[ENTER] would copy the second-previous *result*.

Once an expression is on the new-entry line at the bottom of the home screen, it can be edited in various ways. Text can be deleted (using [←] to delete the character before the cursor, or [◆][←] to delete the character after the cursor). New text can be inserted (see the previous section). One can even highlight text by pressing and holding [↑] together with an arrow key, then cut ([◆][2nd]) or copy ([◆][↑]) it to a "clipboard," so that it can be pasted ([◆][ESC]) somewhere else in the expression.

7 Menus

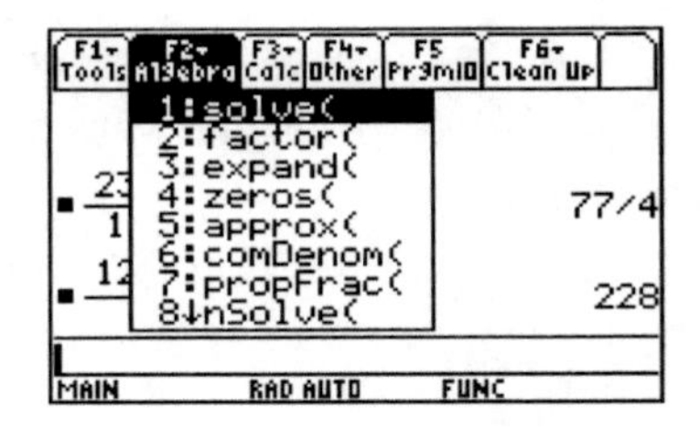

Certain keys and key combinations bring up a menu in a window with a variety of options. Shown is the menu produced by pressing [F2] (Algebra) from the home screen. The arrow next to "8" means that there are more options available (which can be seen by pressing ⏶ or ⏷). To select one of these options (and paste the corresponding command on the entry line), simply press the number (or letter) next to the option. Alternatively, use ⏶ and ⏷ to highlight the desired option and press [ENTER].

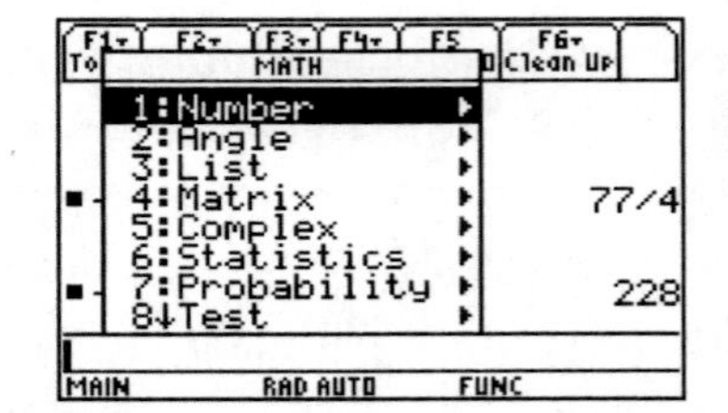

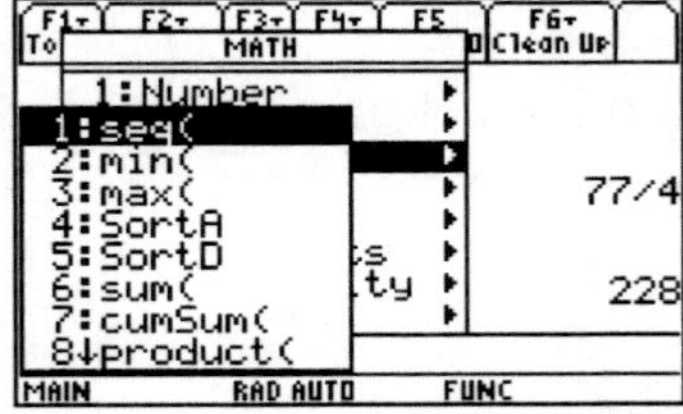

[2nd][5] brings up the MATH menu shown on the right (above). For each of these menu options, the triangle ("▸") on the right side indicates that selecting that option brings up a sub-menu. Below on the right is the List sub-menu (option 3 of the MATH menu). Note that the status line contains a reminder of how to use these menus; [ESC] can be used to exit from one level of a menu (and [2nd] [ESC] would remove all menus and return to the home screen).

The various commands in these menus are too numerous to be listed here. They will be mentioned as needed in the examples.

One last comment is worthwhile, however. Some functions that may be used frequently are buried several levels deep in the menus, and may take many keystrokes to access. Worse, the location of the function might be forgotten (is it in the Algebra or MATH:Number menu?), necessitating a search through the menus. It is useful to remember three things:

- Any command can be typed one letter at a time, in either upper- or lowercase; e.g., [⬆][ALPHA][=][(] [3][ALPHA][(]will type the letters "ABS(", which has the same effect as [2nd][5][1][2].

- Any command can be found in the [CATALOG] menu. Since the commands appear in alphabetical order, it may take some time to locate the desired function. Pressing any letter key (it is not necessary to press [alpha] first) brings up commands starting with that letter ; e.g., pressing [9] brings up the list on the right, while pressing [−] shows commands starting with "O."

- An alternate home-screen menu bar can be found by pressing [2nd] [HOME] (CUSTOM). Pressing [2nd][HOME] again toggles back to the "standard" home-screen menu bar. It is possible to change the commands listed in this CUSTOM menu, but the process is somewhat tedious. Full details can be found in the TI-89 manual, but as an example: The command

```
Custom:Title "Fns":Item "abs(":Item "log(":Item "ln(":EndCustm
```

would result in a CUSTOM menu bar for which [F1] would produce the menu shown on the right.

8 Variables

The letters A through Z (upper- or lowercase), and also sequences of letters (like "High" or "count") can be used as variables (or "memory") to store numerical values. To store a value, type the number (or an expression) followed by [STO▸], then a letter or letters (pressing [alpha] if necessary), then [ENTER]. That variable name can then be used in the same way as a number, as demonstrated at right. These variable names are *not* case-sensitive, so "A" is the same as "a," and "CoUnT" is the same as "count."

Note: The TI-89 interprets 2a as "2 times a"—the "*" symbol is not required (this is consistent with how we interpret mathematical notation). As for order of operations, this kind of multiplication is treated the same as "*" multiplication (see the screen above).

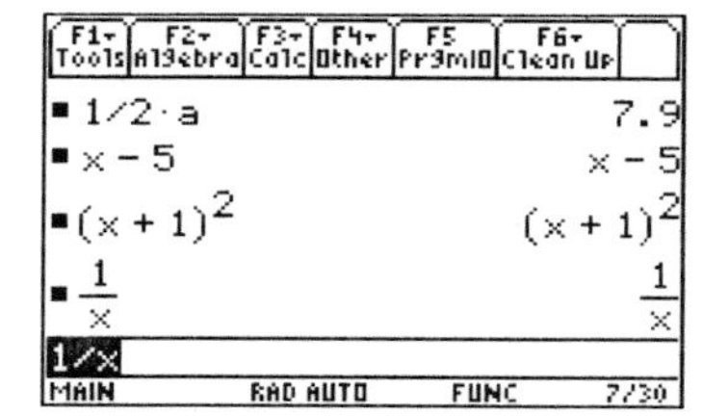

If a variable is used for which no value has been assigned, it is treated as an unknown value, and expressions involving it remain unevaluated. In the screen on the right, the variable x has no assigned value. Any variable's assigned value can be erased ("forgotten") by issuing the command DelVar ([F4][4] from the home screen menubar) followed by the variable name. All one-letter variables can be cleared by choosing [2nd][F1] (Clean Up) [1], or using the NewProb command ([2nd][F1][2]), which also clears the history area.

9 Setting the modes

By pressing [MODE], one can change many aspects of how the calculator behaves. For most of the examples in this manual, the MODE settings will be as shown on the three screens below (although in some cases the settings are not crucial). Some (not all) of these options are described below; consult the TI-89 manual for more details. Changes in the settings are made using the arrows keys and [ENTER].

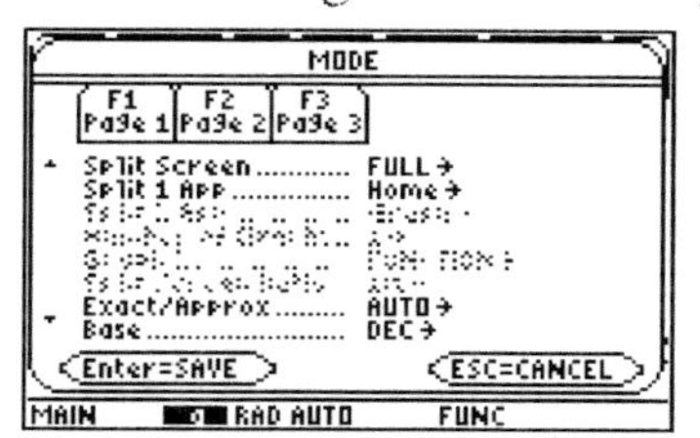

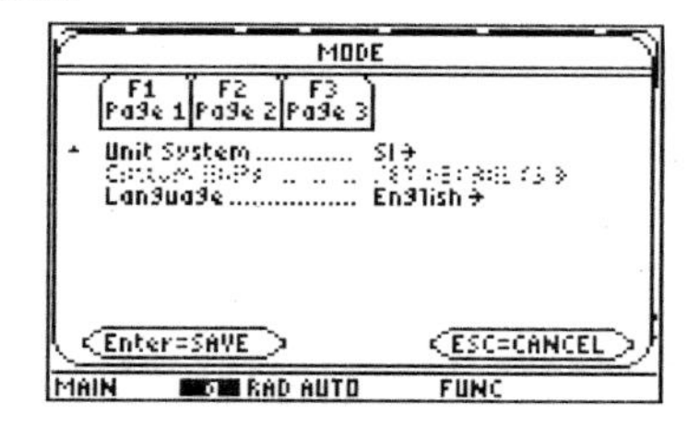

The Graph setting can be either FUNCTION, PARAMETRIC, POLAR, SEQUENCE, 3D, or DIFF EQUATIONS. This setting determines whether formulas to be graphed are functions (y as a function of x), parametric equations (x and y as functions of t), polar equations (r as a function of θ), or sequences (u as a function of n); the other two settings are beyond the scope of this manual. The text accompanying this manual uses FUNCTION, PARAMETRIC, and SEQUENCE modes. The current value of this setting is indicated in the status line at the bottom of the calculator screen—FUNC, PAR, etc.

The Display Digits setting can be FLOAT, FIX n, or FLOAT n, where n is an integer from 1 to 12. This specifies how numbers should be displayed: FLOAT means that the TI-89 should display all non-zero digits (up to a maximum of 12), while (e.g.) FLOAT 4 means that a total of 4 digits will be displayed. Meanwhile, FIX 4 means that the TI-89 will attempt to display 4 digits beyond the decimal point.

`Angle` can be either `RADIAN` or `DEGREE`, indicating whether angle measurements should be assumed to be in radians or degrees. (A right angle measures $\frac{\pi}{2}$ radians, which is equivalent to 90°.) The current value of this setting is indicated in the status line at the bottom of the calculator screen—`RAD` or `DEG`. This text does not refer to angle measurement.

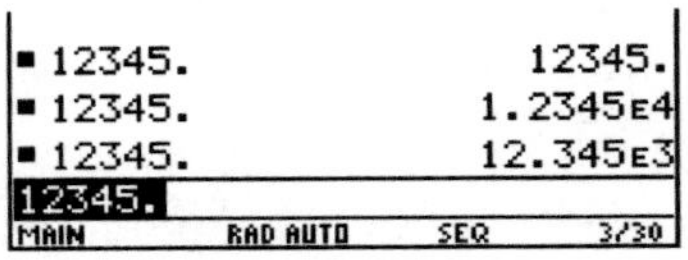

The `Exponential Format` setting is either `NORMAL`, `SCIENTIFIC`, or `ENGINEERING`; this specifies how numbers should be displayed. The screen on the right shows the number "`12345.`" displayed in each of these modes: `NORMAL` mode displays numbers in the range ±999, 999, 999, 999 with no exponents, `SCIENTIFIC` mode displays all numbers in scientific notation, and `ENGINEERING` mode uses only exponents that are multiples of 3. Note: "`E`" is short for "times 10 to the power," so 1.2345E4 $= 1.2345 \times 10^4 = 1.2345 \times 10000 = 12345$.

```
■ 12345.                12345.
■ 12345.              1.2345E4
■ 12345.              12.345E3
12345.
MAIN        RAD AUTO    SEQ   3/30
```

The `Complex Format` is either `REAL`, `RECTANGULAR`, or `POLAR`, and specifies the display mode for complex numbers. `REAL` means that the TI-89 will produce an error if an expression requires the computation of (e.g.) a square root of a negative number, and the other two settings determine whether complex results should be displayed in rectangular or polar format. These two formats are essentially the same as the two used by the textbook. **Note:** The text prefers the term "trigonometric format" rather than "polar format." More information about complex numbers can be found beginning on page 117 (Example 1 from Section 3.1) of this manual.

The `Vector Format` setting (`RECTANGULAR`, `CYLINDRICAL`, or `SPHERICAL`) indicates the default display format for vectors (not used in this text).

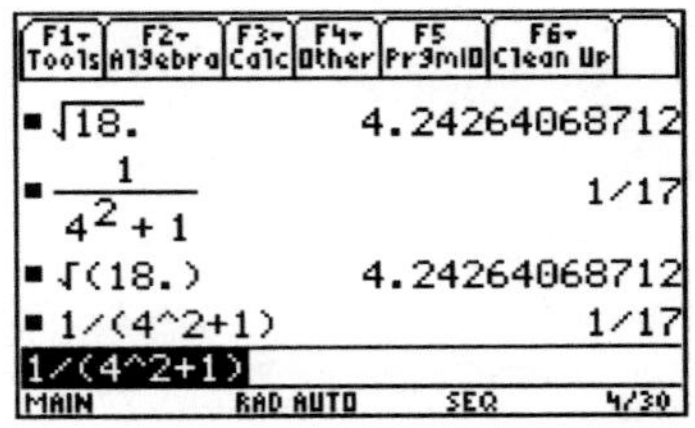

`Pretty Print` (`ON` or `OFF`) determines how expressions (input and output) should be displayed. The first two entries on the right were performed with Pretty Print on, and the other two were done with Pretty Print off.

```
F1▾   F2▾     F3▾  F4▾   F5      F6▾
Tools Algebra Calc Other PrgmIO Clean Up
■ √18.           4.24264068712
■   1
  ------                  1/17
  4² + 1
■ √(18.)         4.24264068712
■ 1/(4^2+1)               1/17
1/(4^2+1)
MAIN        RAD AUTO    SEQ   4/30
```

The `Exact/Approx` setting (`AUTO`, `EXACT`, or `APPROXIMATE`) determines whether results should be considered to be exact or approximate. `EXACT` means that all decimals are converted to fractions; e.g., `.9` is displayed as `9/10`, and `√(2.5)` is $\sqrt{10}/2$. `APPROXIMATE` means that everything is converted to decimal form; for example, `√(2.5)` produces 1.5811.... With the `AUTO` setting, the TI-89 decides whether to display a result as exact or approximate based on whether there is a decimal point in the entry—for example, `√(5/2)` yields $\sqrt{10}/2$, while `√(2.5)` yields 1.5811.... **Note:** Regardless of this setting, pressing [◆][ENTER] instead of [ENTER] to process an entry will cause the TI-89 to show a decimal (approximate) result. (The current value of this setting is indicated in the status line at the bottom of the calculator screen—`AUTO`, `EXACT`, or `APPROX`.)

The other mode settings deal with issues that are beyond the scope of the textbook, and are not discussed here.

10 Setting the graph window

Pressing [◆][F2] brings up the WINDOW settings. The exact contents of the WINDOW menu vary depending on the Graph mode setting; below are six examples showing this menu in each possible Graph modes. (The last two are not used in this manual, but are shown here for reference.)

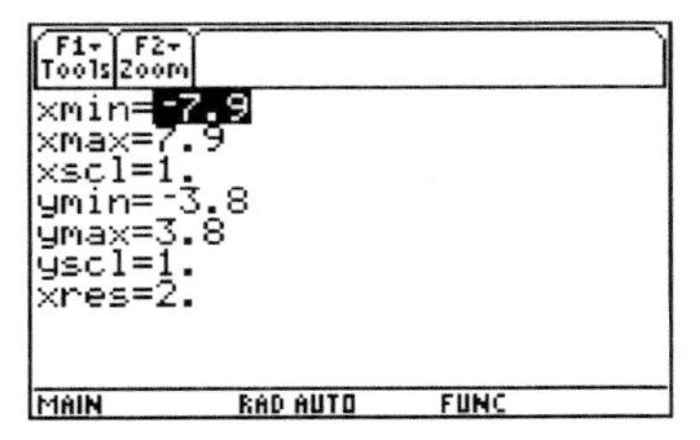

Function mode

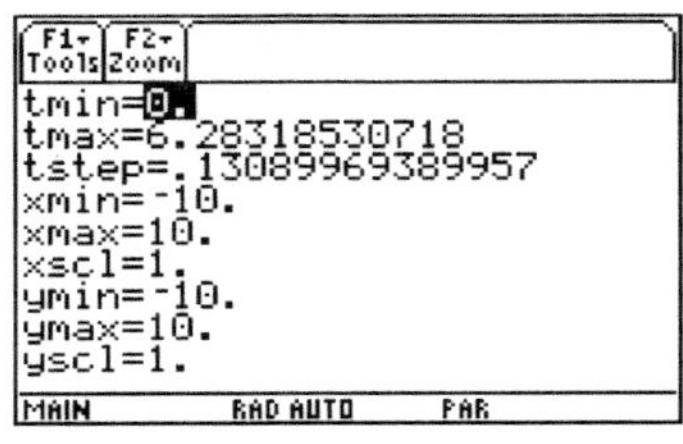

Parametric mode

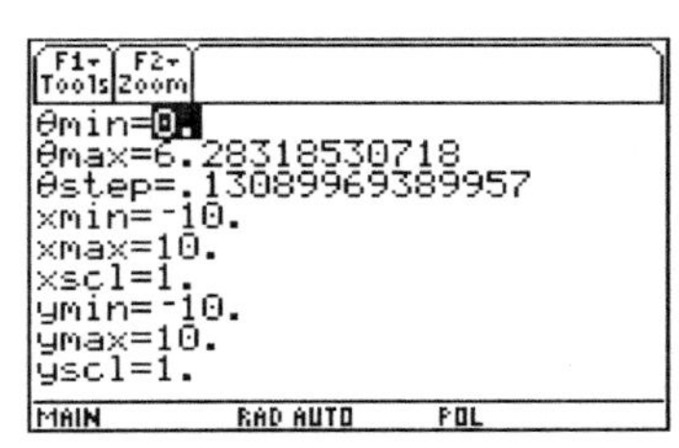

Polar mode

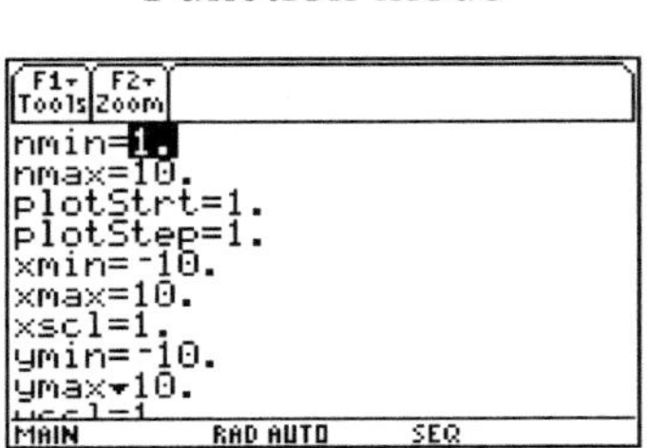

Sequence mode

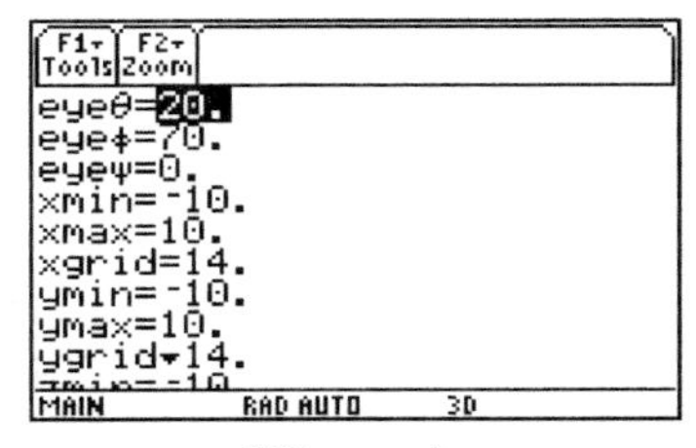

3D mode

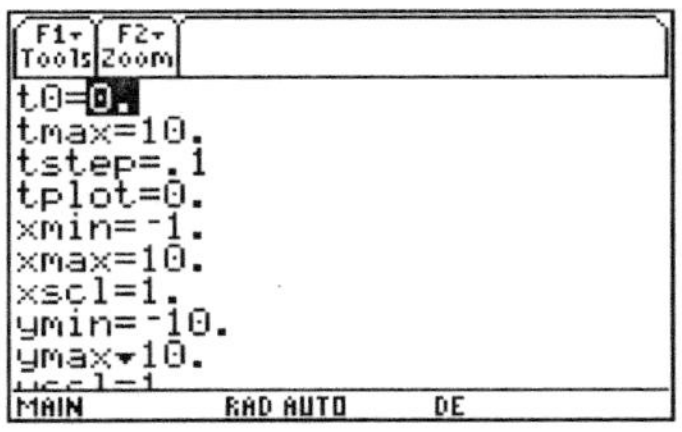

Diff Equations mode

All these menus include the values `xmin`, `xmax`, `xscl`, `ymin`, `ymax`, and `yscl`. When [◆][F3] (GRAPH) is pressed, the TI-89 will show a portion of the Cartesian (x-y) plane determined by these values. In function mode, this menu also includes `xres`, the behavior of which is described in section 12 of this manual (page 105). The other settings in this screen allow specification of the smallest, largest, and step values of t (for parametric mode) or θ (for polar mode), or initial conditions for sequence mode.

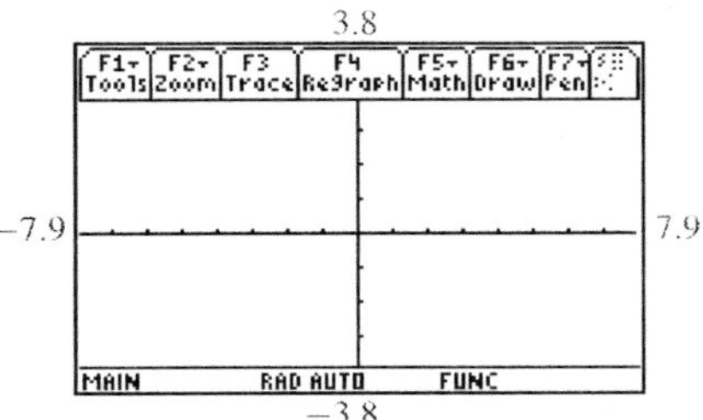

With settings as in the "Function mode" screen shown above, the TI-89 would display the screen at right: x values from -7.9 to 7.9 (that is, from `xmin` to `xmax`), and y values between -3.8 to 3.8 (`ymin` to `ymax`). Since `xscl` $=$ `yscl` $= 1$, the TI-89 places tick marks on both axes every 1 unit; thus the x-axis ticks are at -7, $-6, \ldots, 6$, and 7, and the y-axis ticks fall on the integers from -3 to 3. This window is called the "decimal" window, and is most quickly set by pressing [◆][F2][F2][4] (Zoom:ZoomDec).

Below are three more sets of window settings, and the graph screens they produce. Note that the first graph on the left has tick marks every 10 units on both axes. The second window is called the "standard" viewing window, and is most quickly set by pressing [◆][F2][F2][6] (Zoom:ZoomStd). The setting `yscl` $= 0$ in the final graph means that no tick marks are placed on the y-axis.

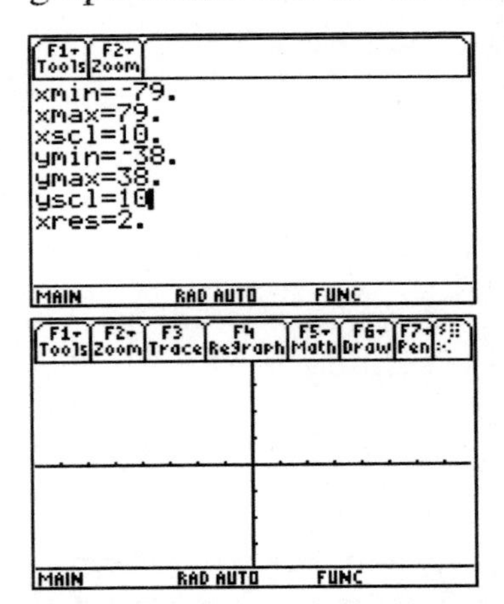

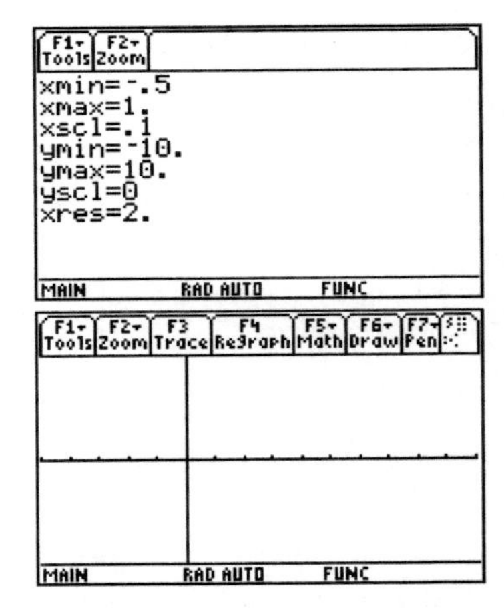

11 The graph screen

The TI-89's graph screen—that is, the portion of the screen used to display graphs, below the menu bar and above the status line—is made up of an array of rectangular dots (pixels) with 77 rows and 159 columns. All the pixels in the leftmost column have x-coordinate `xmin`, while those in the rightmost column have x-coordinate `xmax`. The x-coordinate changes steadily across the screen from left to right, which means that the coordinate for the nth column (counting the leftmost column as column 0) must be `xmin` $+ n\Delta$x, where Δx $=$ (`xmax` $-$ `xmin`)/158. Similarly, the nth row of the screen (counting up from the bottom row, which is row 0) has y-coordinate `ymin` $+ n\Delta$y, where Δy $=$ (`ymax` $-$ `ymin`)/76.

It is not necessary to memorize the formulas for Δx and Δy. Should they be needed, they can be determined by pressing [◆][F3] (GRAPH) and then the arrow keys. When pressing ▶ or ◀ successively, the displayed x-coordinate changes by Δx; meanwhile, when pressing ▲ or ▼, the y-coordinate changes by Δy. Alternatively, the values can be found by typing "Δx" and "Δy" on the home screen; this is most easily done by pressing [2nd][+][1][5] to access the CHAR:Greek menu and type the "Δ" character, then press [X] or [Y]. This produces results like those shown on the right; the values of Δx and Δy there are those for the standard viewing window.

In the decimal window `xmin` $= -7.9$, `xmax` $= 7.9$, `ymin` $= -3.8$, `ymax` $= 3.8$, note that Δx $= 0.1$ and Δy $= 0.1$. Thus, the individual pixels on the screen represent x-coordinates $-7.9, -7.8, -7.7, \ldots, 7.7, 7.8, 7.9$ and y-coordinates $-3.8, -3.7, -3.6, \ldots, 3.6, 3.7, 3.8$. This is where the decimal window gets its name.

Windows for which Δx $=$ Δy, such as the decimal window, are called square windows. Since there are just over twice as many columns as rows on the graph screen, this means that square windows should have `xmax` $-$ `xmin` just over twice as big as `ymax` $-$ `ymin`. Any window can be made square be pressing [◆][F2][F2][5] (Zoom:ZoomSqr). To see the effect of a square window, observe the two pairs of graphs below. In each pair, the first graph is on the standard window, and the second is on a square window (after pressing [◆][F2][F2][5]). (This changes `xmin` and `xmax` to about -20.8 and 20.8, respectively, while `ymin` and `ymax` remain unchanged at -10 and 10.) The first pair shows the line $y = x$; on the square window, this line (correctly) appears to make a 45° angle with the x- and y-axes. The second pair shows the lines $y = 2x - 3$

and $y = 3 - \frac{1}{2}x$; note that on the square window, these lines look perpendicular (as they should). Finally, the last pair shows a circle centered at the origin with a radius of 8. On the standard window, this looks like an oval since the screen is wider than it is tall. (The reason for the gaps in the circle will be addressed in the next section.)

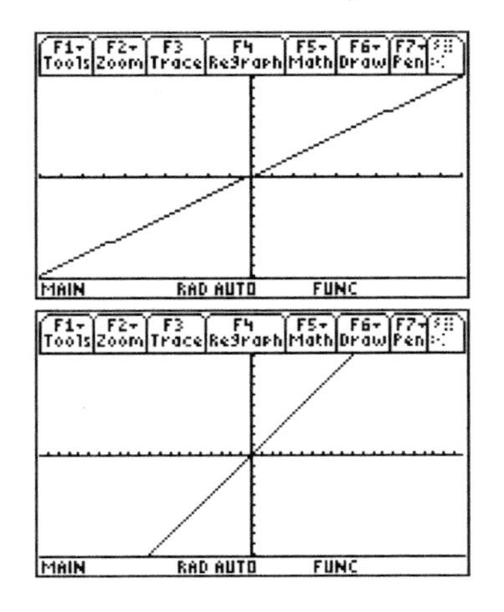

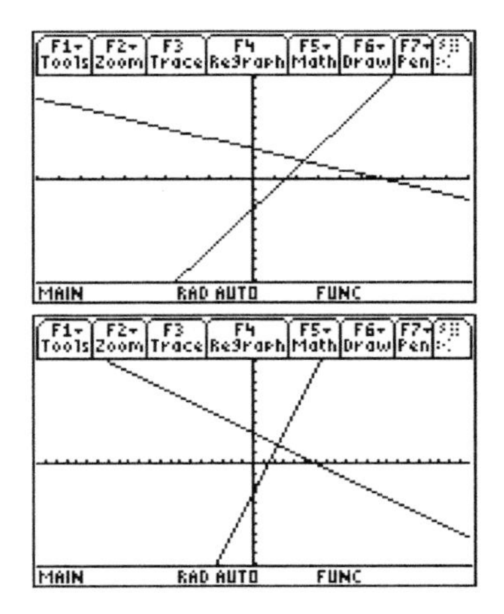

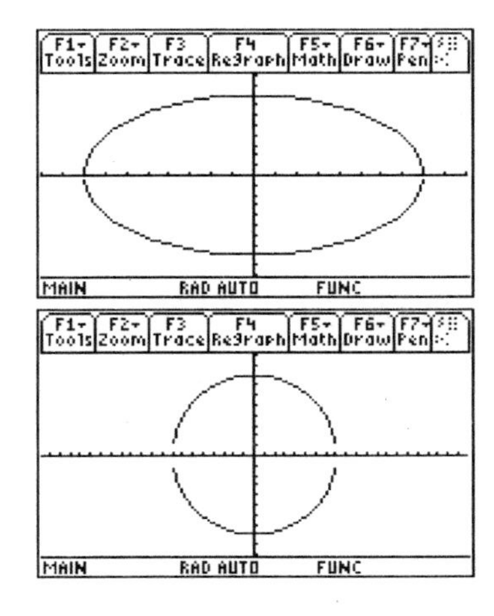

12 Graphing a function

This introductory section only addresses creating graphs in function mode. Procedures for creating parametric and polar graphs are very similar, and are described in this manual in the material related to the examples from the text.

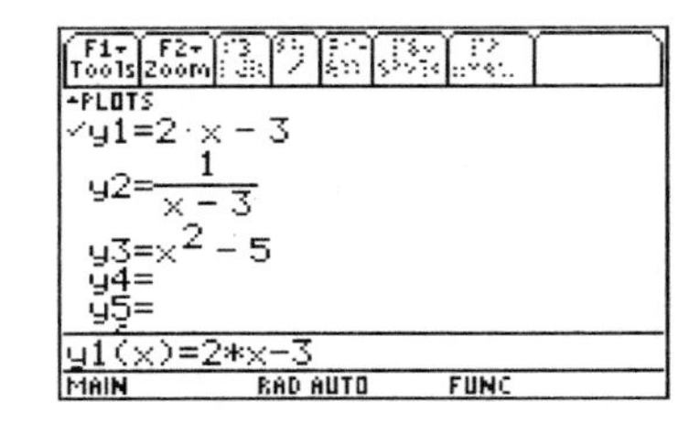

To see the graph of $y = 2x - 3$, begin by entering the formula into the calculator. This is done by pressing [◆][F1] to access the "y equals" screen of the calculator. Enter the formula as `y1` (or any other `y`n). (If `y1` already has a formula, press [ENTER], [F3], or [CLEAR] first, then type the new formula.) If another y variable has a formula, position the cursor on that line and press either [CLEAR] (to delete the formula) or [F4]. The latter has the effect of toggling the check mark for that line; which tells the TI-89 whether or not to graph that formula. In the screen on the right, only `y1` will be graphed.

The next step is to choose a viewing window. See the previous section for more details on this. This example uses the standard window ([F2][6]).

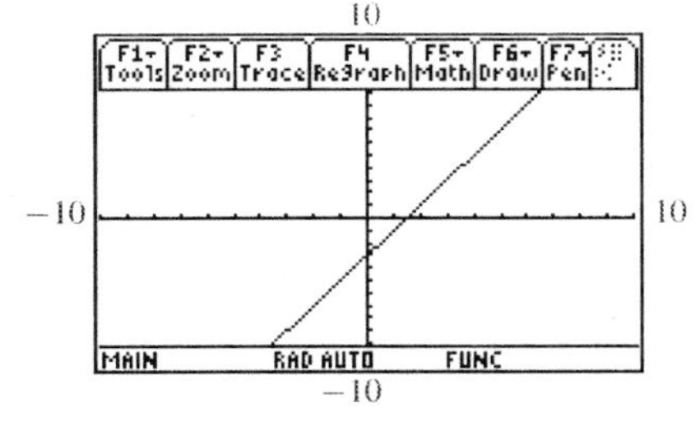

If the graph has not been displayed, press [◆][F3], and the line should be drawn. In order to produce this graph, the TI-89 considers 159 values of x, ranging from `xmin` to `xmax` in steps of Δx (assuming that `xres` = 1; see below for other possibilities). For each value of x, it computes the corresponding value of y, then plots that point (x, y) and draws a line between this point and the previous one. (See also the information about graph styles later in this section.)

If `xres` is set to 2, the TI-89 will only compute y for every other x value; that is, it uses a step size of 2Δx. Similarly, if `xres` is 3, the step size will be 3Δx, and so on. Setting `xres` higher causes graphs to appear faster (since fewer points are plotted), but for some functions, the graph may look "choppy" if `xres` is too large, since detail is sacrificed for speed.

Note: If the line does not appear, or the TI-89 reports an error, double-check all the previous steps. Also, check the mode settings (discussed in section 9, page 101).

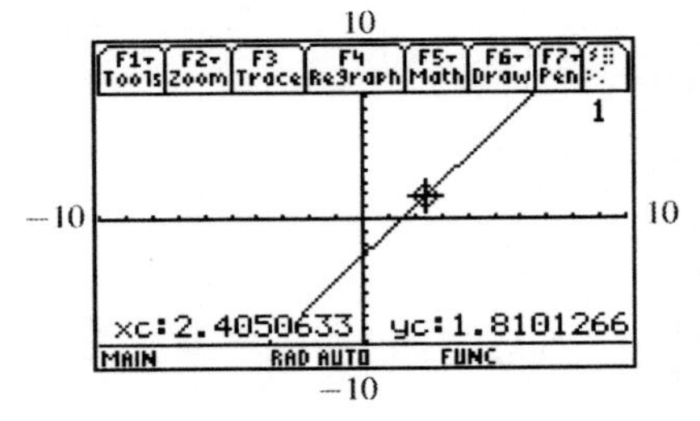

Once the graph is visible, the window can be changed using [◆][F2] (WINDOW) or [F2] (ZOOM). Pressing [F3] (TRACE) brings up the "trace cursor," and displays the x- and y-coordinates for various points on the line as the ◀ and ▶ keys are pressed. (These variables—`xc` and `yc`—can also be referenced from the home screen; that is, typing `xc` [ENTER] on the home screen would show the value 2.40506. . . .) Tracing beyond the left or right columns causes the TI-89 to adjust the values of `xmin` and `xmax` and redraw the graph.

To graph the function

$$y = \frac{1}{x-3},$$

enter that formula into the "y equals" screen (note the use of parentheses on the entry line). As before, this example uses the standard viewing window.

For this function, the TI-89 produces the graph shown on the right. This illustrates one of the pitfalls of the connect-the-dots method used by the calculator: The nearly-vertical line segment drawn at $x = 3$ *should not be there*, but it is drawn because the calculator connects the points

$$x \approx 2.911,\ y \approx -11.286 \ \text{ and } \ x \approx 3.165,\ y \approx 6.077.$$

Calculator users must learn to recognize these flaws in calculator-produced graphs.

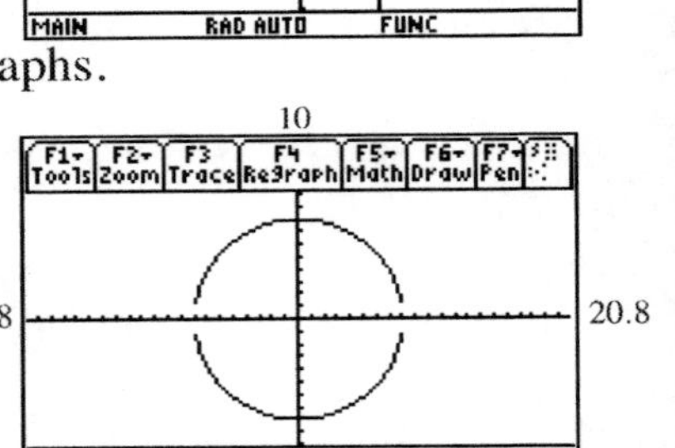

The graph of a circle centered at the origin with radius 8 (shown on a square window, with `xres` = 1) shows another problem that arises from connecting the dots. When $x = -8.157895$, y is undefined, so no point is plotted (that is, there is no point on this circle that has x-coordinate less than -8, or greater than 8). The next point plotted on the upper half of the circle is $x = -7.894737$ and $y = 1.2934953$; since no point had been plotted for the previous x-coordinate, this is not connected to anything, so there appears to be a gap between the circle and the x-axis. The calculator is not "smart" enough to know that the graph should extend from -8 to 8.

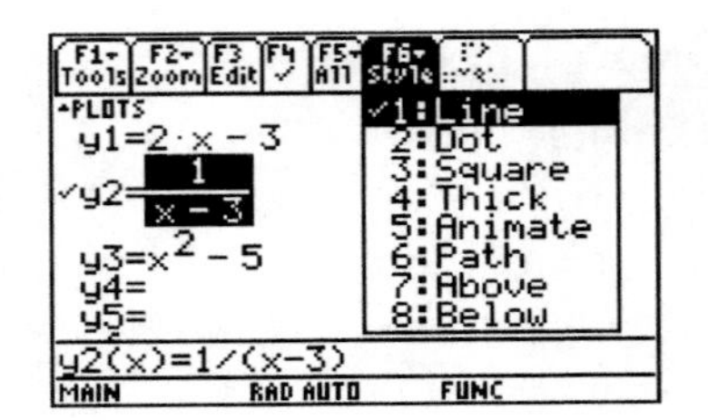

One additional feature of graphing with the TI-89 is that each function can have a "style" assigned to its graph. To see this style, go to the "y equals" screen and press [2nd][F1] (Style); the check mark indicates which style applies to the current function. These options are shown on the right; "`Line`," the default, means that the calculator should draw lines between the plotted points. More information can be found in the examples, and complete details are in the TI-89 manual.

13 Adding programs to the TI-89

The TI-89's capabilities can be extended by downloading or entering programs into the calculator's memory. Instructions for writing a program are beyond the scope of this manual, but programs written by others and downloaded from the Internet (or obtained as printouts) can be transferred to the calculator in one of three ways:

1. If one TI-89 already has a program, it can be transferred to another using the calculator-to-calculator link cable. To do this, first make sure the cable is firmly inserted in both calculators. On both calculators, press [2nd][−] (VAR-LINK). On the sending calculator, use the arrow keys and [F4] to place check marks next to the programs to be transferred. On the receiving calculator, press [F3][2] (Link:Receive), then on the sending calculator, press [F3][1] (Link:Send to TI-89/TI-92 Plus).

2. If a computer with the TI-Graph Link is available, and the program file is on that computer (e.g., after having been downloaded from the Internet), the program can be transferred to the calculator using the TI Connect (or TI Graph Link) software. This transfer is done in a manner similar to the calculator-to-calculator transfer described above; specific instructions can be found in the documentation that accompanies the software. (They are not given here because of slight differences between platforms and software versions.)

3. View a listing of the program and type it in manually. (**Note:** Even if the TI-Graph Link cable is not available, the software can be used to view program listings on a computer.) While this is the most tedious method, studying programs written by others can be a good way to learn programming. To enter a program, start by choosing [APPS][7] [3] (Program Editor:New), then specify whether this is a program or a function, and give it a name (up to eight characters, like "`quadform`" or "`midpoint`")—note that the TI-89 is automatically put into alpha mode while typing this name. Then press [ENTER] (OK), enter the commands in the program or function, and press [2nd][ESC] (QUIT) to return to the home screen when finished.

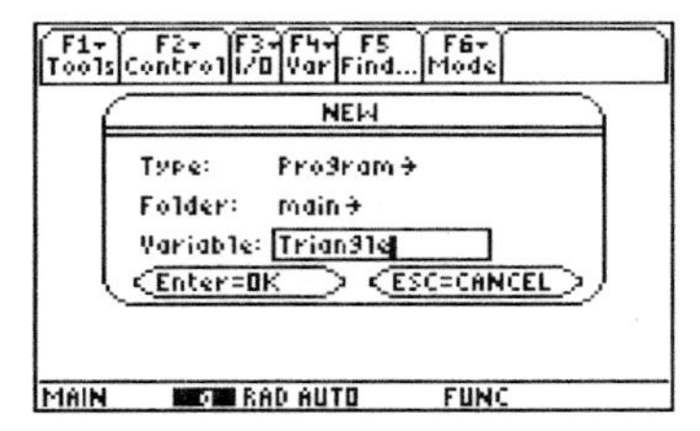

To run the program, make sure there is nothing on the current line of the home screen, then type the name of the program or function (this name is not case sensitive). Follow this with a set of parentheses, containing any required arguments, then press [ENTER]. If the program was entered manually (option 3 above), errors may be reported; in that case, press [ENTER] (GOTO), correct the mistake and try again.

Programs can be found at many places on the Internet, including:

- `http://www.bluffton.edu/~nesterd`—the Web site of the author of this manual;
- `http://tifaq.calc.org`—a "Frequently Asked Questions" page maintained by Ray Kremer; and
- `http://www.ticalc.org`.

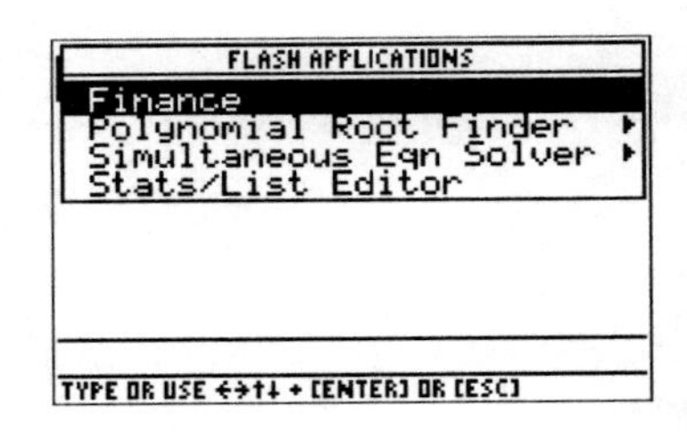

Additionally, one can install a variety of "APPs"—applications (programs) which can extend the capabilities of the calculator in various ways. APPs can be viewed by pressing [APPS][1], or [◆][APPS]; the TI-89 Titanium edition has a number of APPs preloaded, while the standard TI-89 has none. Shown is a TI-89 with four APPs installed, some of which are discussed in this manual. These and other APPs can be downloaded from `education.ti.com`, then installed using a Graph Link cable.

Examples

Here are the details for using the TI-89 for several of the examples from the textbook. Also given are the keystrokes necessary to produce some of the commands shown in the text's examples. In some cases, some suggestions are made for using the calculator more efficiently.

Throughout this section, it is assumed that the textbook is available for reference. The problems from the text are not restated here, and there are frequent references to the calculator screens shown in the text.

Section 1.1 Technology Note (page 3) — Viewing Windows

Information about setting viewing windows is given in section 10 of the introduction, page 103.

Section 1.1 Example 1 (page 5) — Finding Roots on a Calculator

Aside from the square root function √ ([2nd][×]), the TI-89 has no other root functions; specifically, the ³√ and ˣ√ commands shown in Figure 11 are not available. Therefore, expressions like those in this example must be entered in exponential form.

Section 1.2 Technology Note (page 18) — Producing Tables

To use the table features of the TI-89, begin by entering the formula ($y = 9x - 5$, in the example shown in Figure 26 of the text) on the Y= screen, as one would to create a graph. (The check marks determine which formulas will be displayed in the table, just as they do for graphs.)

Next, press [◆][F4] to access the Table Setup screen. The table will display y values for given values of x. The `tblStart` value sets the lowest value of x, while `Δtbl` determines the "step size" for successive values of x. These two values are only used if the `Independent` option is set to `AUTO`—this means, "automatically generate the values of the independent variable (x)." The effect of setting this option to `ASK` is illustrated at the end of this example.

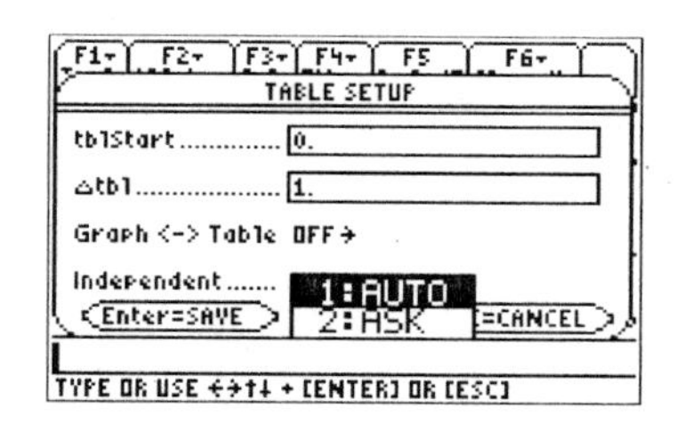

When the Table Setup options are set satisfactorily, press [ENTER] then [◆][F5] to produce the table. The first screen on the right is the table generated based on the settings in the above screen (essentially the same as Figure 26 in the text). By pressing ▼ repeatedly, the x values are increased, and the y values updated. After pressing ▼ nine times, the table looks like the second screen. Similarly, the x values can be decreased by pressing ▲.

x	y1
0.	-5.
1.	4.
2.	13.
3.	22.
4.	31.

x=0.

x	y1
5.	40.
6.	49.
7.	58.
8.	67.
9.	76.

x=9.

Finally, the screen on the right shows the effect of setting `Independent` to `ASK`. The settings of `tblStart` and `Δtbl` are ignored; you enter an x value into the first column, and the corresponding y values will be computed.

x	y1
1.	4.
12.5	107.5
100.	895.

x=-5

x	y1
1.	4.
12.5	107.5
100.	895.
-5.	-50.

x=-5.

Section 1.4 Example 5 (page 41) Using the Slope Relationship for Perpendicular Lines

See page 104 for information about setting square windows, including an illustration of perpendicular lines on square and non-square windows.

Section 1.4 Example 6 (page 42) Modeling Medicare Costs with a Linear Function

Given a set of data pairs (x, y), the TI-89 can produce a scatter diagram (as well as other types of statistical plots).

The first step is to enter the data into the TI-89. This is done by pressing [APPS][6][3] (Data/Matrix Editor), then entering a name for the "data variable" (which will contain all of the numbers for the scatterplot). Alternatively, use an existing data variable, if there is one.

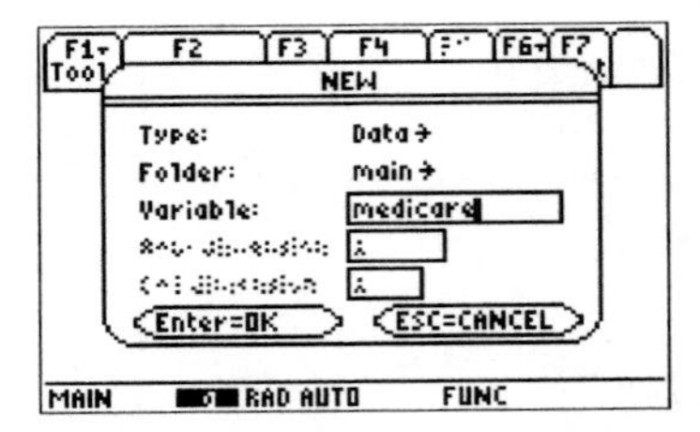

In the spreadsheet-like screen that appears, enter the year values into the first column (c1) and the costs into the second column (c2). If re-using an existing data variable, old data can be cleared out using Tools:Clear Editor, Utils:Delete, or Utils:Clear Column. Make sure that both columns contain the same number of entries. This screen approximately matches the one shown in Figure 59.

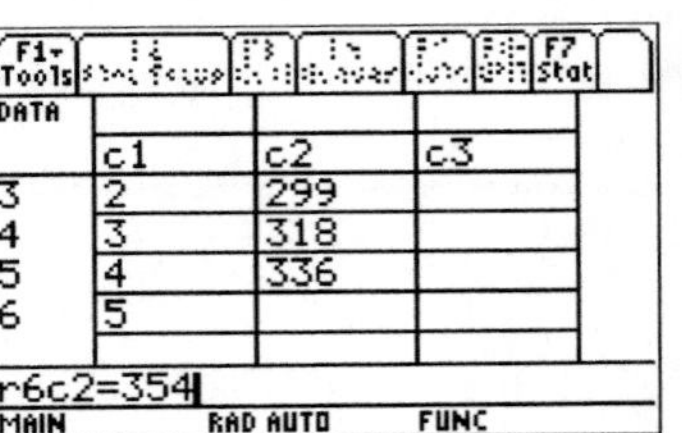

To produce the scatter diagram, press [F2][F1] (Plot Setup:Define), then choose Scatter (the default) for the Plot Type, and set x and y as c1 and c2, respectively.

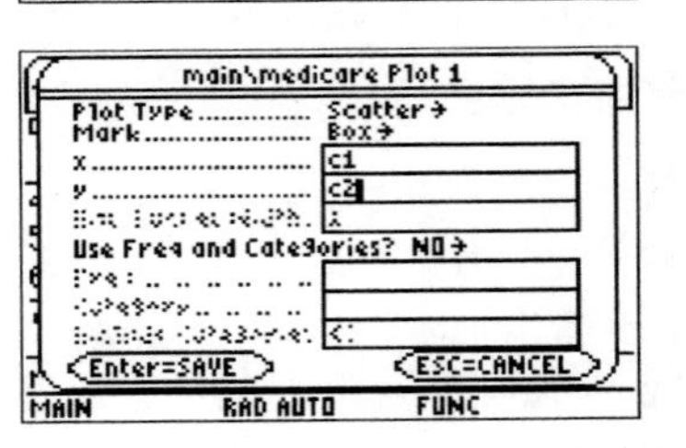

Next, check that nothing else will be plotted: Press [◆][F1] and make sure that only Plot1 has a check mark next to it. Use [F4] to turn off the check marks next to everything else.

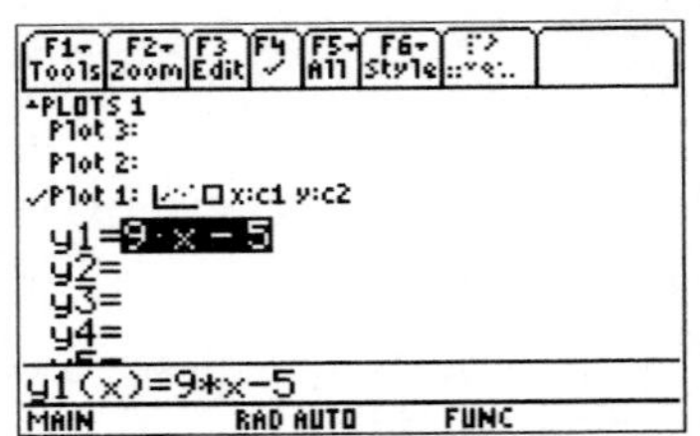

Finally, set up the viewing window as shown in Figure 58 of the text—or press [◆][F2][F2][9] (Zoom:ZoomData), which automatically adjusts the window to show all the data in the plot. (The resulting window does not quite match the one shown in Figure 58.)

Note: When finished with a statistics plot like this one, it is a good idea to turn it off so that the TI-89 will not attempt to display it the next time [◆][F3] (GRAPH) is pushed. This can be done from the Y= screen using [F4] to un-check the plot, or by pressing [F5][5] (All:Data Plots Off).

Section 1.4 Example 7 (page 43) Finding the Least-Squares Regression Line

Given a set of data pairs (x, y), the TI-89 can find various formulas (including linear, as well as more complex formulas) that approximate the relationship between x and y. These formulas are called "regression formulas."

The first step is to enter the data into the TI-89. See the previous example for a description of this process.

Once the data have been entered (and while still in the data editor), press [F5] (Calc). For `Calculation Type`, choose `LinReg`, and set x and y as `c1` and `c2`, respectively.

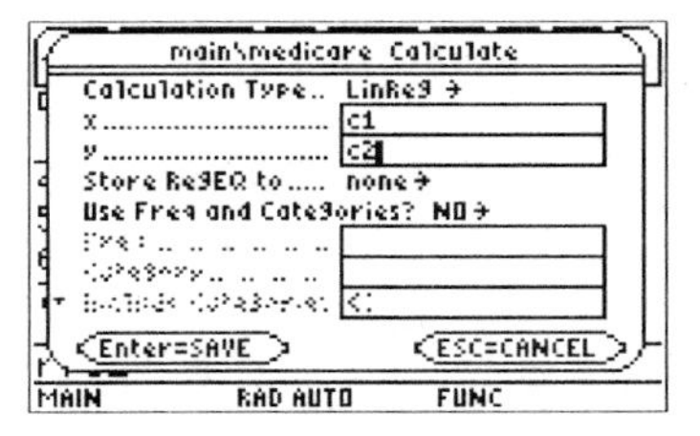

Pressing [ENTER] displays the results of the `LinReg` (shown on the right). These numbers agree (after rounding) with those shown in Figure 61(b) of the text.

Following this example, the text introduces the idea of the correlation coefficient r. When the TI-89 finds a regression formula, it also computes r and r^2, and reports them along with the formula (naming the correlation "`corr`" rather than `r`).

Section 1.5 Example 1 (page 52) Solving a Linear Equation

As is illustrated on the right, the TI-89's `solve` and `nSolve` functions attempt to find solutions to an equation. The format is `solve(`*equation*`,`*variable*`)`, so the first output shown in the history area was from the command `solve(10+3(2x-4)=17-(x+5),x)`. Additionally, the `zeros` function will attempt to find the zeros of an expression; the appropriate entry for this example would be `zeros(10+3(2x-4)-(17-(x+5)),x)`. Full details on how to use these functions (all of which are found in the Algebra menu) can be found in the TI-89 manual.

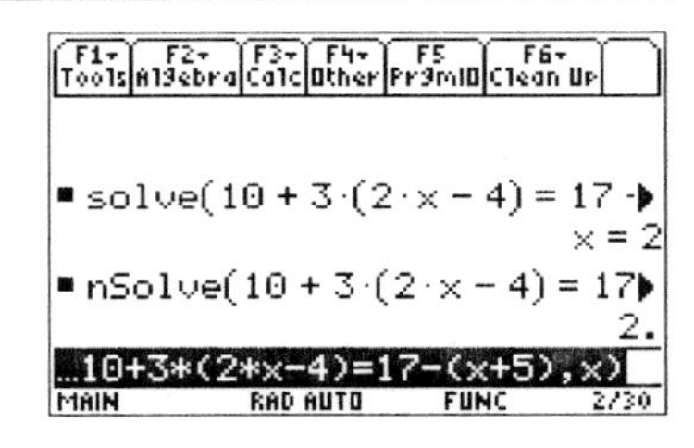

Section 1.5 Example 4 (page 54) Applying the Intersection-of-Graphs Method

We need to solve the equation $f(x) = g(x)$, where $f(x) = 5.91x + 13.7$ and $g(x) = -4.71x + 64.7$. We are looking for an x value that will make the left and right sides of this equation equal to each other, which corresponds to the x-coordinate of the point of intersection of the graphs of $y = f(x)$ and $y = g(x)$.

In order to have the TI-89 locate this intersection, begin by setting up the TI-89 to graph the left side of the equation as `y1`, and the right side as `y2`.

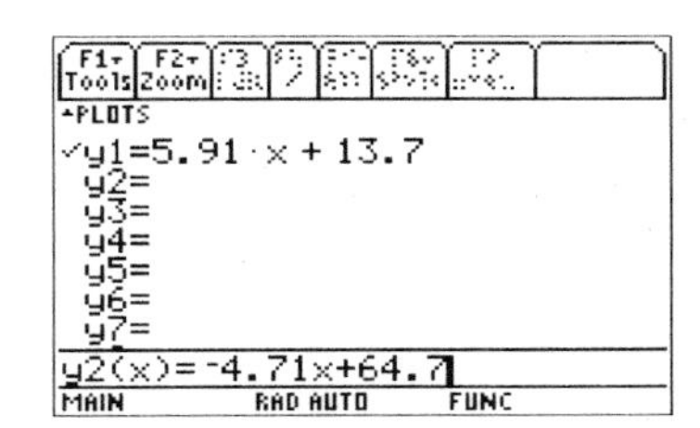

Next, select a viewing window which shows the point of intersection; we use the window shown in Figure 66(a) of the text: [0, 12] × [0, 100]. The TI-89 can automatically locate this point using the GRAPH:Math:Intersection feature. Use ⊙, ⊙ and [ENTER] to specify which two functions to use (in this case, the only two being displayed). The TI-89 then prompts for lower and upper bounds (numbers that are, respectively, less than and greater than the location of the intersection). After pressing [ENTER], the TI-89 will try to find an intersection of the two graphs. The screens below illustrate these steps.

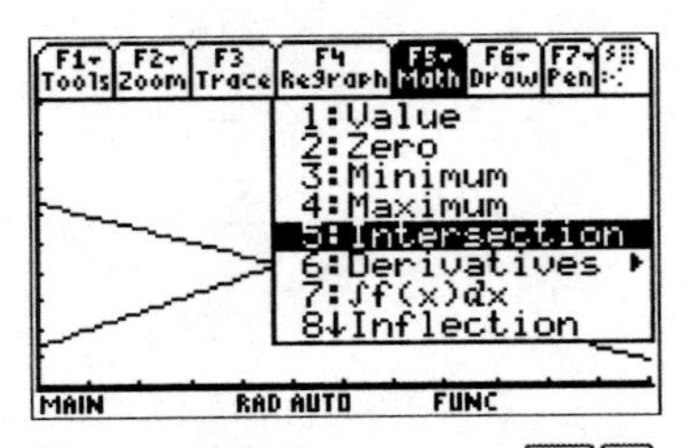

From graph screen: [F5][5]

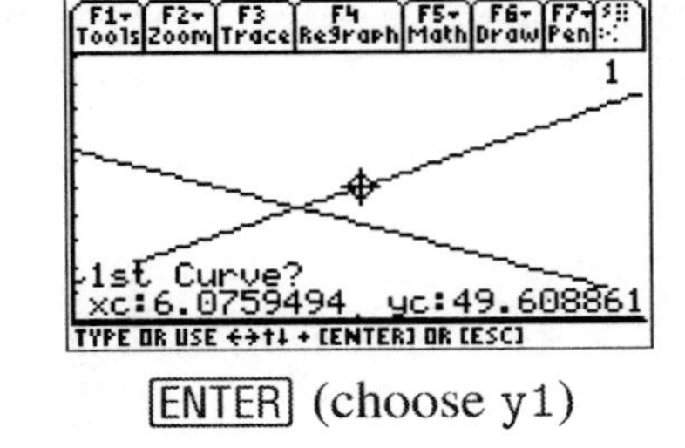

[ENTER] (choose y1)

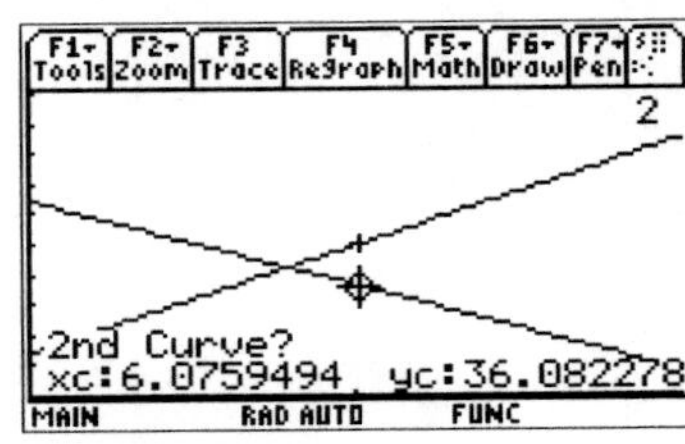

[ENTER] (choose y2)

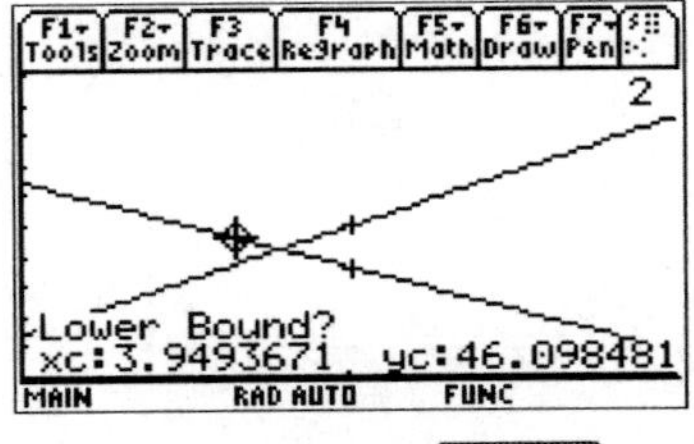

Move cursor [ENTER]

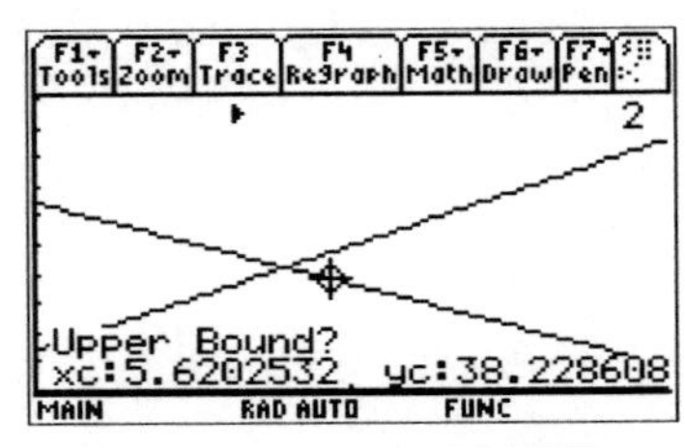

Move cursor [ENTER]

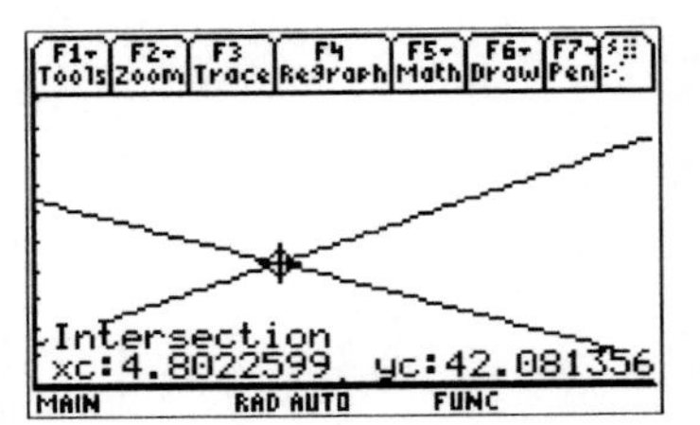

The intersection is found.

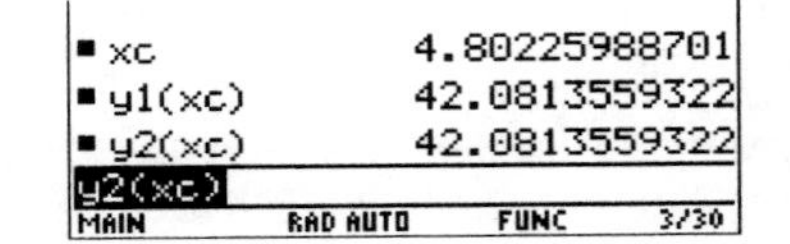

The x-coordinate of this point of intersection is calculated to many digits of accuracy. Also note that following this "intersection" procedure, the calculator variable xc contains the x-coordinate of this intersection. This might be useful for performing computations with the solution; in the screen shown here it is used to confirm that f and g give identical output values at this x value.

Note: An approximation for the point of intersection can be found simply by moving the TRACE cursor as near the intersection as possible. The amount of error can be minimized by "zooming in" on the graph. This is the only method available for graphing calculators such as the TI-81.

See the next example for another graphical approach to solving equations. Additionally, the previous example notes the `solve`, `nSolve`, and `zeros` functions for solving such problems.

Section 1.5 Example 5 (page 55)

Using the x-Intercept Method

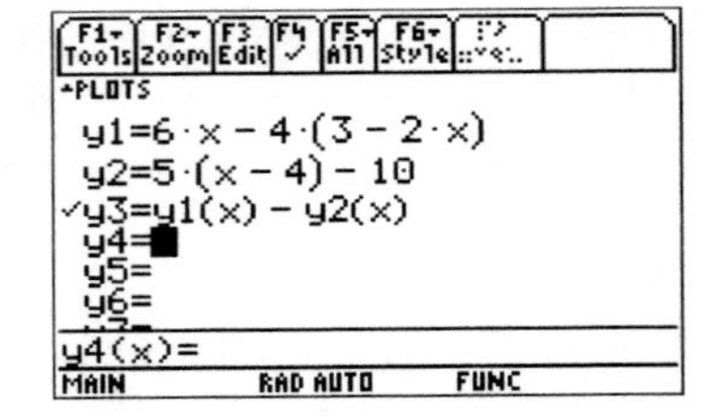

The text suggests graphing `y1=6x-4(3-2x)-(5(x-4)-10)`, noting the importance of placing parentheses around the subtracted expression. An equivalent approach is illustrated on the right: Define y1 and y2 using the expressions on the left and right sides of the equation, and then set `y3=y1(x)-y2(x)`. This method avoids the need for so many (potentially confusing) parentheses, at the cost of a few more keystrokes. Note that y1 and y2 have been "de-selected" (unchecked) so that they will not be graphed (see section 12 of the introduction, page 105).

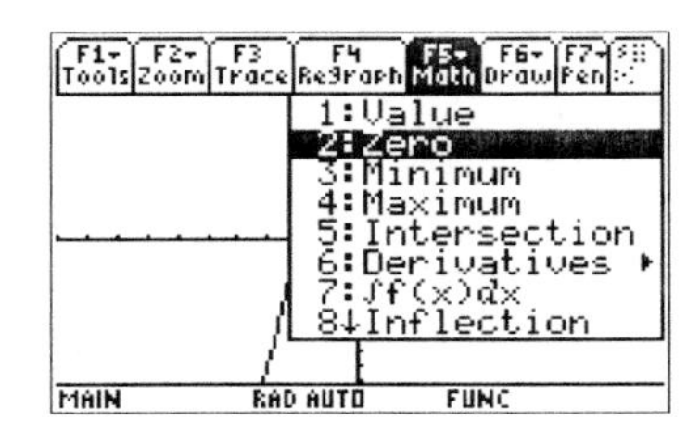

The TI-89 can automatically locate this point with the GRAPH:Math:Zero feature. The TI-89 prompts for lower and upper bounds (numbers that are, respectively, less than and greater than the zero), then attempts to locate the zero between the given bounds. (Provided there is only one zero between the bounds, and the function is "well-behaved"—meaning it has some nice properties like continuity—the calculator will find it.) The screens below illustrate these steps.

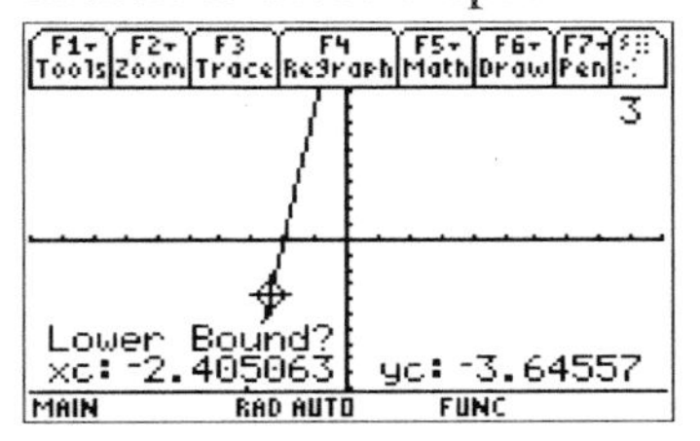

Move cursor to the left of the zero, press [ENTER]

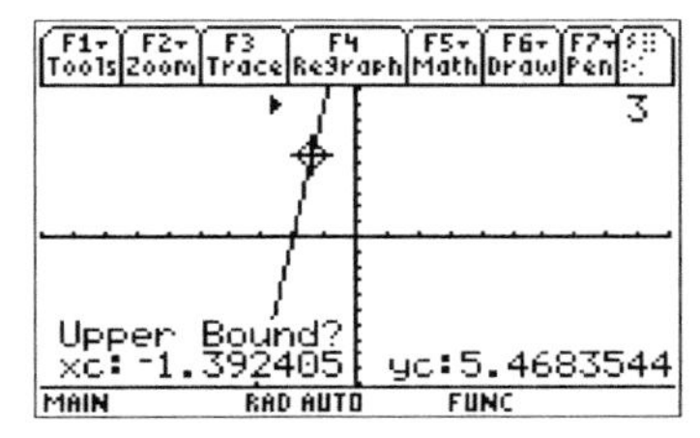

Move cursor to the right of the zero, press [ENTER]

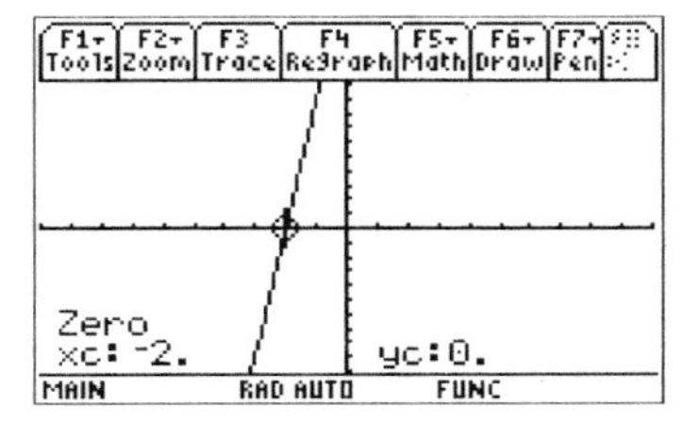

The TI-89 finds the zero.

As with the intersection method, after the TI-89 locates the x-intercept, the calculator variable xc contains the x-coordinate.

Section 1.5 Example 6 (page 58) — Solving a Linear Inequality

Section 1.5 Example 7 (page 58) — Solving Linear Inequalities

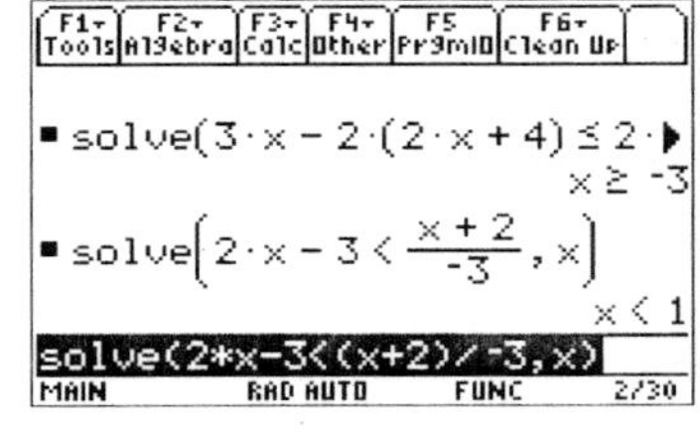

The solve command (see page 111) can also be used to solve inequalities like these. Use [2nd][0] and [2nd][.] to type < and >, and [♦][0] and [♦][.] to type ≤ and ≥. (These symbols can also be found in the MATH:Test menu, [2nd][5] [8].) While this is much faster, students should recognize that there are some benefits to doing problems like this by hand (the "hard" way)—specifically, working through the steps shown in the text gives one insight into what is going on, while using solve lends little to one's understanding of inequalities.

Section 1.5 Example 8 (page 59) — Using the Intersection-of-Graphs Method

In Figure 71, the text illustrates using a graph to support the solution $[-3, \infty)$ to the inequality $3x - 2(2x + 4) \leq 2x + 1$, making the observation that solutions to this inequality correspond to those x-values for which the graph of $y = 3x - 2(2x + 4)$ *intersects or is below* the graph of $y = 2x + 1$. This connection between "<" and "below" (or ">" and "above") is an important one, and students should strive to understand it. However, it can sometimes be confusing, especially when one is just learning it, and the following graphical approach may be useful. (Also note the use of the solve command, illustrated in the previous example.)

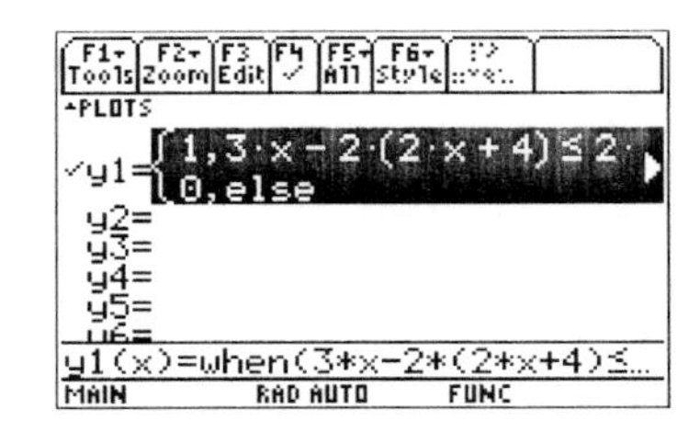

To solve (or confirm the solution of) an inequality like this, enter the formula shown on the right. This was entered as when(3x-2(2x+4)≤2x+1,1,0); note how the TI-89 displays it. The ≤ symbol is [♦][0]. This definition means that y1 will equal 1 for the values of x which satisfy the inequality, and 0 for all other values of x.

With this understanding, one can observe the graph produced (right), and confirm that the solution is $x \geq -3$. (Care must be taken to determine whether or not -3 should be included in the solution set; the graph does not make that clear. This same observation is made in the Technology Note next to this example in the text.)

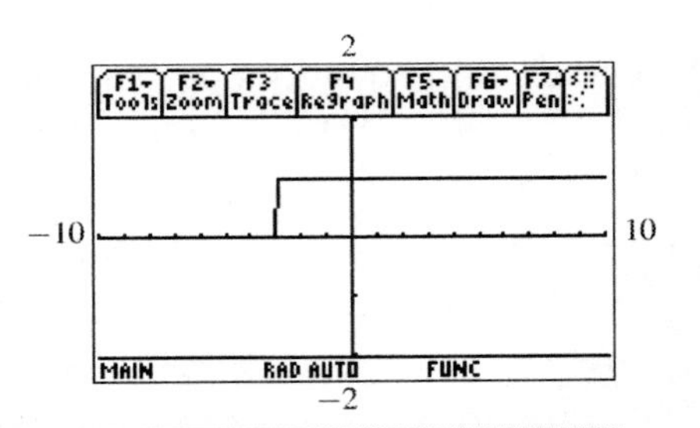

Section 1.5 Technology Note (page 60) Typing Function Variables

For the technique of defining `y3=y1(x)-y2(x)`, see the comments about Example 5 from Section 1.5 (page 112 of this manual).

Section 1.5 Example 11 (page 61) Solving a Three-Part Inequality

The `solve` feature (see page 111) can be used for this problem, but it will only handle one inequality at a time, so we must solve the two pieces separately (as shown) and then (optionally) join them with "`and`" (found in the MATH:Test menu). The final output shown is equivalent to $-\frac{7}{3} < x < 5$.

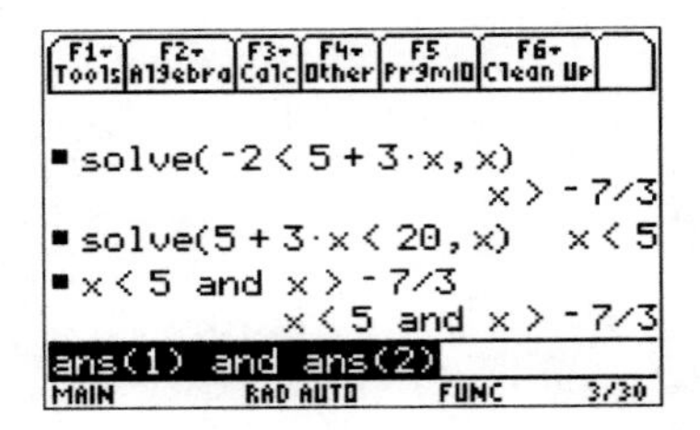

As was mentioned in the discussion of Example 8 from Section 1.5 (page 113 of this manual), the understanding that "<" and ">" go with "below" and "above" (respectively) is very important. The technique mentioned in that discussion can be applied to three-part inequalities, with some slight adjustments. Specifically, we must graph `when(-2<5+3x and 5+3x<20,1,0)`.

Section 2.1 Technology Note (page 96) Rational Exponents

See page 109 of this manual for more information about roots and rational exponents.

Section 2.1 Technology Note (page 96) Absolute Values

The `abs(` function of the TI-89 is in the MATH:Number menu ([2nd][5][1][2]). Note that while this function is entered as `abs(`, it is displayed in the history area using the "bar" notation.

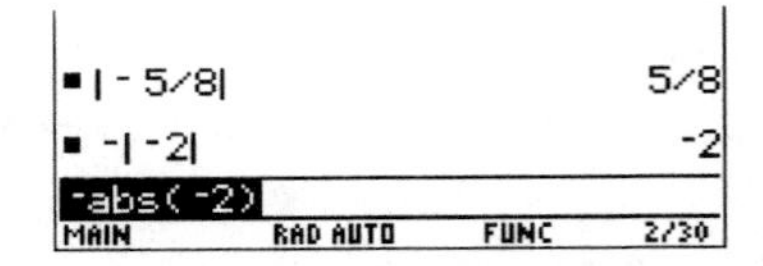

Section 2.1 Technology Note (page 98) Function and Parametric Modes

For information about selecting function and parametric modes, see section 9 of the introduction, page 101.

Section 2.2 Technology Note (page 103) Graphing Groups of Similar Functions
Section 2.3 Technology Note (page 113)
Section 2.3 Technology Note (page 114)
Section 2.3 Technology Note (page 115)
Section 2.3 Technology Note (page 117)

The screen shown on page 103 of the text illustrates one approach to graphing groups of similar functions: Set `y2=y1(x)+3`, `y3=y1(x)-2`, and `y4=y1(x)+5`. This allows an entire "family" to be graphed by simply changing `y1`.

An alternative is to define `y1=x^2+{0,3,-2,5}`. (The curly braces { and } are [2nd][(] and [2nd][)]). When a list (like `{0,3,-2,5}`) appears in a formula, it tells the TI-89 to graph this formula several times, using each value in the list; therefore, this one definition will graph the four functions $y = x^2$, $y = x^2 + 3$, $y = x^2 - 2$, and $y = x^2 + 5$. Different families can be produced simply by changing `x^2` to another function.

This list approach translates nicely to other types of transformations. For horizontal shifts, use, e.g., `y1=(x+{0,-3,-5,4})^2`. For vertical stretches and shrinks, use `y1={1,2,3,4}x^2`; the Technology Note on page 115 of the text illustrates a similar approach. For horizontal stretches and shrinks, use the approach shown in that Technology Note, or something like `y2=y1({2,0.5}x)`.

For reflections (page 117 of the text), use the approach shown in the text, or a variation of the one given above. For example, to graph $y = \sqrt{x}$ and $y = \sqrt{-x}$, for example, define `y1=√({1,-1}x)`.

Note that the usefulness of using lists to accomplish horizontal transformations (horizontal shifts and reflection across the y axis) is limited. For example, the graph of $y = \sqrt{x^2 - 3x}$—or any expression in which x appears more than once—is not conveniently reflected across the y-axis or horizontally shifted by this method. An adaptation of the method shown in the text would work better: E.g., define `y1=√(x^2-3x)` and `y2=y1(x+{2,5,-3})` to see the graph of $f(x) = \sqrt{x^2 - 3x}$, $f(x + 2)$, $f(x + 5)$, and $f(x - 3)$.

Section 2.5 Example 1 (page 139) Finding Function Values for a Piecewise-Defined Function
Section 2.5 Example 2 (page 139) Graphing a Piecewise-Defined Function

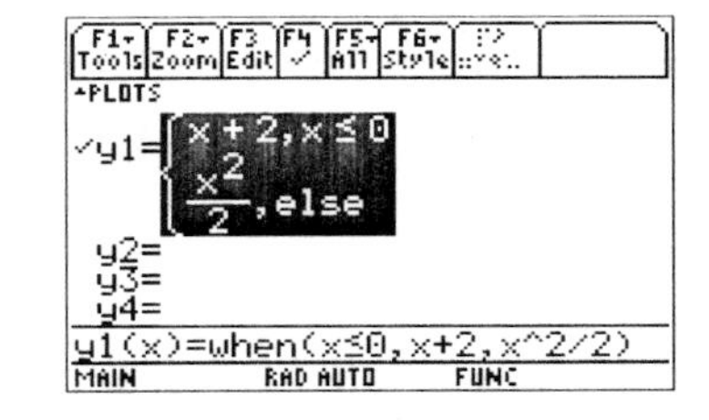

The text shows how to enter piecewise-defined functions on the TI-83; for a TI-89, the appropriate way to enter the function in Example 1 is shown on the right. Note the difference between how the function is entered and how it is displayed; in particular, the display is similar to how the function appears in print. (Use [◆][0] to type "≤".) The use of `Dot` graphing style is not crucial to the graphing process, as long as one remembers that vertical line segments connecting the "pieces" of the graph (in this case, at $x = 0$) are not really part of the graph.

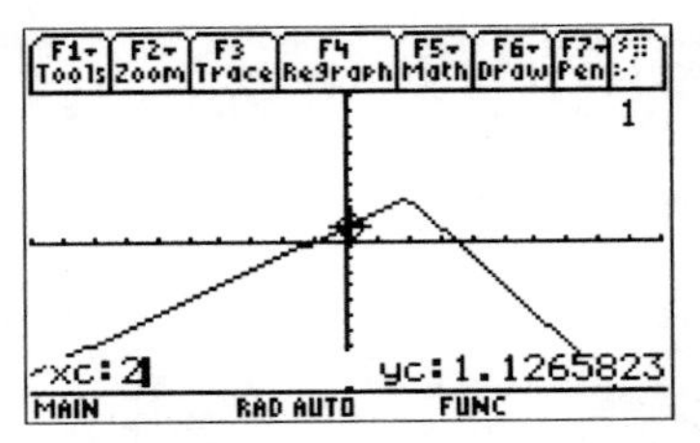

Figure 55 of the text shows that $y = 2$ when $x = 0$—a detail that is not evident from the calculator graph (since the TI-89 does not show open and closed circles as does the graph shown in text). Similarly, the screen in Figure 57(b) shows that $y = 3$ when $x = 2$, using the Trace feature, but note that on the window shown, one cannot get to $x = 2$ by pressing ◀ and ▶ to move the trace cursor. Instead, one must use the the TI-89's "extended" Trace feature, which allows one to trace to any x value between `xmin` and `xmax`. Simply type a number or expression (like `2`, `1/π` or `√(2)`) while in Trace mode. The number appears at the bottom of the window next to the `xc` trace coordinate. Pressing [ENTER] causes the trace cursor to jump to that x-coordinate. This same result can be achieved using the GRAPH:Math:Value command.

Extension: A more complicated piecewise-defined functions, such as

$$f(x) = \begin{cases} 4 - x^2 & \text{if } x < -1 \\ 2 + x & \text{if } -1 \leq x \leq 4 \\ -2 & \text{if } x > 4 \end{cases}$$

is entered as

```
when(x<-1,4-x^2,when(x≤4,2+x,-2))
```

which the TI-89 displays as shown on the right. This tells the TI-89 to compute `y1` as follows: (a) If $x < -1$, use the expression $4 - x^2$. (b) If $x \geq -1$ and $x \leq 4$, use $2 + x$. (c) Otherwise, use -2.

Section 2.5 Example 4 (page 141) Evaluating $[\![x]\!]$

On the TI-89, the greatest integer function is called `floor` or `int`; the former is found in the MATH:Number menu ([2nd][5][1][6]), while the latter is in the [CATALOG] (or either can be typed one letter at a time). In Figure 59, note the use of lists to find $[\![x]\!]$ for all five values with one entry.

Do not confuse `floor` and `int` with the `iPart` ("integer part") function, which is slightly different—specifically, `iPart(-6.5)` returns -6 while `int(-6.5)` gives -7.

Section 2.6 Example 1 (page 149) Using the Operations on Functions
Section 2.6 Example 2 (page 150) Using the Operations on Functions
Section 2.6 Example 4 (page 151) Finding the Difference Quotient

In Example 1, the screens in the text (Figure 64) show how to evaluate function sums, differences, products, and quotients using the TI-83's function notation. Similar entries on a TI-89 will produce the same results as those shown in the text.

The TI-89 can also give the appropriate answers for expressions like those in Examples 2 and 4, as the first two screens below show. However, it only produces correct results if the variables `x` and `h` are undefined. As we see in the third screen, if these variables have values, that value is used to evaluate $f(x)$ and $f(x+h)$, leading to a numeric result. (It may also produce an error if a variable contains something other than a number.) The simplest way to fix this is to type DelVar x, or press [2nd][F1][1][ENTER] (Clean up:Clear a-z).

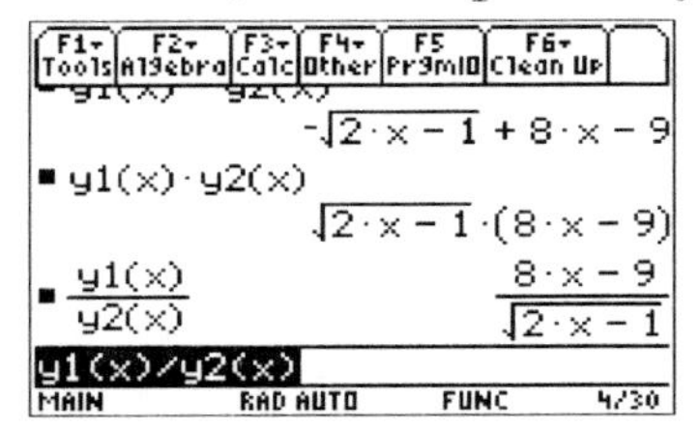

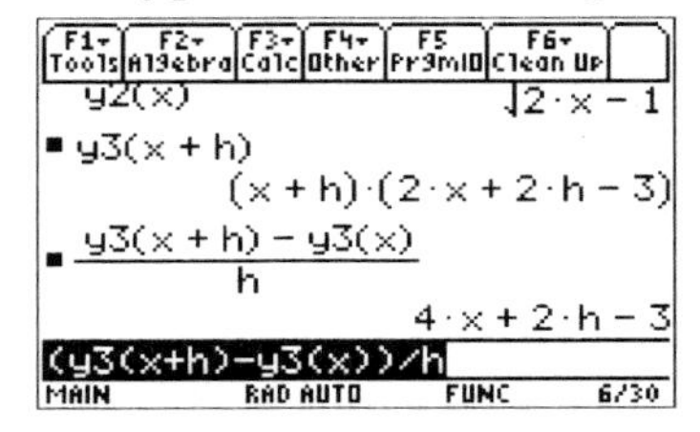

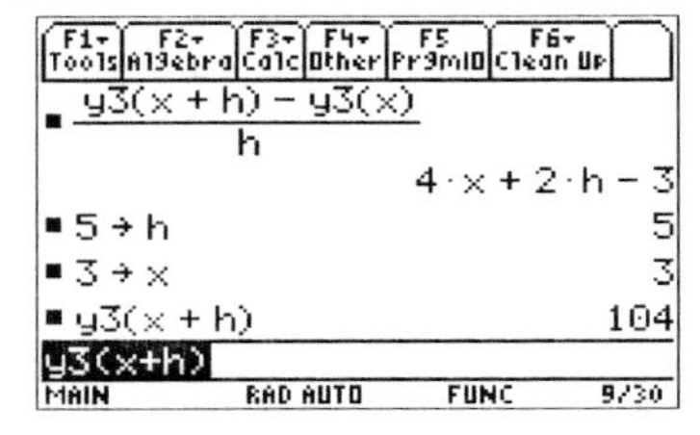

Section 2.6 Example 5 (page 153) Evaluating Composite Functions

Section 2.6 Example 6 (page 153) Finding Composite Functions

The screens shown in Figure 68 illustrate how to evaluate composite functions at specific input values; the TI-89's approach is essentially the same. Note that compositions of three or more functions can be accomplished just as simply: If y1, y2, and y3 are defined as the functions f, g, and h, one can evaluate $(f \circ g \circ h)(3)$ by typing y1(y2(y3(3))).

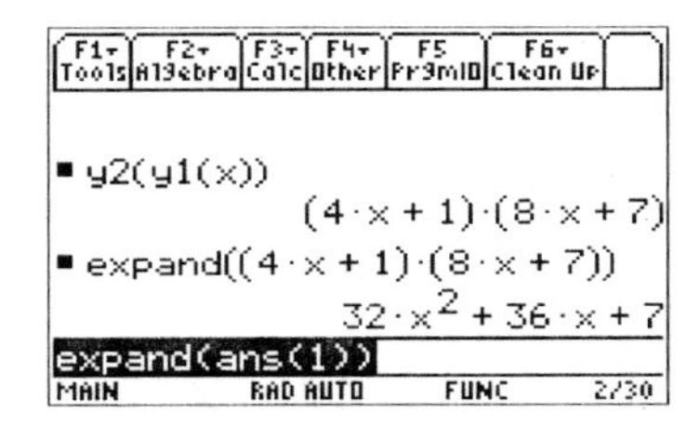

The TI-89 can also give symbolic expressions like those required in Example 6, provided the variable x is undefined (see the previous example). Note that the expand command forces the TI-89 to multiply out the result for Example 6(a), which it first gave in factored form.

Section 3.1 Technology Note (page 174) Complex Number Mode

See page 102 for more information about putting the TI-89 in RECTANGULAR mode (the equivalent of the TI-83's a+bi mode).

Section 3.1 Example 1 (page 175) Writing $\sqrt{-a}$ as $i\sqrt{a}$

Section 3.1 Example 2 (page 176) Finding Products and Quotients Involving $\sqrt{-a}$

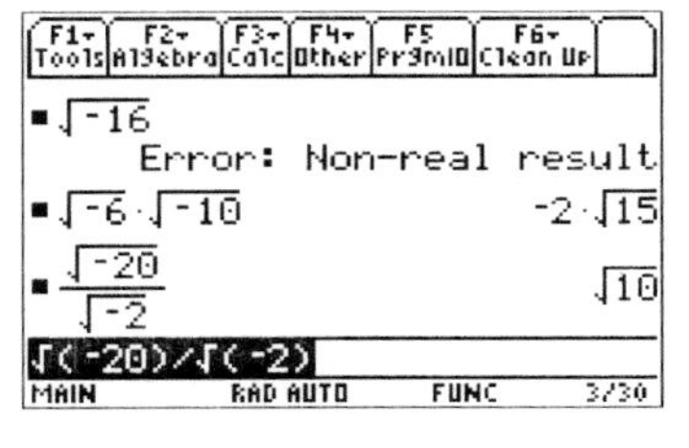

Note that the TI-89's Complex Format mode should be RECTANGULAR (see page 102). The screen on the right shows results from performing some computations in REAL mode; note that this *will* work for (a), (b), and (c) in Example 2, for which the final result is a real number, but an error occurs if the result is complex.

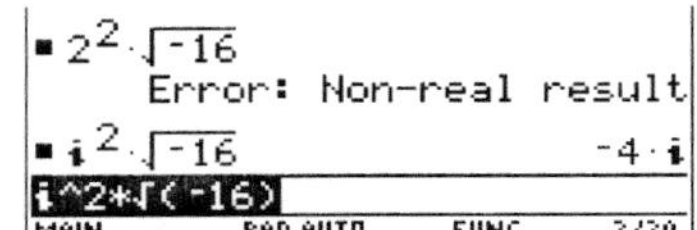

The screen on the right (also produced in REAL mode) illustrates an exception to this rule: A complex result produces no error in REAL mode if the entered expression included the character "i" ([2nd][CATALOG]—this is different from the "regular" lowercase i, [alpha][9]).

Section 3.1 Example 3 (page 176) Adding and Subtracting Complex Numbers
Section 3.1 Example 4 (page 177) Multiplying Complex Numbers
Section 3.1 Example 5 (page 178) Simplifying Powers of *i*
Section 3.1 Example 6 (page 179) Dividing Complex Numbers

Use [2nd][CATALOG] for the "i" character. Note that the TI-89's `Complex Format` mode should be `REAL` or `RECTANGULAR` (see page 102) to produce output similar to that shown in the text. The command `▸Frac` ("to fraction"), shown in the screen accompanying Example 6, is not necessary (nor is it available) on a TI-89; provided it is set to either `EXACT` or `AUTO` mode, results will be displayed as exact fractions.

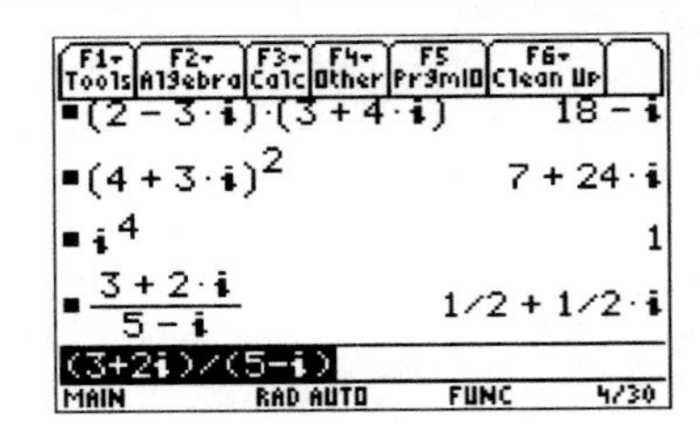

Section 3.2 Example 3 (page 185) Using the Vertex Formula

The text identifies the vertex of the graph of $f(x) = -0.65x^2 + \sqrt{2}\,x + 4$ as a maximum at $(\sqrt{2}/1.3, 4 + 1/1.3) \approx (1.09, 4.77)$, and shows calculator screens that support those values. The TI-89 can automatically locate extreme values ("hills" and "valleys") in a graph using the Minimum and Maximum options in the GRAPH:Math menu ([F5] from the graph screen).

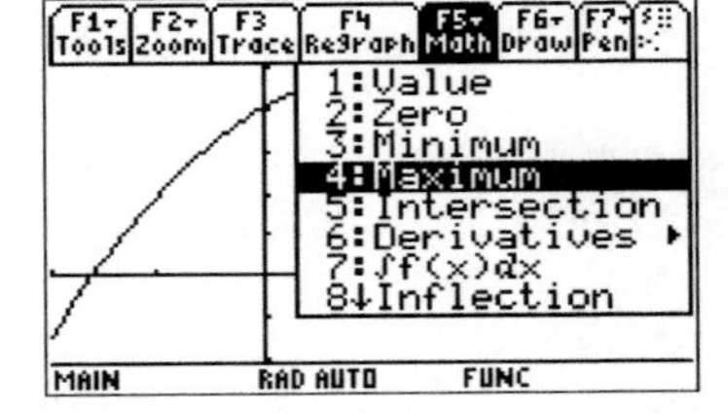

Enter the function in `y1`, and graph in a window that shows the extreme value (such as the window shown in the text: $[-2.4] \times [-2.5]$). Since the coefficient of x^2 is negative, this is a parabola that opens down, and the extreme point is a maximum value. Press [F5][4], then use the arrow keys and [ENTER] to define lower and upper bounds, as was done previously with the Math:Zero and Math:Intersection commands—see the discussion on Examples 4 and 5 from Section 1.5 beginning on page 111 of this manual. When finished, the TI-89 shows a graph similar to Figure 12 in the text.

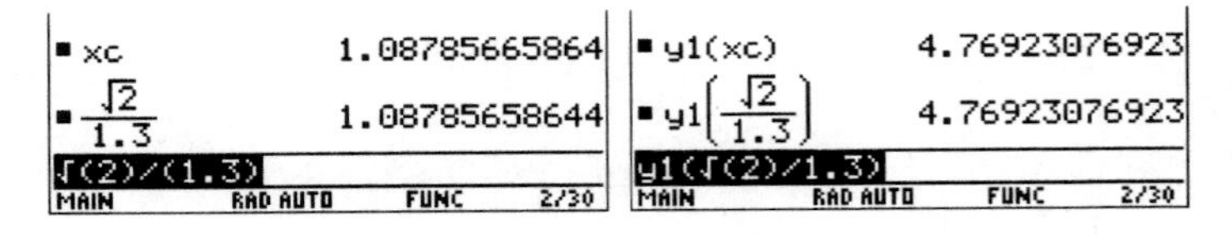

After going through the process of locating a maximum or minimum, the calculator variables `xc` and `yc` contain the coordinates of the point. Depending on the window and the specified bounds, the x value may be off a bit from the exact answer, as is the case here. A limitation of the technology is that the calculating algorithms are programmed to stop within a certain degree of accuracy. (In other words, the TI-89 looks for the vertex until it decides that it is "close enough"; it will not always find exact answers.) In this case, the value of `y1` is nearly identical for both the exact answer and the TI-89's approximate value. It is important for the user to recognize this limitation for two reasons: First, do not report all digits displayed by the calculator, as they are not all reliable. Second, if the calculator reports a result of (say) 1.49999956, it is reasonable to guess that the exact answer might be 1.5.

Section 3.2 Example 5 (page 186) — Identifying Extreme Points and Extreme Values

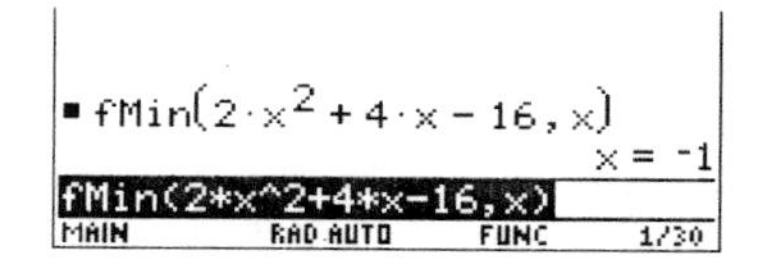

Figure 15 of the text illustrates the use of the TI-83's `fMin` operation. The TI-89 has this function (found in the Calc menu), as well as the similar `fMax` function, but they behave differently. The TI-83 requires a range of numbers between which it will seek a minimum (or maximum); Figure 15(b) uses the range −20 to 20. The TI-89 will attempt to find the locations of *all* minimum values for a given function.

Section 3.2 Example 6 (page 187) — Modeling Hospital Spending with a Quadratic Function

See page 110 for instructions on creating scatter diagrams on the TI-89. The TI-89's `QuadReg` (quadratic regression) feature—option 9 in the Data/Matrix Editor:Calc menu—can be used to find a quadratic function to approximate a set of data. The procedures for doing this are similar to those for a linear regression, described on page 111 of this manual. (Note, though, that the quadratic regression formula is not the same as the approximating function given in the text.)

Section 3.3 Technology Note (page 198) — Calculator Programs

See section 13 of the introduction (page 107) for information about installing programs in the TI-89.

Section 3.3 Example 4 (page 199) — Using the Quadratic Formula
Section 3.3 Example 5 (page 200) — Using the Quadratic Formula

The `solve` and `zeros` functions, previously described on page 111, can be used for many types of equations, including quadratics like these. For equations with complex solutions, use `cSolve` or `cZeros` rather than `solve` and `zeros`.

■ solve(x·(x − 2) = 2·x − 2, x)
x = −(√2 − 2) or x = √2 + 2
■ cSolve(2·x² − x + 4 = 0, x)
x = 1/4 + (√31/4)·i or x = 1/4 ▸
cSolve(2x^2-x+4=0,x)
MAIN RAD AUTO FUNC 2/30

Section 3.3 Example 6 (page 202) — Solving a Quadratic Inequality
Section 3.3 Example 7 (page 203) — Solving a Quadratic Inequality

As the screen on the right shows, `solve` fails to give useful results for inequalities like these.

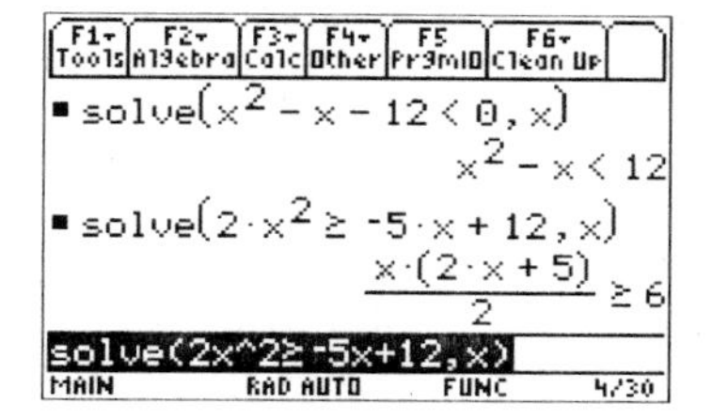

Section 3.6 Example 2 (page 232) — Dividing a Polynomial by a Binomial

The TI-89's `expand(` command (found in the Algebra menu, [F2][3]) nicely shows the results of this division: the quotient is the polynomial, and the remainder is the numerator of the rational expression.

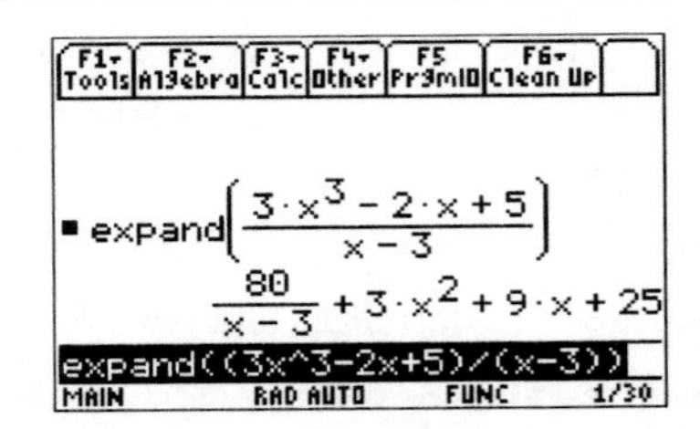

Section 3.7 Example 2 (page 242) — Finding All Zeros of a Polynomial Function

The `cZeros` command produces a list of all the zeros (real and complex) of the polynomial, confirming the answers given in the text.

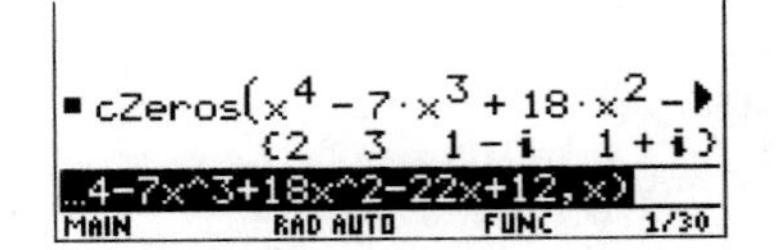

Additionally, if installed on your calculator (see section 13 of the introduction, page 107), the "Polynomial Root Finder" APP can find all zeros (real and complex) of a polynomial. The screens below illustrate the process.

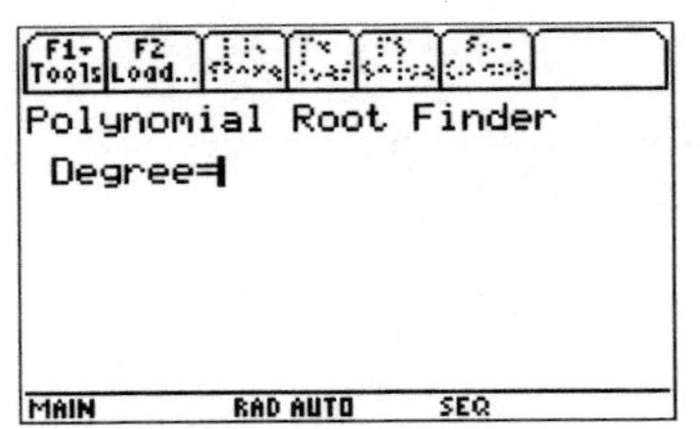

Give the degree of the polynomial.

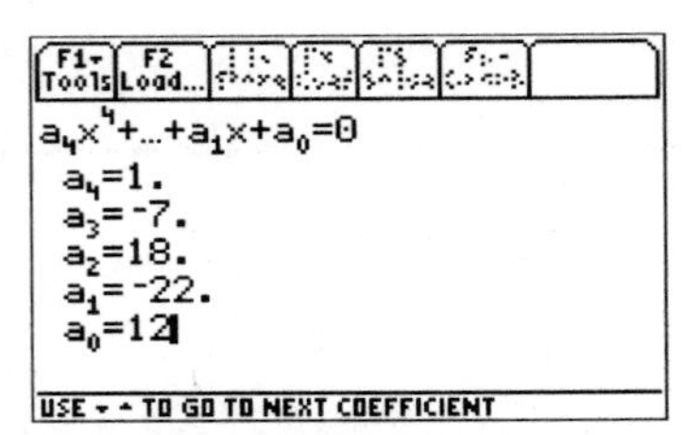

Enter coefficients on this screen.

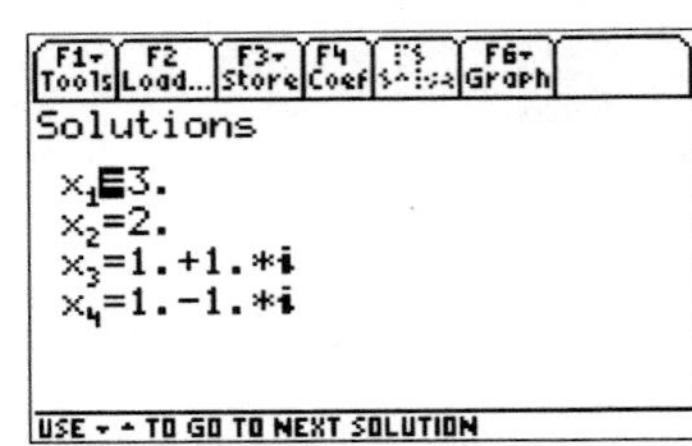

Press [F5] to find the zeros.

Section 3.8 Example 7 (page 255) — Examining Polynomial Models for Debit Card Use

The procedures for creating scatter diagrams and performing regressions are described on pages 110 and 111 of this manual.

Section 4.1 Example 2 (page 273) — Graphing a Rational Function

Note that this function is entered as `y1=2/(x+1)`, **not** `y1=2/x+1`. (The TI-89's "pretty print" display makes this sort of mistake unlikely.)

The issue of incorrectly drawn asymptotes is also addressed in Section 12 of the introduction to this chapter (page 105). Changing the window to `xmin` $= -5$ and `xmax` $= 3$ (or any choice of `xmin` and `xmax` which has -1 halfway between them) eliminates this vertical line because it forces the TI-89 to attempt to evaluate the function at $x = -1$. Since this function is not defined at -1, it cannot plot a point there, and as a result, it does not attempt to connect the dots across the "break" in the graph.

Section 4.2 Example 8 (page 285)

Graphing a Rational Function Defined by an Expression That Is Not in Lowest Terms

Figure 23 shows the graph of `y1=(x^2-4)/(x-2)` on a variation of the decimal window, on which the "hole" in the graph can be seen. There are many possible windows on which the hole is visible; any window for which $x = 2$ is halfway between `xmin` and `xmax` would work. Likewise, there are many windows for which the hole would not be visible.

It is important to realize that some holes cannot be made visible. For example, take the graph of the function `y1=(x^4-4)/(x^2-2)`—which looks like the function $y = x^2 + 2$, except at $x = \pm\sqrt{2}$. It is difficult (if not impossible) to find a window showing the holes at $x = \pm\sqrt{2}$.

Section 4.3 Example 1 (page 291)

Solving a Rational Equation

Section 4.3 Example 2 (page 292)

Solving a Rational Equation

The `solve` function (see page 111) works nicely for rational equations; note that the extraneous solution $x = 2$ for Example 2 is not reported.

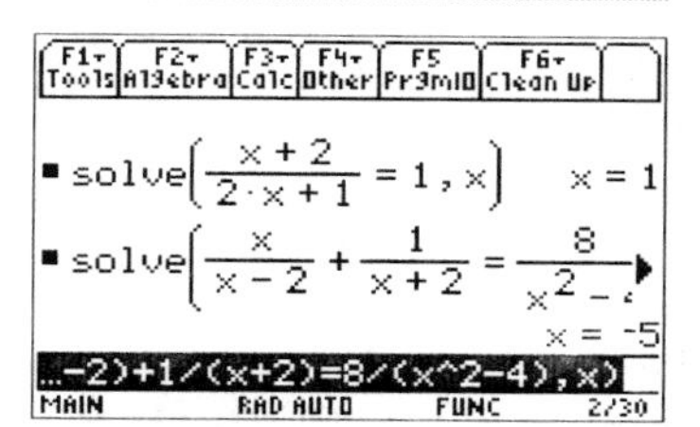

Section 4.3 Example 3 (page 292)

Solving a Rational Inequality

As the screen on the right shows, `solve` fails to give useful results for inequalities like these.

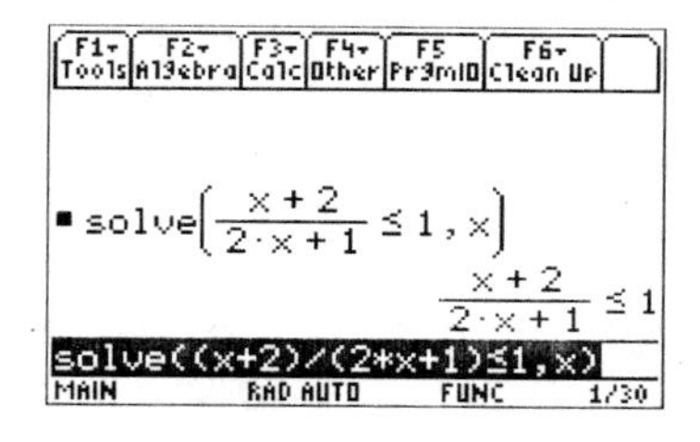

Section 4.4 Example 5 (page 309)

Modeling the Period of Planetary Orbits

Section 4.4 Example 6 (page 309)

Modeling the Length of a Bird's Wing

The procedures for creating scatter diagrams are covered on page 110 of this manual. The "power regression" illustrated in Example 4 (Figures 47&48) is performed in a manner similar to linear regression (see page 111).

Here is the output for the power regression performed on the data in Example 3. Note that this gives further confirmation that the formula $f(x) = x^{1.5}$ does a good job of modeling the relationship between average distance from the sun x and period of revolution y.

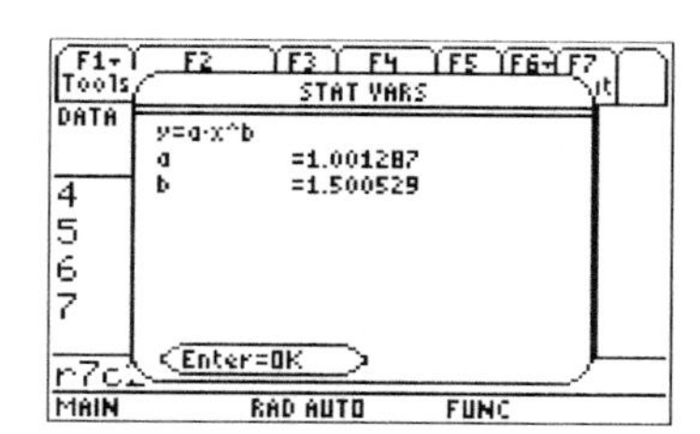

Section 4.4 Example 9 (page 313) Graphing a Circle

The text suggests graphing `y1=√(4-x^2)` and `y2=-y1`. Here are two alternative ways to graph this circle; these approaches might be useful in other situations where one wishes to graph to or more complicated (but similar) functions:

- After typing the formula in `y1`, move the cursor to `y2`, press [(-)], then press [2nd][STO▸] (RCL), then type "`y1`" and press [ENTER]. This will "recall" the formula of `y1`, placing it on the entry line for `y2`. This takes a few additional keystrokes, but can be a useful approach in cases where the second function to be graphed is similar to the first, but cannot easily be written in terms of `y1`.
- Enter the single formula `y1={-1,1}√(4-x^2)`. (The curly braces { and } are [2nd][(] and [2nd][)]). See page 115 for more information about this approach.

Section 5.1 Example 6 (page 345) Finding the Inverse of a Function with a Restricted Domain

The graph in Figure 8 shows the functions $f(x) = \sqrt{x+5}$ and $f^{-1}(x) = x^2 - 5,\ x \geq 0$. To produce a similar graph on the TI-89, enter the second function as a piecewise-defined function with only one "piece": Enter `y2=when(x≥0,x^2-5,1/0)`. This function is undefined (because of division by 0) whenever $x < 0$.

Section 5.2 Example 5 (page 355) Using Graphs to Evaluate Exponential Expressions

Graphs like those shown in Figure 20 can be produced using the "extended trace" features of the TI-89, mentioned previously on page 115 of this manual. In Trace mode, simply type a number or expression (like `√(6)` or `-√(2)`). The number appears at the bottom of the window next to the `xc` trace coordinate. Pressing [ENTER] causes the trace cursor to jump to that x-coordinate. This same result can be achieved using the GRAPH:Math:Value command.

Section 5.3 Technology Note (page 365) Logarithms of Nonpositive Numbers

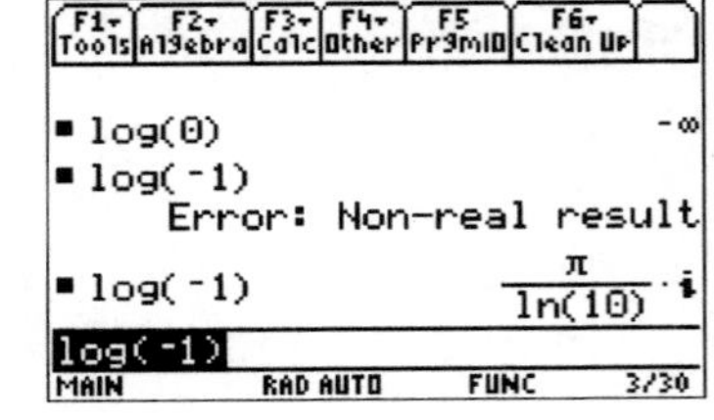

To type the `log` function, use the CATALOG or type it one letter at a time. This Technology Note states that (under some circumstances) a calculator will give an error message when asked to compute (e.g.) `log(0)` or `log(-1)`. In fact, the TI-89 reports the value of `log(0)` as $-\infty$. `log(-1)` gives a `Non-real result` error when the TI-89 is in REAL mode, but reports complex results if in RECTANGULAR or POLAR complex mode. The third computation on the right shows log(−1) in rectangular format. In general, if $x > 0$, then $\log(-x) = \log(x) + \log(-1)$, as the product rule for logarithms would suggest.

Section 5.3 Example 3 (page 365) Finding pH and [H_3O^+]

The natural logarithm function (ln) is [2nd][X]. For the common (base-10) logarithm, use the CATALOG or type `log` one letter at a time.

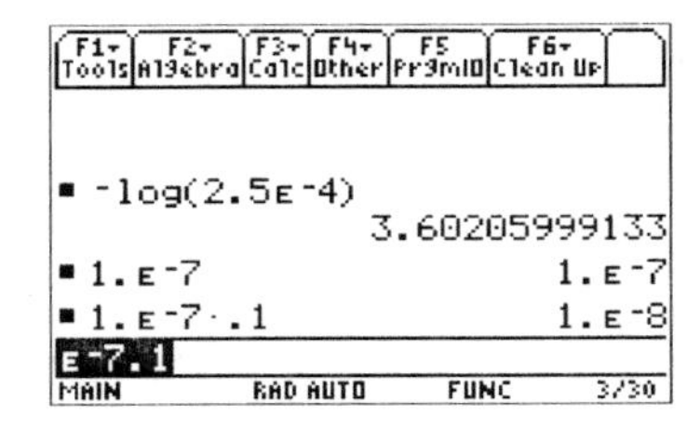

For (a), the text shows `-log(2.5*10^(-4))`. The TI-89 has a "10^" function buried in the [CATALOG], but [1][0][^] is faster to type and produces the same results. This could also be entered as shown on the first line of the screen on the right, since "E" (produced with [EE]) and "`*10^`" are nearly equivalent. The two are not completely interchangeable, however; in particular, in part (b), `10^(-7.1)` **cannot** be replaced with `E-7.1`, because "E" is only valid when followed by an *integer*. That is, `E-7` produces the same result as `10^-7`, but `E-7.1` is interpreted as `(1E-7)*0.1`.

Section 5.3 Example 6 (page 368) — Using the Properties of Logarithms

Section 5.3 Example 7 (page 369) — Using the Properties of Logarithms

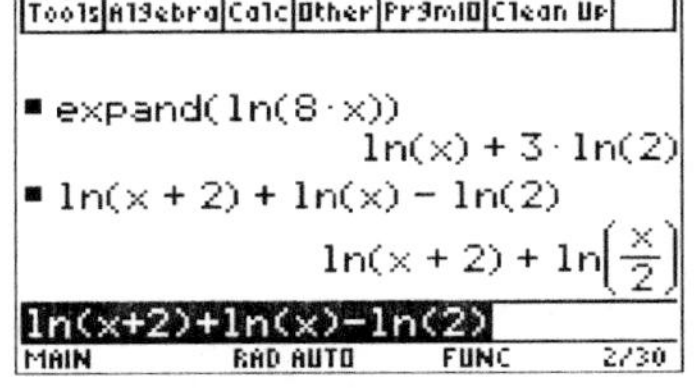

The TI-89 "knows" the properties of logarithms, but can really only apply them usefully to the natural logarithm `ln`; for other bases, it uses the change-of-base formula to write the expression in terms of `ln`. Shown are the TI-89's results for Example 6(a) and Example 7(a), as rewritten with natural logs. Note that they are equivalent to the expressions given in the text.

Section 5.4 Technology Note (page 375) — Asymptotes in Logarithmic Graphs

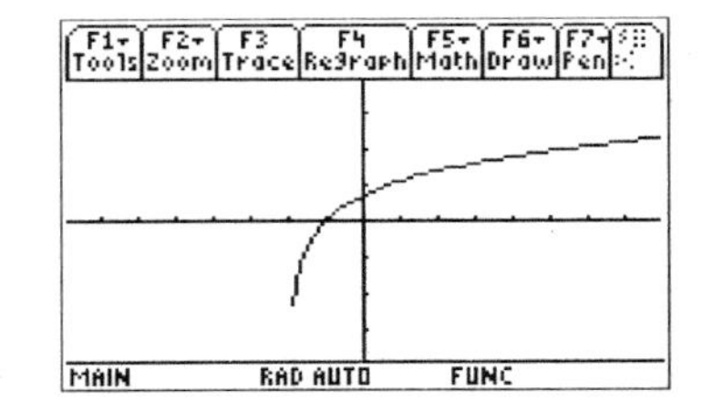

The calculator screen behavior described here is more evident when accompanied by a horizontal shift. Shown here (on the decimal window) is the graph of $y = \ln(x+2)$; note that the graph seems to have an endpoint at about $(-1.9, -2.3)$. In reality, the graph has a vertical asymptote at $x = -2$, and that portion of the graph "goes down to $-\infty$"—that is, as $x \to -2$ from the right, $y \to -\infty$.

Section 5.6 Technology Note (page 398) — Financial Calculations

Section 5.6 Example 6 (page 399) — Using Amortization to Finance an Automobile

To access the finance menus shown in this Technology Note, you must use the Finance APP. This is pre-installed on the TI-89 Titanium edition; for a standard TI-89, it can be downloaded from `education.ti.com` and installed over a Graph Link cable.

This package makes many financial functions available on your TI-89, including "TVM" ("time value of money") functions, which are useful for solving the problem posed in this Example. With this package installed, the TI-89's TVM features are nearly equivalent to those found in the TI-83; see page 23 of this manual for information about how they work.

Section 5.6 Example 8 (page 400) — Modeling Atmospheric CO_2 Concentrations

Section 5.6 Example 9 (page 401) — Modeling Interest Rates

The procedures for creating scatter diagrams and performing regressions are described on pages 110 and 111 of this manual.

Section 6.1 Example 3 (page 420) — Graphing a Circle

For part (b), the text suggests graphing `y1=4+√(36-(x+3)^2)` and `y2=4-√(36-(x+3)^2)`. Here are three options to speed up entering these formulas (see also page 122):

- After typing the formula in `y1`, move the cursor to y2, press [ENTER], then press [2nd][STO▸] (RCL), then type "`y1`" and press [ENTER]. This will "recall" the formula of `y1`, placing it on the entry line for y2. Now edit this formula, changing the first "+" to a "−."
- After typing the formula in `y1`, define `y2=8-y1(x)`. This produces the desired results, since $8 - \texttt{y1} = 8 - (4 + \sqrt{36 - (x+3)^2}) = 8 - 4 - \sqrt{36 - (x+3)^2} = 4 - \sqrt{36 - (x+3)^2}$.
- Enter the single formula `y1=4+{-1,1}√(36-(x+3)^2)`. (The curly braces { and } are [2nd][(] and [2nd][)], respectively). When a list (like `{-1,1}`) appears in a formula, it tells the TI-89 to graph this formula several times, using each value in the list.

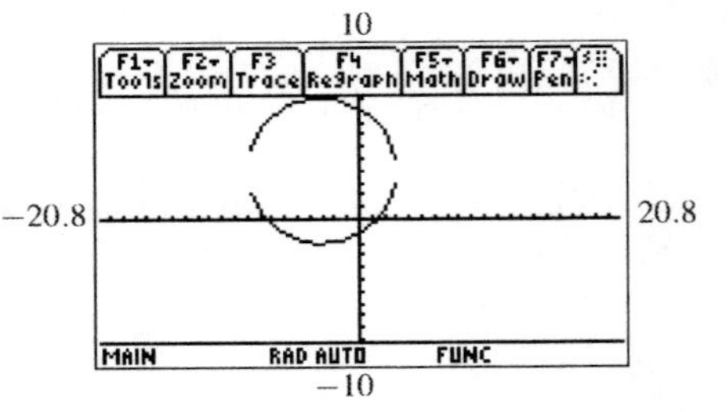

The window chosen in Figure 6 of the text is a square window (see section 11 of this chapter's introduction), so that the graph looks like a circle. (On a non-square window, the graph would look like an ellipse—that is, a distorted circle.) Note, however, that on the TI-89, this window is not square. Other square windows would also produce a "true" circle, but some will leave gaps similar to those circles shown in Sections 11 and 12 of the introduction (pages 104–105). Shown is the same circle on the square window produced by "squaring up" the standard window. The Technology Note next to this example shows a similar graph for the circle in part (a).

Section 6.4 Technology Note (page 454) — Parametric Mode

See the next example, as well as section 9 of the introduction (page 101), for information about selecting Parametric mode. The process of creating a graph in this mode is described in the next example.

Section 6.4 Example 1 (page 454) — Graphing a Plane Curve Defined Parametrically

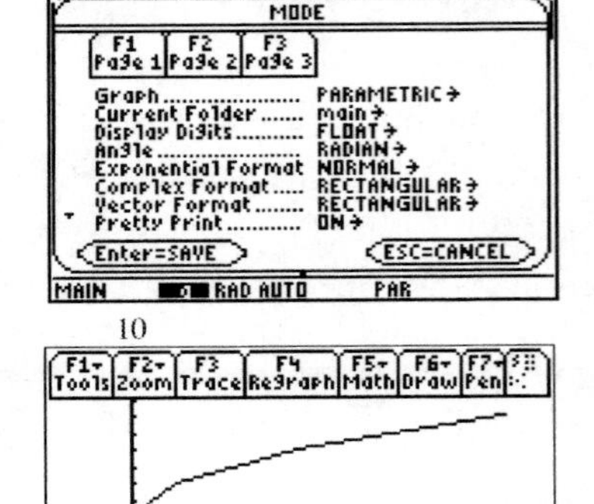

Place the TI-89 in Parametric mode, as the screen on the right shows. In this mode, [◆][F1] allows entry of up to 99 pairs of parametric equations (x and y as functions of t). The TI-89 will graph any pair of equations for which at least one of x and y is selected (has a check mark next to it).

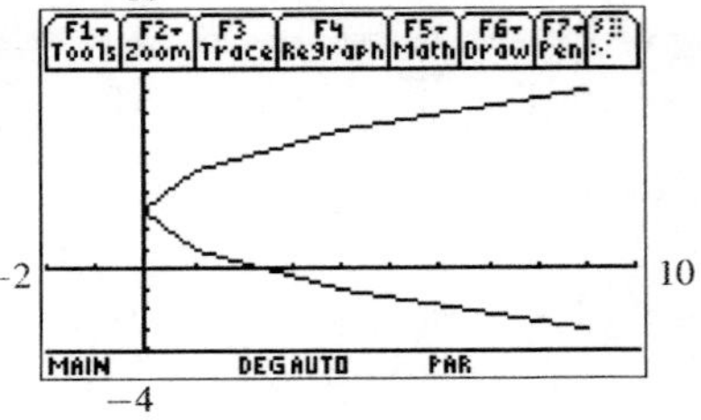

The value of `tstep` does not need to be 0.05, although that choice works well for this graph. Too large a choice of `tstep` produces a less-smooth graph, like the one shown on the right (drawn with `tstep=1`). Setting `tstep` too small may produce a smooth graph, but it might be drawn very slowly. Sometimes it may be necessary to try different values of `tstep` to choose a good one.

One can trace on a parametric graph, just as on a function-mode graph. The screen shown here is what appears when one first enters Trace mode: The trace cursor begins at the (x, y) coordinate corresponding to `tmin`, and pressing ⊙ increases the value of t (and likewise, ⊙ decreases t). This can be somewhat disorienting, since for this graph, pressing ⊙ moves the cursor to the *left*.

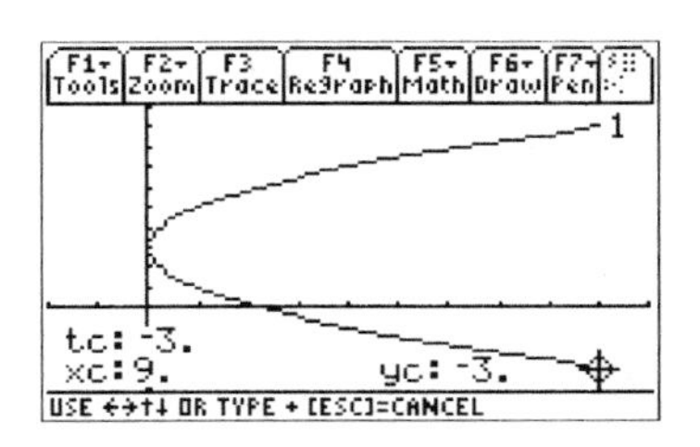

Section 7.1 Example 1 (page 468) Solving a System by Substitution

Section 7.1 Example 2 (page 468) Solving a System by Elimination

See page 111 for a description of the TI-89's intersection-locating procedure.

Section 7.1 Example 6 (page 472) Solving a Nonlinear System by Elimination

See the discussion related to Example 9 from Section 4.4 (page 122 of this manual) for tips on entering formulas like these.

In Figure 7, the text shows the intersections as found by the procedure built in to the calculator (described on page 111 of this manual). This will not work on the TI-89, because every other x coordinate besides -2 and 2 is outside the domain of either the circle or the hyperbola. Therefore, any set of lower and upper bounds will include points where at least one equation is undefined.

The TI-89's `solve` function can handle systems like this: Enter

```
solve(x^2+y^2=4 and 2x^2-y^2=8,{x,y})
```

and the result is "`x=2 and y=0 or x=-2 and y=0`."

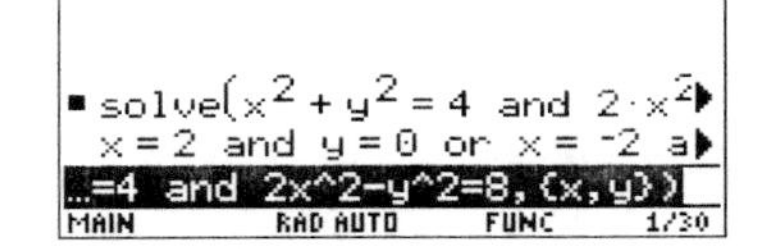

Section 7.2 Example 1 (page 481) Solving a System of Three Equations in Three Variables

If installed on your calculator (see section 13 of the introduction, page 107), the "Simultaneous Equation Solver" APP allows you to solve systems of linear equations. The screens below illustrate the process. When all coefficients have been entered (in the form of an *augmented matrix*, discussed in Section 7.3 of the text), pressing F5 solves for the three unknowns (which the TI-89 calls x_1, x_2, and x_3, rather than x, y, and z).

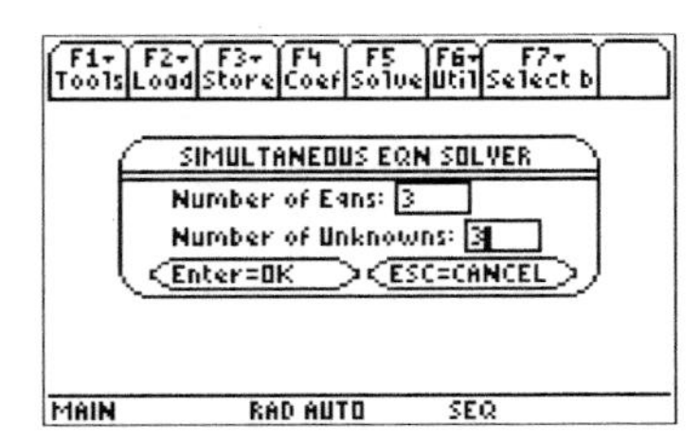

Give the number of equations and unknowns.

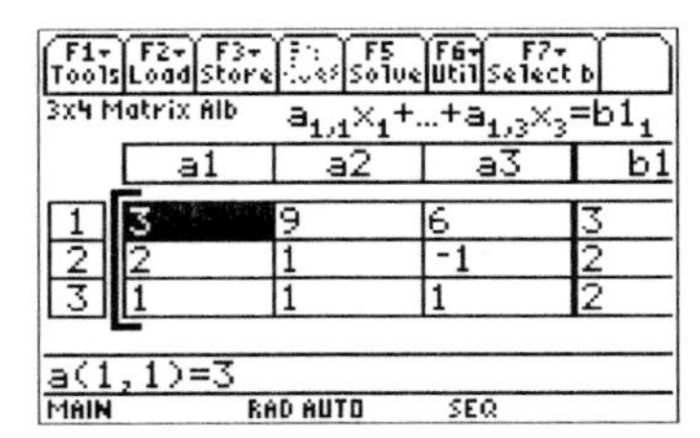

Enter coefficients on this screen.

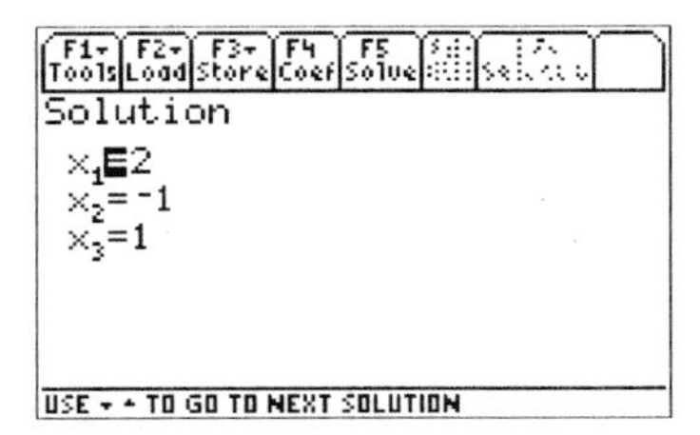

Press F5 to solve the system.

Section 7.2 Example 5 (page 484)

Using a System to Fit a Parabola to Three Data Points

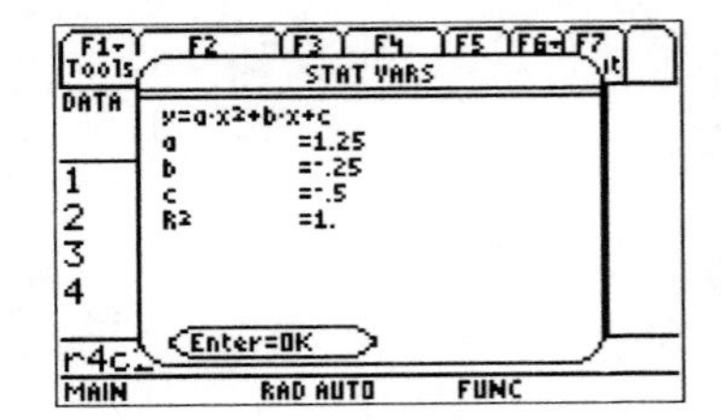

Note that the equation found in this example by algebraic methods can also be found very quickly on the TI-89 by performing a quadratic regression (`QuadReg`—see page 110) on the three given points. These agree with the coefficients found in the text.

It is worth noting that the `CubicReg` and `QuartReg` procedures could likewise be used to find functions to exactly fit sets of four or five data points (much more easily than solving the related systems of equations). Furthermore, the `LinReg` procedure can be used with a pair of points to find the equation of the line through those points.

Section 7.3 Technology Note (page 488)

Entering Matrices

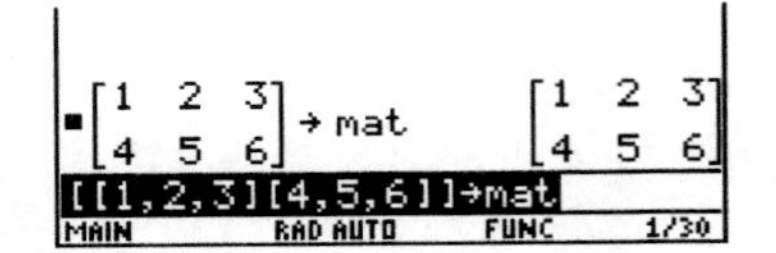

The matrices shown in the calculator screens in the text have the names "`[A]`" and "`[B]`." On the TI-82/83/84, all matrices have names of the form [*letter*]; on the TI-89, matrices can have any variable name. (In fact, variable names *cannot* include the bracket characters on the TI-89.) One way to set the values in a matrix is by "storing" the contents on the home screen; an example is shown on the right. (Note the difference between how this is *entered* and how it is *displayed* in the history area.) The square brackets are [2nd][,] and [2nd][÷].

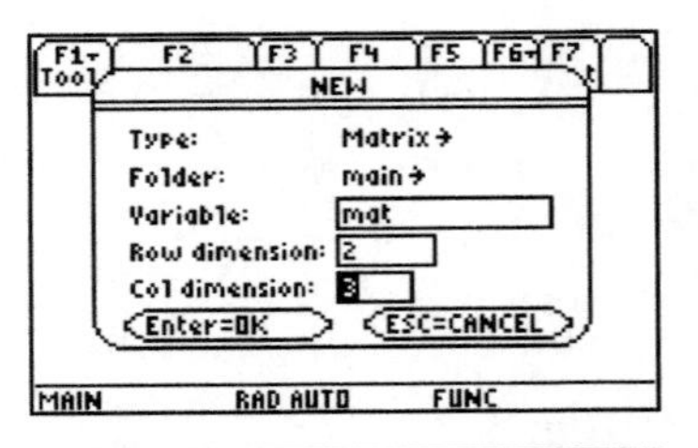

Alternatively, press [APPS][6] (Data/Matrix editor) and either create a new matrix, or open an existing one. If creating a new matrix, you will be prompted for a name and the size of the matrix, as the screen on the right illustrates.

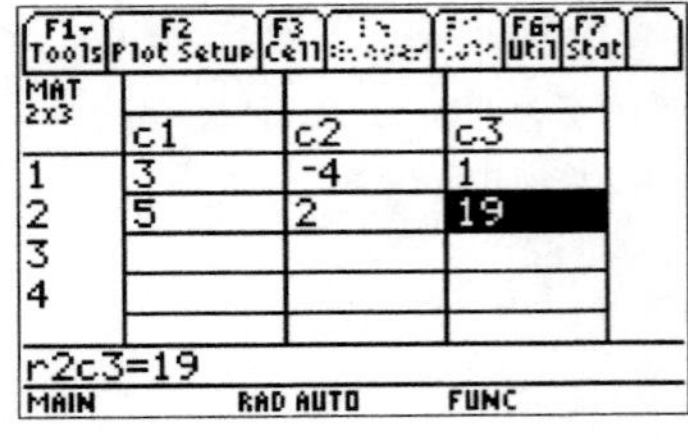

The matrix editor resembles a spreadsheet. Individual values of the matrix can be entered, or the matrix can be resized or sorted using the Util options ([2nd][F1]). Press [2nd][ESC] (QUIT) when finished.

These two methods are equivalent to those illustrated in Figure 13 of the text.

Section 7.3 Technology Note (page 489)

Matrix Row Transformations

Section 7.3 Example 1 (page 489)

Using Row Transformations

Here are the formats for the matrix row-operation commands ([2nd][5][4][alpha][×]), along with examples of their effect on the matrix `mat`$=\begin{bmatrix} 1 & 4 & 7 \\ 2 & 5 & 8 \\ 3 & 6 & 9 \end{bmatrix}$:

- `rowSwap(`*matrix*`,`*A*`,`*B*`)` produces a new matrix that has row *A* and row *B* swapped.

  Input: `rowSwap(mat,1,2)` Output: $\begin{bmatrix} 2 & 5 & 8 \\ 1 & 4 & 7 \\ 3 & 6 & 9 \end{bmatrix}$

- `rowAdd(`*matrix*`,`*A*`,`*B*`)` produces a new matrix with row *A* added to row *B*.

  Input: `rowAdd(mat,1,2)` Output: $\begin{bmatrix} 1 & 4 & 7 \\ 3 & 9 & 15 \\ 3 & 6 & 9 \end{bmatrix}$

- `mRow(`*number*`,`*matrix*`,`*A*`)` produces a new matrix with row *A* multiplied by *number*.

  Input: `mRow(-4,mat,1)` Output: $\begin{bmatrix} -4 & -16 & -28 \\ 2 & 5 & 8 \\ 3 & 6 & 9 \end{bmatrix}$

- `mRowAdd(`*number*`,`*matrix*`,`*A*`,`*B*`)` produces a new matrix with row *A* multiplied by *number* and added to row *B* (row *A* is unchanged).

  Input: `mRowAdd(2,mat,1,3)` Output: $\begin{bmatrix} 1 & 4 & 7 \\ 2 & 5 & 8 \\ 5 & 14 & 23 \end{bmatrix}$

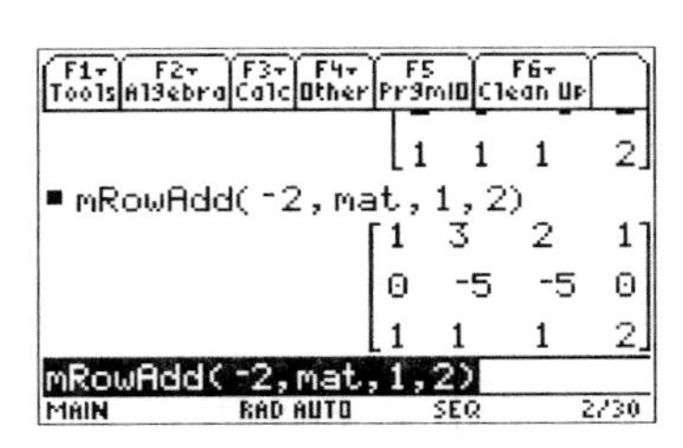

Keep in mind that these row operations leave the matrix `mat` untouched. To perform a sequence of row operations, each result must either be stored in a matrix, or use the result variable `ans(1)` as the matrix. For example, with `mat` equal to the augmented matrix for the system given just before this Technology Note and Example, the command shown on the right shows the appropriate first step in solving the system, displaying the resulting matrix on the screen—but `mat` is exactly as it was before. The three screens below illustrate the next steps in solving the system; note the use of `ans(1)`.

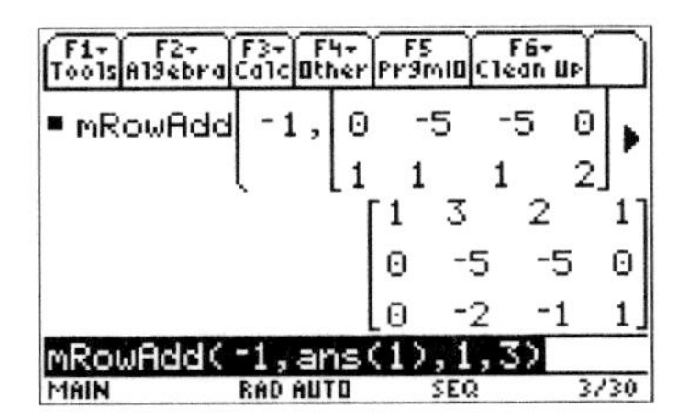

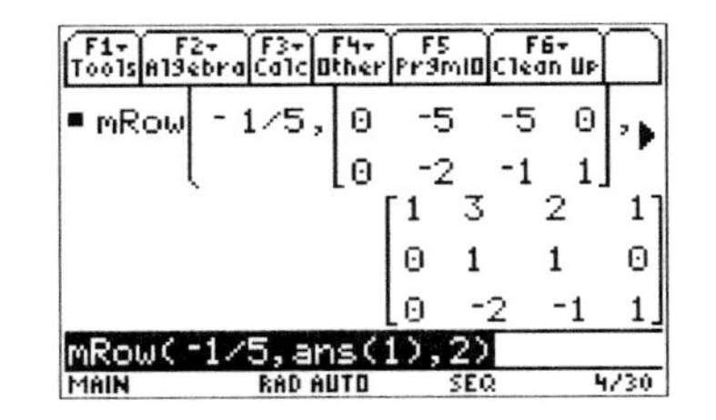

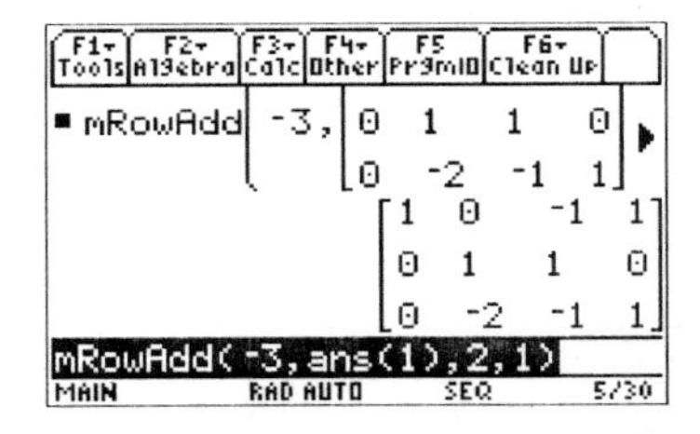

Section 7.3 Example 2 (page 490) Solving a System by the Row Echelon Method

Section 7.3 Example 5 (page 493) Solving a System by the Reduced Row Echelon Method

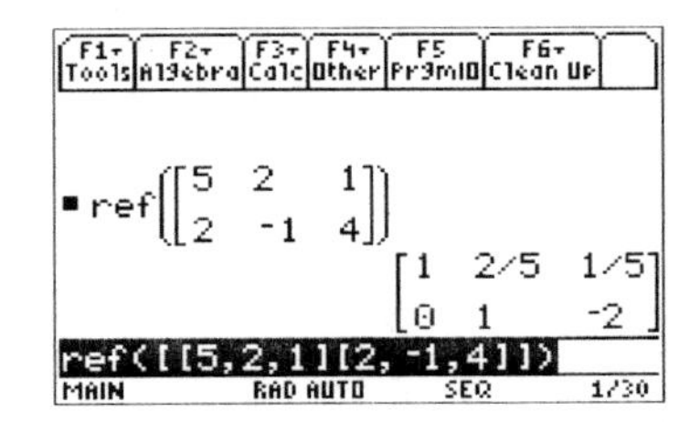

The `ref` and `rref` commands are options 3 and 4 in the MATRX:MATH menu ([2nd][5][4]). These will do all the necessary row operations at once, making these individual steps seem tedious. However, doing the whole process step-by-step can be helpful in understanding how it works.

Section 7.4 Example 1 (page 501) Classifying Matrices by Dimension

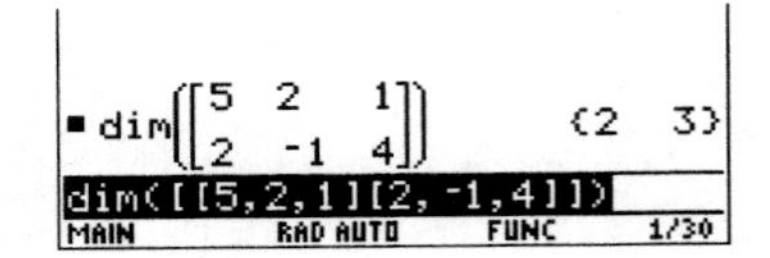

If a matrix has been entered in the TI-89, the dimensions can be found using the `dim` command, which is buried deep in the MATH:Matrix menu. (It is easier to find in the [CATALOG], or to type it one letter at a time.) This function returns a list containing two numbers: { row count, column count }. Of course, counting rows and columns is arguably much simpler than using the `dim` command.

Section 7.4 Example 2 (page 501) Determining Equality of Matrices

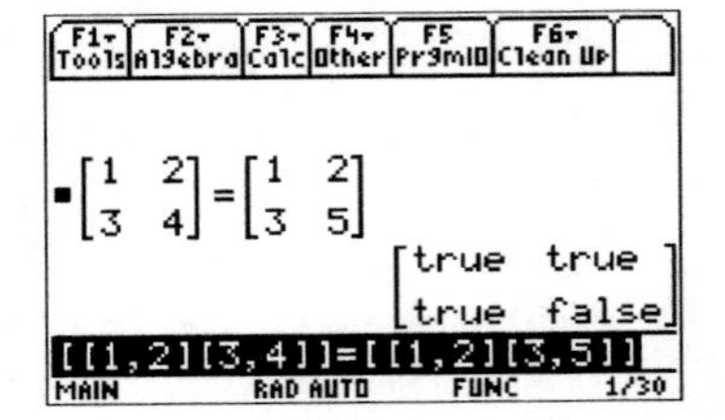

The TI-89 can compare two matrices directly simply by typing (e.g.) `mat1=mat2`. If both matrices have the same dimensions, then the result is a matrix containing "true" and "false" for each entry (as is shown on the right). If the two matrices do not have the same dimensions, the test simply reports "false."

Section 7.4 Example 3 (page 502) Adding Matrices
Section 7.4 Example 4 (page 504) Subtracting Matrices
Section 7.4 Example 5 (page 505) Multiplying Matrices by Scalars
Section 7.4 Example 8 (page 508) Multiplying Matrices

As the screens in the text indicate, basic arithmetic with matrices is relatively straightforward. To perform these computations on a TI-89, the entries are essentially identical to those shown in the text, except for the names of the matrices.

Section 7.4 Example 9 (page 508) Multiplying Square Matrices

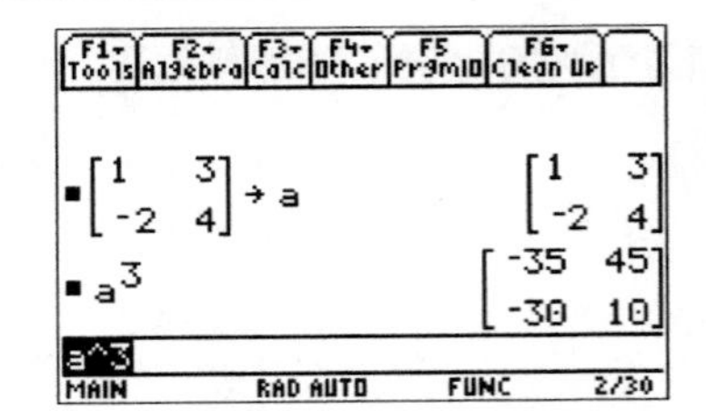

One additional observation about multiplying square matrices: One can raise a square matrix to any power using the [^] key. (This is most useful for integer powers, but the TI-89 will report results for other powers, too.)

Section 7.5 Example 1 (page 514) Evaluating the Determinant of a 2×2 Matrix

Section 7.5 Example 4 (page 517) Evaluating the Determinant of a 3×3 Matrix

Section 7.5 Example 5 (page 518) Evaluating the Determinant of a 4×4 Matrix

The determinant of any square matrix is easy to find with a TI-89. (At least, it is easy for the user; the calculator is doing all the work!) Simply choose the command `det(` from the MATH:Matrix menu ([2nd][5][4][2]), then type the matrix or the name of matrix variable. Note that trying to find the determinant of a non-square matrix (for example, a 3×4 matrix) results in a `Dimension` error.

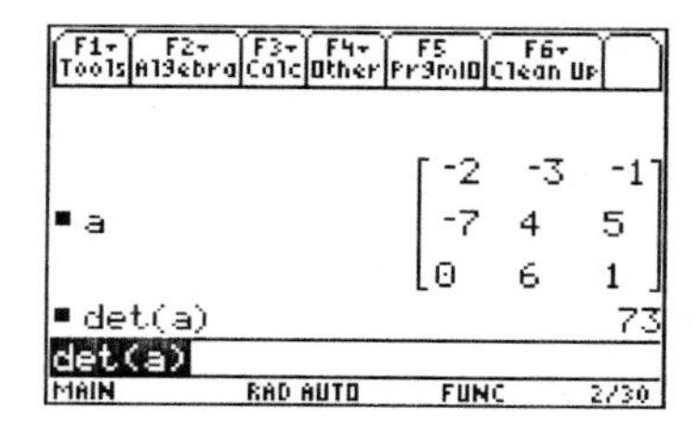

Section 7.6 Example 1 (page 525) Using the 2×2 Identity Matrix

On the TI-89, the function to produce identity matrices is `identity(`, located in the MATH:Matrix menu ([2nd][5][4]). The screen on the right shows the 3×3 identity matrix.

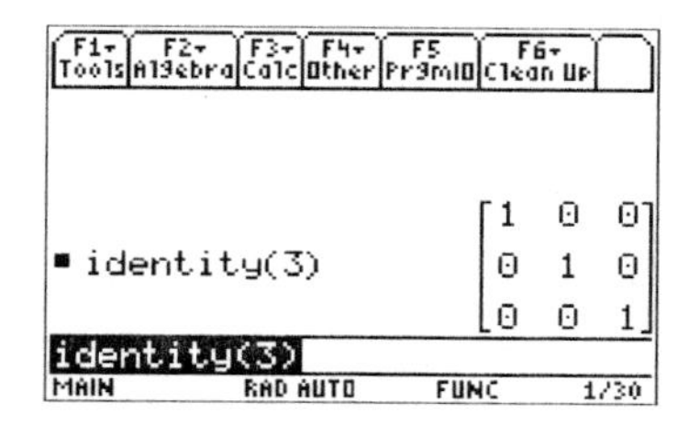

Section 7.6 Example 3 (page 528) Finding the Inverse of a 3×3 Matrix

Section 7.6 Example 4 (page 530) Finding the Inverse of a 2×2 Matrix

The TI-89 will quickly find the inverse of a square matrix (if it exists) by simply raising that matrix to the power -1 (that is, type the matrix, followed by [^][(-)][1]). An example is shown. Attempting this with a non-square matrix results in a `Dimension` error. A `Singular matrix` error occurs when we try to find the inverse of this matrix. ("Singular" means noninvertible.)

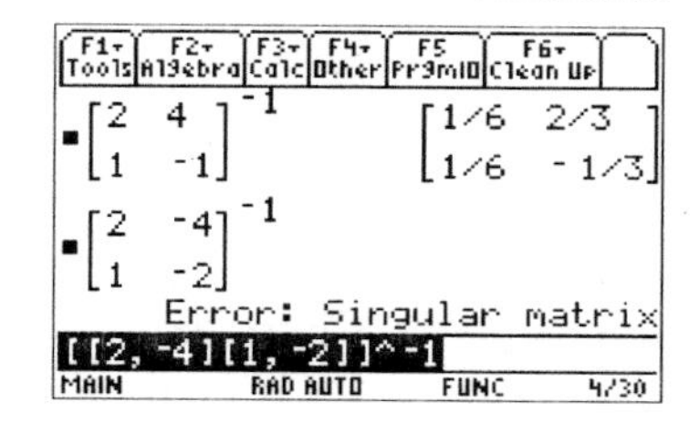

Section 7.7 Example 1 (page 538) Graphing a Linear Inequality

With the TI-89, there are two ways to shade above or below a function. The simpler way is to use the `Above` and `Below` graph styles (see page 106), which mean "draw the graph of this function and shade the region above/below it." This style produces the graph shown below on the right. Note that the TI-89 is not capable of showing the detail that the line is "dashed."

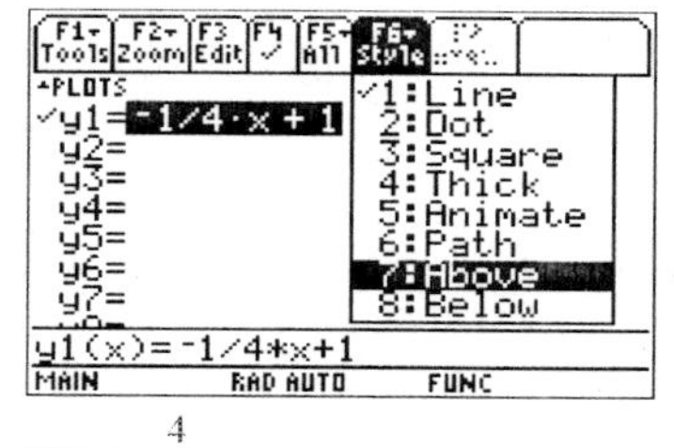

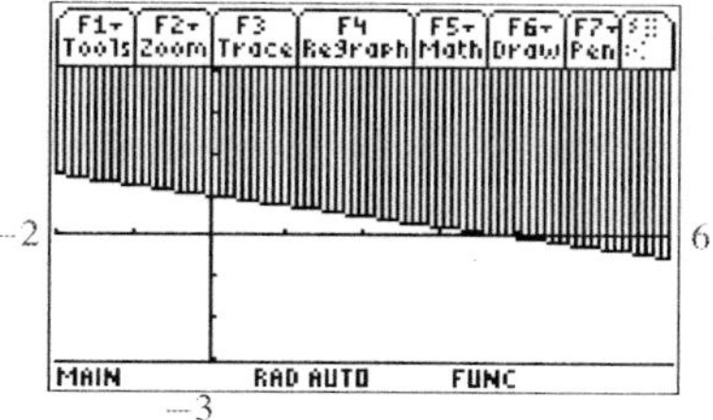

The other way to shade is the `Shade` command, accessed in the [CATALOG]. The format is

`Shade` *lower*`,`*upper*`,`*min x*`,`*max x*`,`*pattern*`,`*resolution*

Here *lower* and *upper* are the functions between which the TI-89 will draw the shading (above *lower* and below *upper*). The last four options can be omitted. *min x* and *max x* specify the starting and ending x values for the shading. If omitted, the TI-89 uses `xmin` and `xmax`.

The last two options specify how the shading should look. *pattern* determines the direction of the shading: 1 (vertical—the default), 2 (horizontal), 3 (negative-slope 45°—that is, upper left to lower right), or 4 (positive-slope 45°—that is, lower left to upper right). *resolution* is an integer from 1 to 10 which specifies how dense the shading should be (1 = shade every column of pixels, 2 = shade every other column, 3 = shade every third column, etc.). If omitted, the TI-89 shades every other column; i.e., it uses *resolution* = 2.

To produce the graph shown above, the appropriate command (typed on the home screen) would be something like

```
Shade -(1/4)x+1,10.
```

The use of 10 for the *upper* function simply tells the TI-89 to shade up as high as necessary; this could be replaced by any number greater than 4 (the value of `ymax` for the viewing window shown). Also, if the function `y1` had previously been defined as `-(1/4)x+1`, this command could be typed as `Shade y1(x),10`.

One more useful piece of information: Suppose one makes a mistake in typing the `Shade` command (e.g., switching *upper* and *lower*, or using the wrong value of *resolution*), resulting in the wrong shading. The screen on the right, for example, arose from typing `Shade -(1/4)x+1,1`. In order to achieve the desired results, the mistake must first be erased by pressing [F4] (Regraph) or [2nd][F1][1](Draw:ClrDraw). Then return to the home screen, correct the mistake in the `Shade` command, and try again.

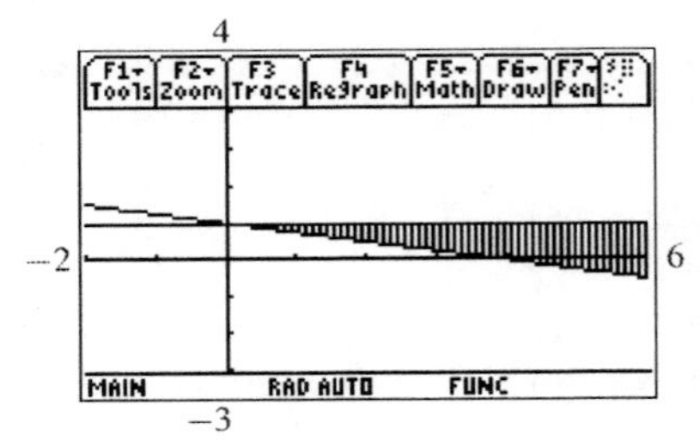

Section 7.7 Example 2 (page 539)

Graphing a System of Two Inequalities

The easiest way to produce (essentially) the same graph as that shown in the text is to use the "shade above/below" graph style (see page 106). The screen on the right (above) shows the style for y2 being set to `Above`; the style for `y1` is also `Above`. The results are shown on the graph below. When more than one function is graphed with shading, the TI-89 rotates through the four shading patterns (see the previous example); that is, it graphs the first with vertical shading, the second with horizontal, and so on. All shading is done with a resolution of 2 (every other pixel).

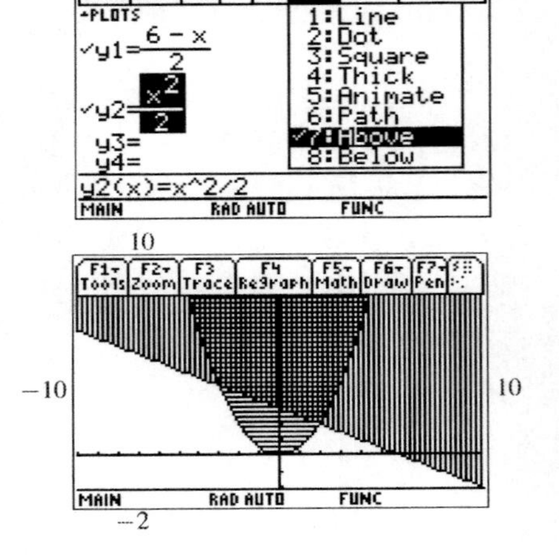

The `Shade` command (see the previous example) can be used to produce this from the home screen. If `y1=(6-x)/2` and `y2=x^2/2`, the commands at right produce the graph shown above.

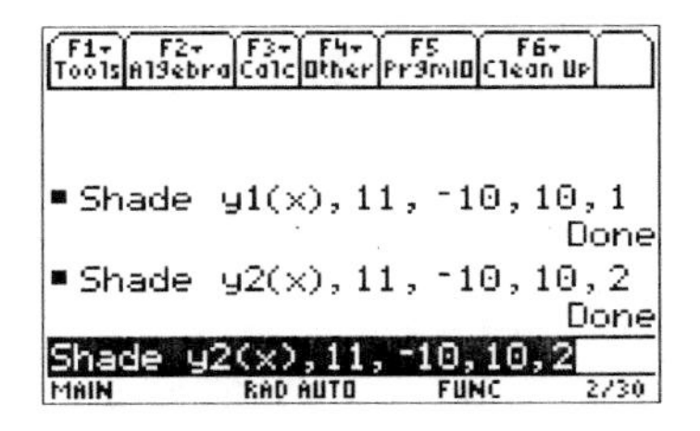

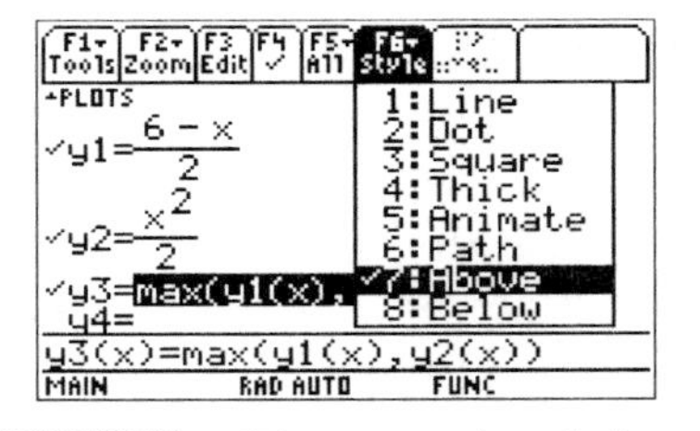

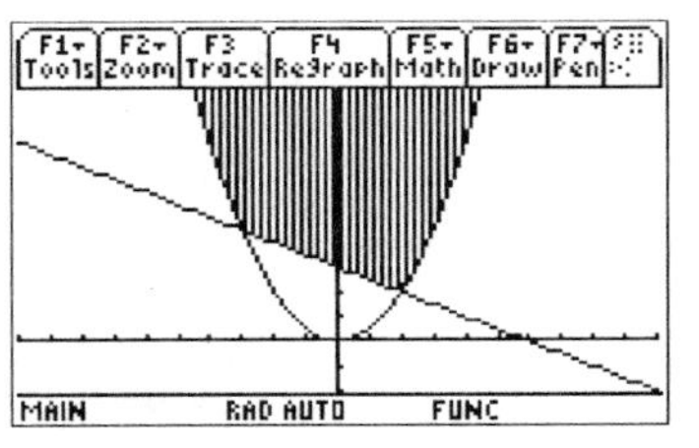

A nicer picture can be created, with a little more work, by making the observation that if y is greater than both $x^2/2$ and $(6-x)/2$, then for any x, y must be greater than the larger of these two expressions. The TI-89 provides a convenient way to find the larger of two numbers with the `max(` function, located in the MATH:List menu ([2nd][5][3]). Then `max(y1(x),y2(x))` will return the larger of `y1` and `y2`, and the Y= screen entries shown will produce the graph on the far right. (Note that `y1` and `y2` have been set to graph as solid curves, while `y3` has the `Above` style.) The home-screen command `Shade max(y1(x),y2(x)),11` would produce similar results.

Extension: The table below shows how (using graph styles or home-screen commands) to shade regions that arise from variations on the inequalities in this example, assuming that `y1=(6-x)/2` and `y2=x^2/2`. (The results of these commands are not shown here.)

For the system...	or equivalently...	the command would be...
$x < 6 - 2y$ $x^2 < 2y$	$y < (6-x)/2$ $y > x^2/2$	shade below y1 and above y2, or enter `Shade y2(x),y1(x)`
$x > 6 - 2y$ $x^2 > 2y$	$y > (6-x)/2$ $y < x^2/2$	shade above y1 and below y2, or enter `Shade y1(x),y2(x)`
$x < 6 - 2y$ $x^2 > 2y$	$y < (6-x)/2$ $y < x^2/2$	shade below y1 and below y2, or enter `Shade -10,min(y1(x),y2(x))`

Section 7.8 Example 1 (page 548) Finding a Partial Fraction Decomposition

Section 7.8 Example 2 (page 549) Finding a Partial Fraction Decomposition

Section 7.8 Example 3 (page 550) Finding a Partial Fraction Decomposition

Section 7.8 Example 4 (page 551) Finding a Partial Fraction Decomposition

The TI-89's `expand(` command (found in the Algebra menu, [F2][3]) automates the task of expanding a rational expression into partial fractions. Shown is the output for Example 1; note that it is not necessary to divide the polynomials first.

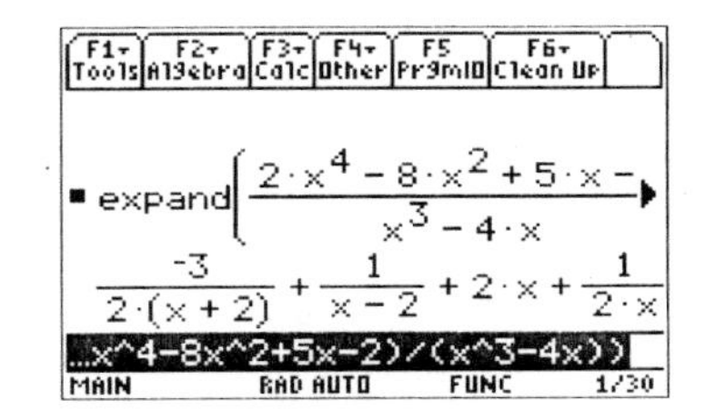

Section 8.1 Technology Note (page 567)

Radian and Degree modes

See section 9 of the introduction (page 101) for information about selecting Degree and Radian modes.

Section 8.1 Example 2 (page 568)

Calculating with Degrees and Minutes

Section 8.1 Example 3 (page 568)

Converting between Decimal Degrees and Degrees, Minutes, Seconds

When working with angles measured in degrees, it is a good idea to select Degree mode (see section 9 of the introduction, page 101). In fact, when in Radian mode, the TI-89 automatically converts angles entered in degrees to radians, which can be confusing.

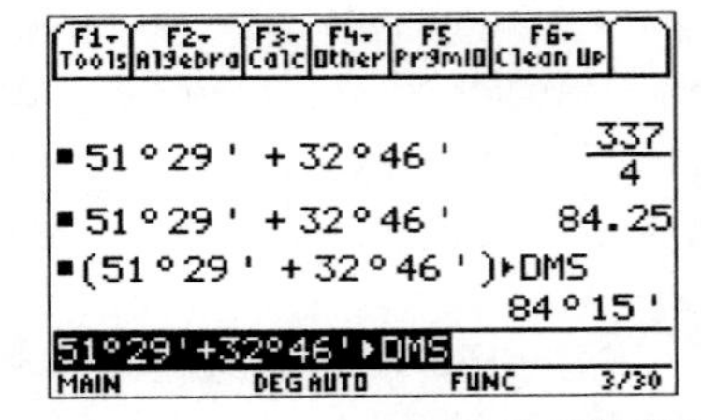

In Degree mode, the conversion to decimal degrees is fairly simple: Enter the angle using [2nd][I], [2nd][=], and [2nd][I] for the degrees, minutes, and seconds symbols. The only potential snag is that, depending on the [MODE] settings, the TI-89 might attempt to display an exact (fractional) representation of the conversion, rather than a decimal (as in the first entry shown in the history area on the right). This can be overcome by either entering one of the quantities (degrees, minutes, or seconds) with a decimal point, or by entering the expression with [◆][ENTER] instead of just [ENTER]. (This produced the second output in the history area.) The third entry makes use of the ▸DMS operator, which causes an angle to be displayed in degrees, minutes, and seconds, rather than as a decimal.

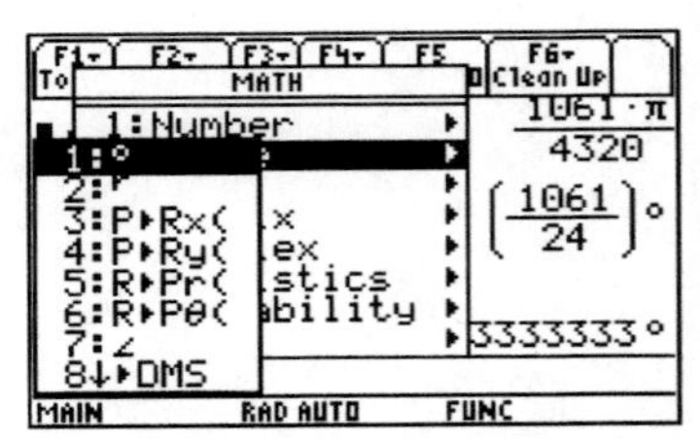

If the TI-89 is left in Radian mode, a little more work is needed. The MATH:Angle menu ([2nd][5][2]) is shown on the right. It includes the degrees symbol, and the ▸DMS operator (mentioned above). Not visible in this menu is option 9: ▸DD (convert to decimal degrees).

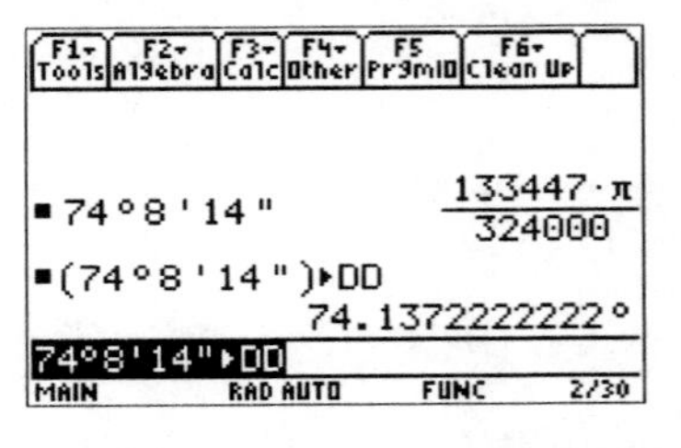

When in Radian mode, simply entering 74°8'14" causes the calculator to first convert interpret this as degrees (and fractions of degrees), then to convert this angle into radians. The top line in the history area shows this (not-too-helpful) result. Appending the ▸DD conversion command to this forces the TI-89 to forego the conversion to radians. The second output arose from pressing [◆][ENTER].

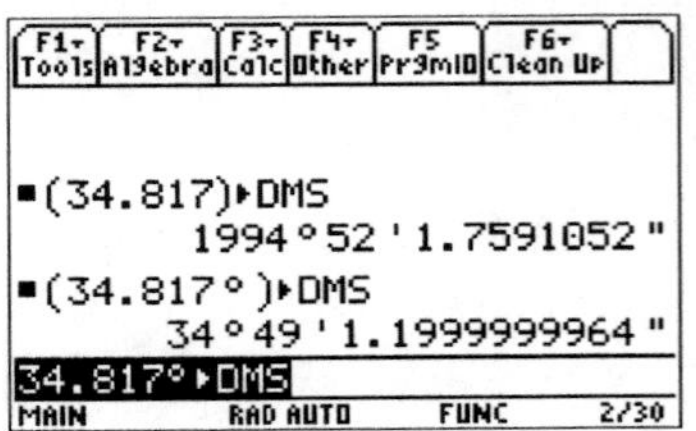

When converting from decimal degrees to DMS while the calculator is in Radian mode, be sure to enter the angle with the degree symbol ([2nd][I]); otherwise, the TI-89 assumes that the given angle is in radians, so that it converts radians to degrees before splitting the result into degrees, minutes, and seconds. Note that some rounding error was introduced in the process, causing the seconds portion of the result to be slightly different from 1.2.

Section 8.1 Example 5 (page 571) Converting Degrees to Radians

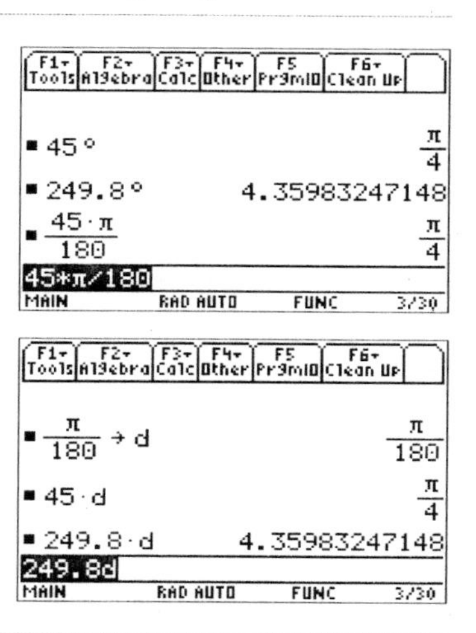

The number π is available as [2nd][^], and the degree symbol is [2nd][|]. With the calculator in Radian mode (see page 101), entering 45° causes the TI-89 to automatically convert to radians. The screen on the right shows the two ways of performing the conversion: Letting the TI-89 do the work by using the degrees symbol, or simply multiplying by $\pi/180$.

An alternative to using the degree symbol is to store $\pi/180$ in the calculator variable `d` (see page 101). Then typing, for example, `45d` [ENTER] will multiply 45 by $\pi/180$. This approach will work regardless of whether the calculator is in Degree or Radian mode. (A value stored in a variable will remain there until it is replaced by a new value, or the variable is deleted or cleared.)

Section 8.1 Example 6 (page 571) Converting Radians to Degrees

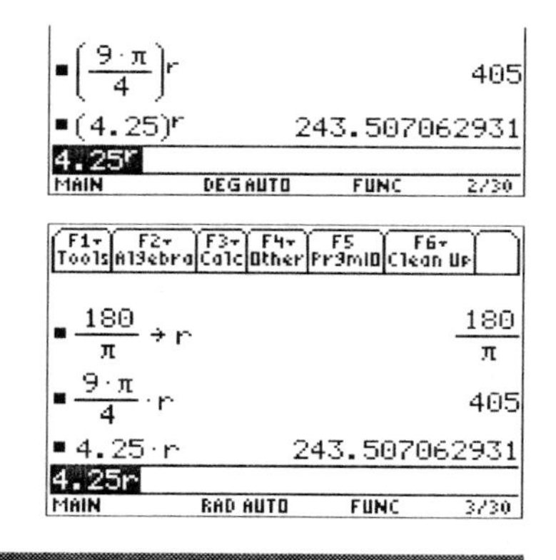

With the TI-89 in Degree mode (see page 101), the radian symbol (a superscripted `r`), produced with [2nd][5][2][2], will automatically change a radian angle measurement to degrees.

Alternatively, with the value $180/\pi$ stored in the calculator variable `r` (see page 101), typing $(9\pi/4)$`r` [ENTER] will convert from radians to degrees regardless of whether the calculator is in Degree or Radian mode. (The same result can be achieved by *dividing by* the calculator variable `d` as defined in the previous example.)

Section 8.2 Example 3 (page 584) Finding Function Values of an Angle

To graph part of a function, like the line $x + 2y = 0$, $x \geq 0$ shown in the text, enter `y1=when(x≥0,-1/2*x,1/0)`. The division by zero means that the function is not graphed when $x < 0$. "≥" is found in the MATH:Test menu, [2nd][5][8].

Section 8.2 Example 4 (page 585) Finding Function Values of Quadrantal Angles

The alternative to putting the calculator in Degree mode is to use the degree symbol ([2nd][|]) following each angle measure; e.g., enter `sin(90°)` rather than just `sin(90)`.

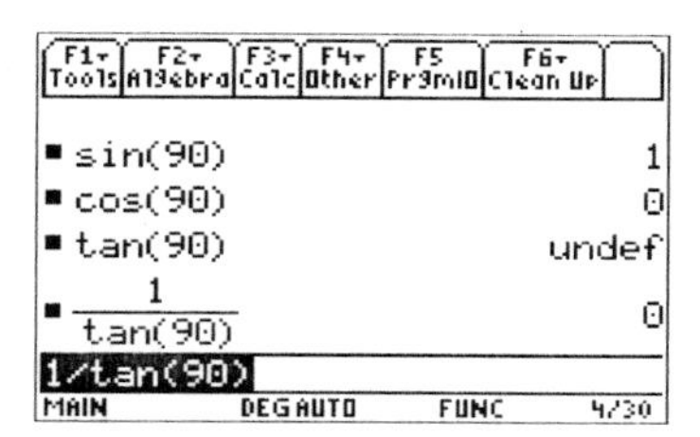

Unlike the TI-83 (used for the screens in the text), the TI-89 reports "undef" (undefined) when asked to compute $\tan 90^\circ$. Since the cotangent, secant, and cosecant functions are the reciprocals of the tangent, cosine, and sine, they can be entered as (e.g.) `1/sin(90)`. This reports the correct value (0) for $\cot 90^\circ$, even though $\tan 90^\circ$ is undefined, and also correctly reports that $\sec 90^\circ$ is undefined.

One might guess that the other three trigonometric functions are accessed with [◆] followed by [Y], [Z], or [T] (which produce, e.g., $\sin^{-1}$). This is **not** what these functions do; in this case, the exponent -1 does not mean "reciprocal," but instead indicates that these are inverse functions (which are discussed in detail in Chapter 9 of the text).

Section 8.2 Technology Note (page 589) Powers of trigonometric functions

Because the TI-89 automatically includes an opening parenthesis on the trigonometric functions, one cannot enter (e.g.) `sin^2`. To compute $sin^2 30°$ (with the TI-89 in Degree mode), one should type either `sin(30)^2` or `(sin(30))^2`—the extra parentheses are not needed but do not hurt. Typing `sin(30^2)` would compute the sine of 900°, and typing `sin(^230)` produces (not too surprisingly) a syntax error.

Section 8.3 Technology Note (page 598) Decimal approximations and exact values

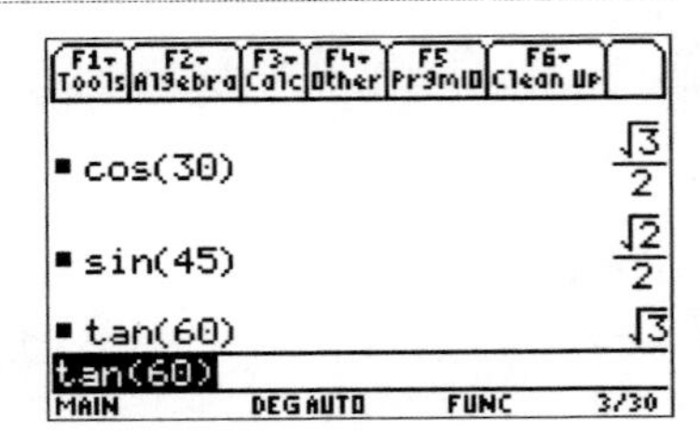

As the comment at the end of this note points out, the TI-89 *can* find exact values for many trigonometric functions, provided it is in EXACT or AUTO mode. Shown are the outputs for several of the special angles shown in the table next to this Technology Note. The TI-89 will report decimal approximations if it is in APPROX mode, if an angle is entered with a decimal point—e.g., `cos(30.)`—or if the expression is entered with [◆][ENTER] rather than just [ENTER].

Section 8.3 Example 7 (page 601) Approximating Trigonometric Function Values with a Calculator

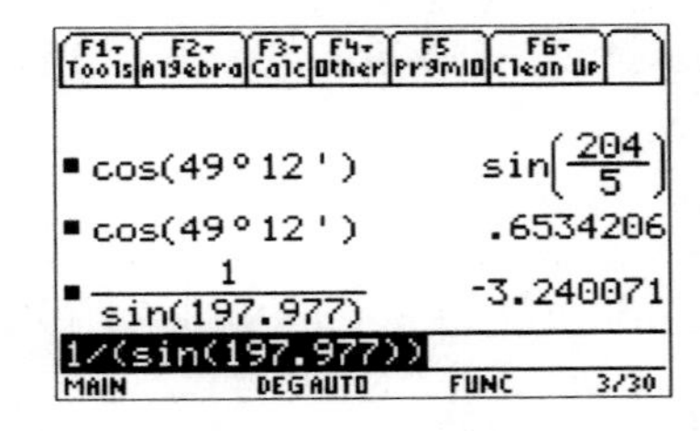

Note that the cosine for part (a) gives an exact value (the first output in the history area) if in EXACT or AUTO mode; to get a decimal approximation, either include a decimal point in the entry, or press [◆][ENTER] instead of [ENTER]. This cosine will be computed correctly in either Degree or Radian mode, but for the cosecant in part (b), we would have to include the degree symbol ([2nd][I]) after the angle measure if the TI-89 were in Radian mode.

Section 8.3 Technology Note (page 602) Inverse versus reciprocal functions

The inverse trigonometric functions were mentioned earlier, and are covered in the next example; more details are given in Chapter 9 of the text.

Section 8.3 Example 8 (page 602) Using Inverse Trigonometric Functions

Section 8.3 Example 10 (page 603) Finding Angle Measures

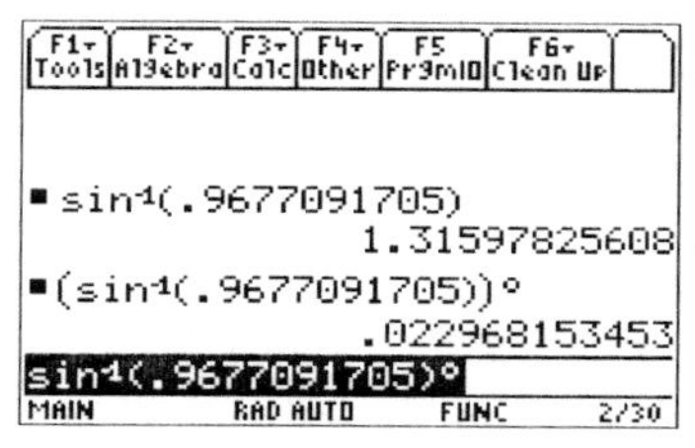

The `sin⁻¹` ("inverse sine," or "arcsine") function is [◆][Y], while `cos⁻¹` is [◆][Z] and `tan⁻¹` is [◆][T]. Note that the first output shown in text Figure 55 was produced in Degree mode, and the second was produced in Radian mode. None of the options available in the MATH:Angle menu can be used to avoid changing the mode; for example, when in Degree mode, the inverse trigonometric functions will always give an angle measure in degrees. The screen on the right shows the result for Example 8(a) when done in Radian mode. Note that in the second entry, an attempt was made to get the TI-89 to report the result in degrees (by placing the degree symbol at the end of the entry), but this does not have the desired result.

Section 8.4 Technology Note (page 609) Programs to solve right triangles

See section 13 of the introduction (page 107) for information about installing and running programs on the TI-89. The text notes that one must "consider the various cases"; there are five such cases: two legs, leg and hypotenuse, angle and hypotenuse, angle and adjacent leg, angle and opposite leg.

Section 8.4 Example 7 (page 613) Solving a Problem Involving the Angle of Elevation

The TI-89 can automatically locate the intersection of two graphs using the GRAPH:Math:Intersection feature. This was previously illustrated on page 111, but we repeat the description here: With the two functions graphed, press [F5] and choose option 5 (Intersection). Use ⊖, ⊖ and [ENTER] to specify which two functions to use (in this case, the only two being displayed). The TI-89 then prompts for lower and upper bounds (numbers that are, respectively, less than and greater than the location of the intersection). After pressing [ENTER], the TI-89 will try to find an intersection of the two graphs. The screens below illustrate these steps.

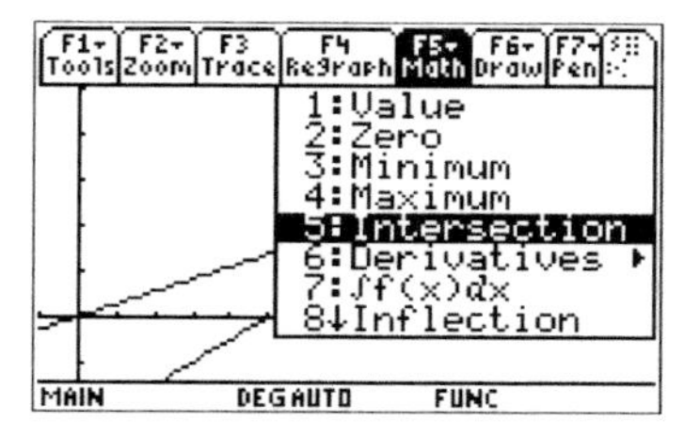

From the graph screen: [F5][5]

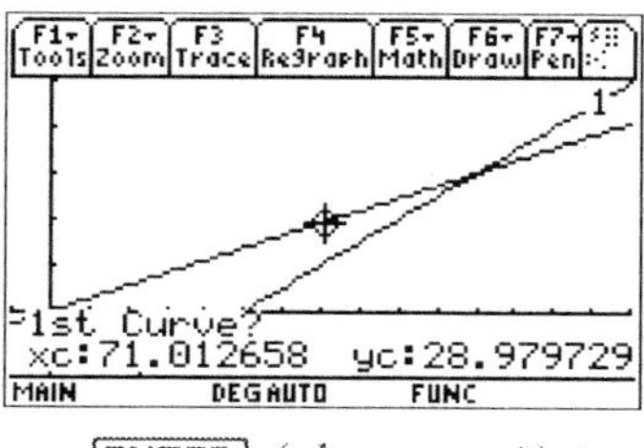

[ENTER] (choose y1)

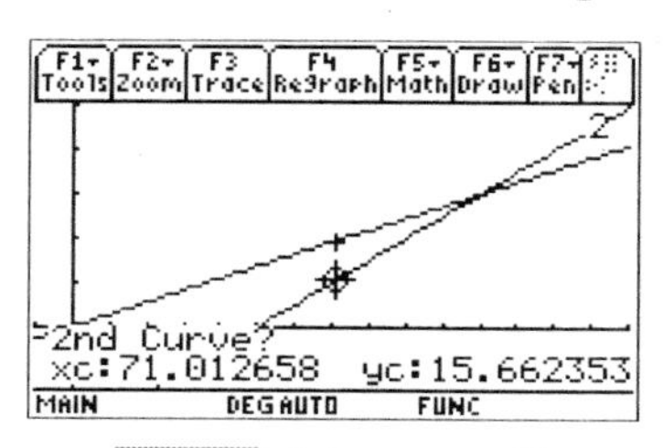

[ENTER] (choose y2)

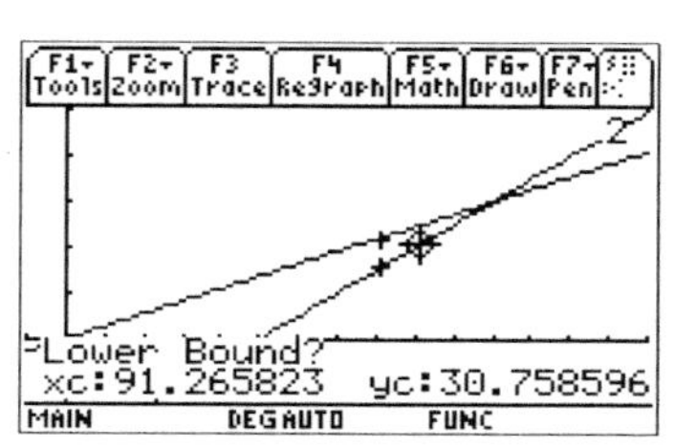

Move to the left of the intersection, press [ENTER]

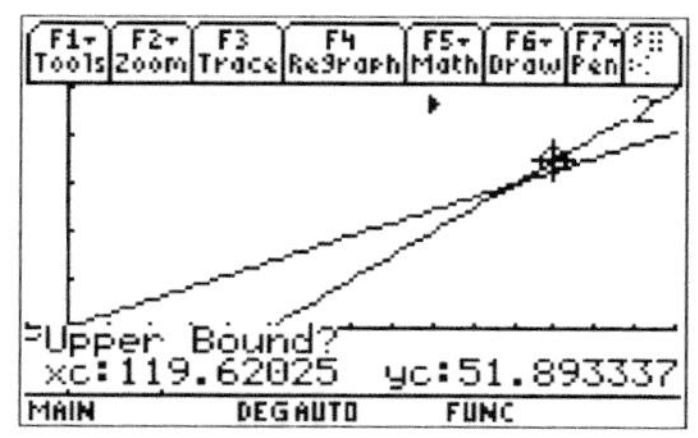

Move to the right of the intersection, press [ENTER]

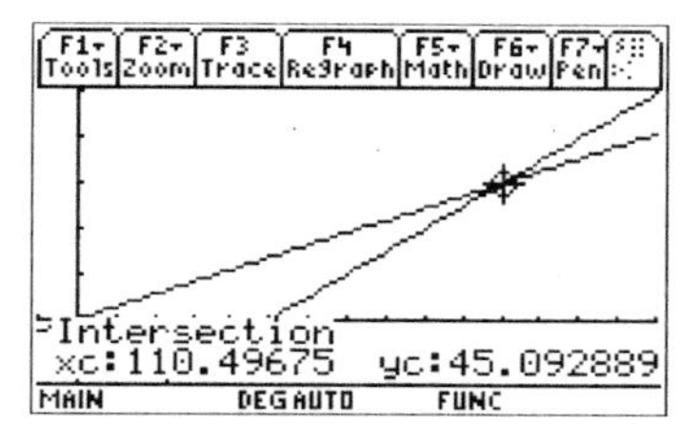

The intersection is found (as in Figure 71)

Section 8.6 Technology Note (page 630) — The trig viewing window

The trig viewing window is set by pressing [◆][F2][F2][7]. On the TI-89, the values of `xmin` and `xmax` are On the TI-89, the values of `xmin` and `xmax` are about $\pm 3.3\pi$ rather than $\pm 2\pi$. These values are chosen so that Δx (see page 101) equals $\pi/24$.

If the TI-89 is in Degree mode, `xmin` and `xmax` will be ± 592.5, so $\Delta x = 7.5$.

Section 8.6 Example 1 (page 631) — Graphing $y = a \sin x$

Section 8.6 Example 2 (page 633) — Graphing $y = \sin bx$

The TI-89's graph styles can produce screens like the one shown in Figure 89 or the graph screen accompanying Example 2. See page 106 for information on setting the thickness of a graph. Note that the TI-89 must be in Radian mode in order to produce the correct graph.

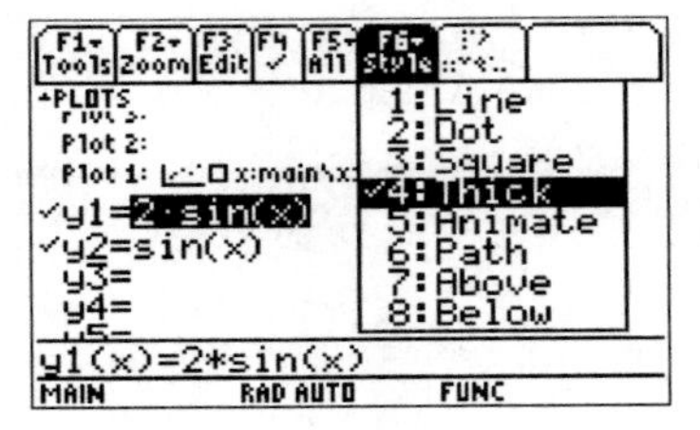

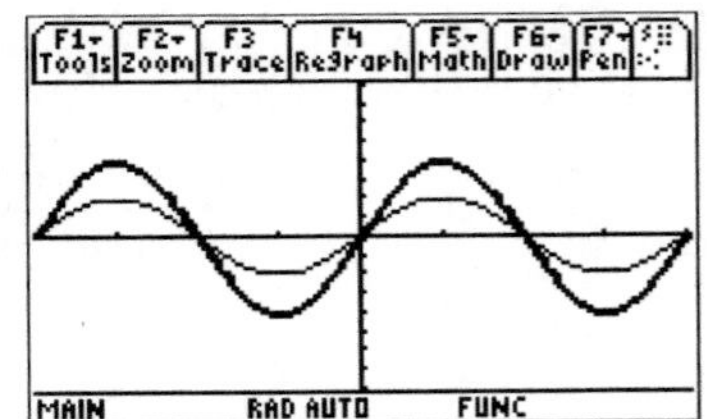

It is possible to distinguish between the two graphs without having them drawn using different styles by using the TRACE feature. On the right, the trace cursor is on graph 2—that is, the graph of $y_2 = \sin x$.

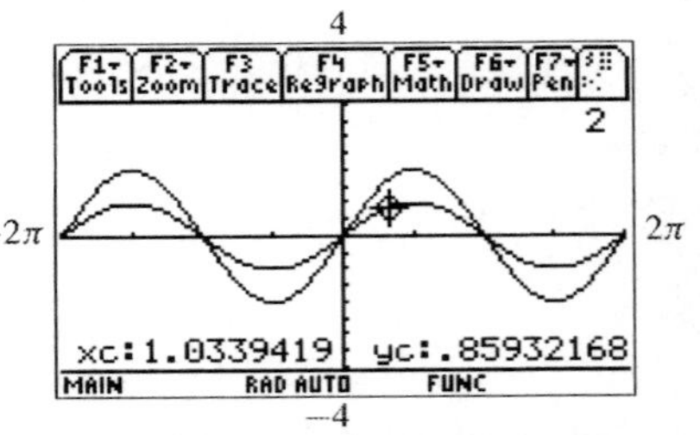

Section 8.6 Example 8 (page 638) — Modeling Temperature with a Sine Function

The "sine regression" illustrated at the end of this example is a built-in feature of the TI-89, found in the Calc menu ([F5]) of the data editor. Also shown here is the output of this command; note that the calculator takes several seconds to perform this computation. Also note that the reported values are not exactly the same as those given by the TI-83. (The values shown in the text are all rounded to two decimal places; however, these values do not round quite the same.) See page 110 for information about using the TI-89 for regression computations.

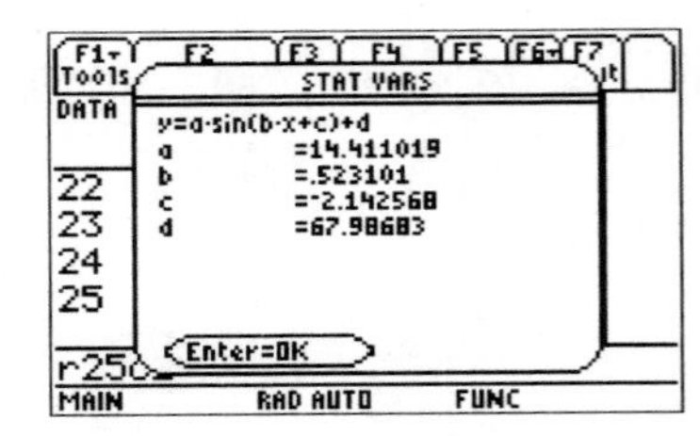

Section 8.7 Technology Note (page 647) — Cosecant and secant functions

Section 8.7 Example 1 (page 649) — Graphing $y = a \sec bx$

Section 8.7 Example 2 (page 650) — Graphing $y = a \csc (x - d)$

See section 12 of the introduction (page 105) for information about graphing functions in the `Dot` graph style. To graph the cosecant function, the actual entry on a TI-89 would be

`y1=1/sin(x)` or `y1=sin(x)^-1`.

The function in Example 1 can be entered as `y1=2(cos(x/2))`$^{-1}$ (or as shown in the text). The function in Example 2 can be entered as `y1=3/(2sin(x-π/2))` (or as shown in the text). A reminder: `sin`$^{-1}$ ([◆][Y]) is *not* the cosecant function.

Section 9.1	Example 2	(page 674)	Rewriting an Expression in Terms of Sine and Cosine
Section 9.1	Example 3	(page 676)	Verifying an Identity (Working with One Side)
Section 9.1	Example 4	(page 677)	Verifying an Identity (Working with One Side)
Section 9.1	Example 5	(page 677)	Verifying an Identity (Working with One Side)
Section 9.1	Example 6	(page 678)	Verifying an Identity (Working with Both Sides)

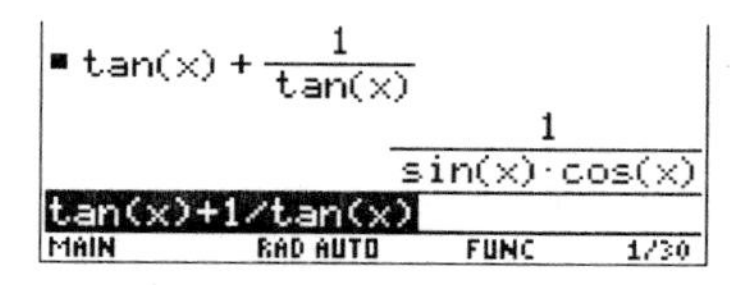

The TI-89 automatically attempts to rewrite many trigonometric expressions in terms of sine and cosine, as the screen here shows. (This will only work correctly if x has not been assigned a numerical value.)

As an alternative to graphing these two functions, the TI-89's table feature (see page 109 of this manual) can be used: If the y values are the same for a reasonably large sample of x values, one can be fairly sure (though not certain) that the two expressions are equal. To make this approach more reliable, be sure to choose x values that are not, for example, all multiples of π. This method is used in Example 4.

The fact that the two graphs are identical on the calculator screen does provide strong support for the identity, especially when confirmed by tracing, as described in the Technology Note next to Example 3 in the text. The tables shown in Figures 3 and 4 also show that the two expressions produce identical output—at least to the number of digits visible in the table. The exception is Figure 3 when x equals $\pi/2 \approx 1.5708$, for which y1 shows ERROR while y2 equals 1. In fact, $\cot(\pi/2) + 1 = 1$, but the value of y1 is reported as ERROR since $\tan(\pi/2)$ is undefined. (If this table is reproduced on the TI-89, the value 1 is shown for both y1 and y2, because the TI-89 evaluates $1/\tan x$ as 0 when $x = \pi/2$.)

The footnote on page 676 points out that the graph (or the table) cannot be used to prove the identity. In particular, the graph in Figure 2 only plots points for values of x that are $\Delta x = \pi/24$ units apart, while the table in Figure 3 shows outputs for input values spaced $\pi/8$ units apart. For example, the function `y3=cos(x)/sin(x)+cos(48x)` would look identical to y1 and y2 if graphed on the trig viewing window, although this function is different from these two functions at any point other than those shown in the graph.

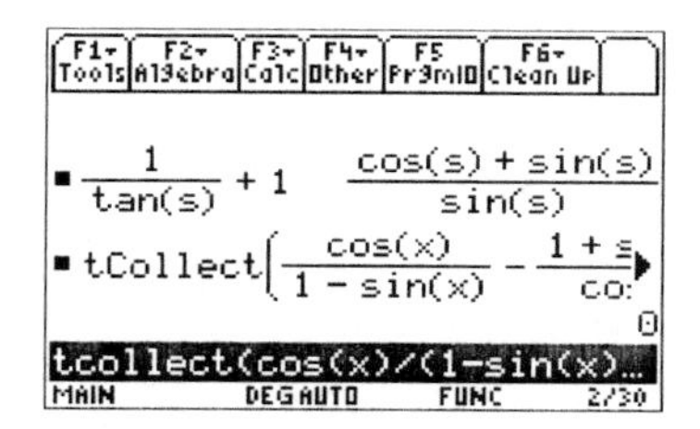

Note, though, that with the TI-89, graphical verification of identities is not usually necessary, because the TI-89 can simplify many such expressions, provided that all trig functions are entered in terms of the sine, cosine, and tangent functions. The screen on the right shows confirmation of the identities in Examples 3 and 5. For the first, the left side of the equation was entered, and the output is equivalent to the right side. The second uses the TI-89's `tCollect` function, found in the MATH:Algebra:Trig menu ([2nd][5][9][9]). This can be used to simplify some trigonometric expressions; note that it reports that the difference between the left and right sides simplifies to 0.

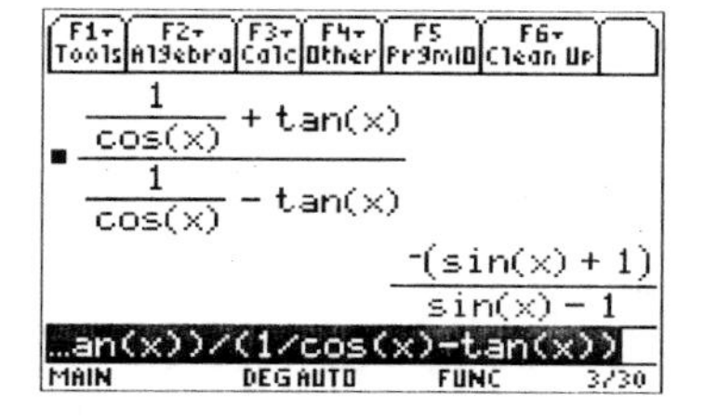

The TI-89's methods of expansion do not always give the same results as those shown in these examples. Here is the output for the left side of the equation in Example 6 (which agrees with the expression found in the text). However, the TI-89 does not simplify the right side to the same expression. (With `tCollect`, though, it does confirm that the difference between the left and right sides is 0.)

Section 9.2 Example 1 (page 685) Finding Exact Cosine Function Values

Section 9.2 Example 3 (page 687) Finding Exact Sine and Tangent Function Values

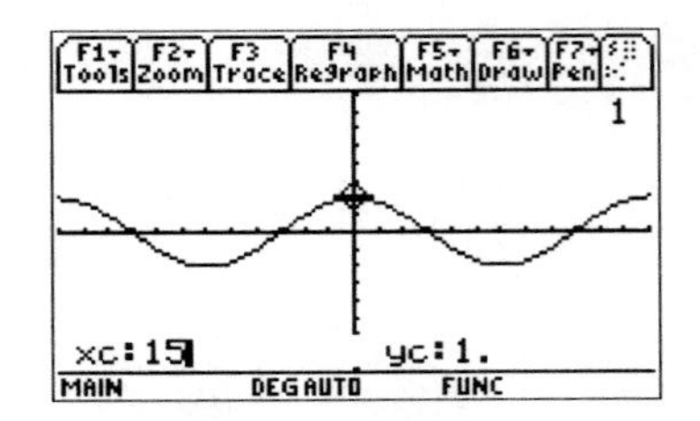

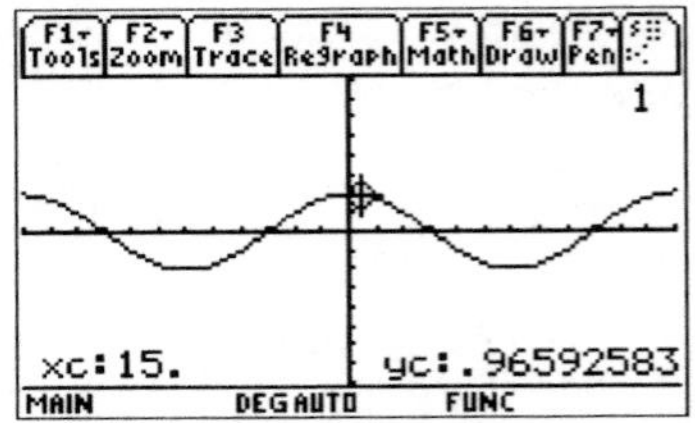

The TI-89 can graphically and numerically support exact value computations such as $\cos 15^\circ = \frac{\sqrt{6}+\sqrt{2}}{4}$. Starting with a graph of `y1=cos(x)`, the TI-89 makes it possible to trace to any real number value for x between `xmin` and `xmax`, using the GRAPH:Trace feature or the GRAPH:Math:Value command. Simply type a number and press [ENTER]; the cursor then jumps to that x-coordinate, and shows a decimal approximation of y. (This feature was previously discussed on page 116.)

Alternatively, a table of values like those shown here can be used to find the value of $\cos 15^\circ$.

x	y1
11.	.98163
12.	.97815
13.	.97437
14.	.9703
15.	.96593

y1(x)=.96592582628907

From the home screen, we see agreement to 12 decimal places for the values of `yc` (which contains the y coordinate from the GRAPH:Trace operation), $\cos 15^\circ$, and $\frac{\sqrt{6}+\sqrt{3}}{4}$. (Decimal points were placed in the second and third entries to get a decimal result.)

```
■ yc                 .965925826289
■ cos(15.)           .965925826289
■ (√6 + √2)/4.       .965925826289
(√(6)+√(2))/4.
```

While this graphical confirmation might be more useful for other situations, it is actually unnecessary in this case, because the TI-89 knows how to use many of these identities to find exact values! Assuming that the `Exact/Approx` mode setting is AUTO or EXACT, the TI-89 computes the exact value of $\cos(15^\circ)$, although it reports that exact value in a slightly different form from that given in the text.

```
■ cos(15)            (√3 + 1)·√2 / 4
■ (√6 + √2)/4        (√3 + 1)·√2 / 4
(√(6)+√(2))/4
```

Note that Example 8 in Section 9.3 (page 699) shows that $\cos 15^\circ = \cos \frac{\pi}{12}$ can also be written as $\frac{\sqrt{2+\sqrt{3}}}{2}$.

Section 9.3 Example 2 (page 694) Verifying a Double-Number Identity

Section 9.3 Example 3 (page 694) Simplifying Expressions by Using Double-Number Identities

Section 9.3 Example 4 (page 695) Deriving a Multiple-Number Identity

Section 9.3 Example 6 (page 697) Using a Product-to-Sum Identity

Section 9.3 Example 7 (page 698) Using a Sum-to-Product Identity

Section 9.3 Example 8 (page 699) Using a Half-Number Identity to Find an Exact Value

For Example 4, the table in the text shows `y1=sin(3x)` and `y2=3sin(x)-4sin(x)^3`—recall the proper way to enter this second expression—with Δ`tbl` $= \pi/8$ and `tblStart=` $-3\pi/8$. Note that choosing

x values which are multiples of a fraction of π is somewhat risky, since the periods of $\sin 3x$ and $\sin x$ involve fractions of π. Stronger support can be obtained by trying input values that are not multiples of π —say, $x = 1$, or $x = \sqrt{2}$.

Again, such methods are not necessary on the TI-89, because it can provide more direct verification of identities. The TI-89's `tExpand` function—in the MATH:Algebra:Trig menu—is more-or-less the opposite of `tCollect`; here we see it's expansion of the two sides of the equation in Example 2.
Note: According to the TI-89 manual, "for best results, `tExpand()` should be used in Radian mode." In fact, `tExpand` seems to do *nothing* when the TI-89 is in Degree mode.

`tCollect` works for some of these expressions, as well; here are it's results for Example 3.

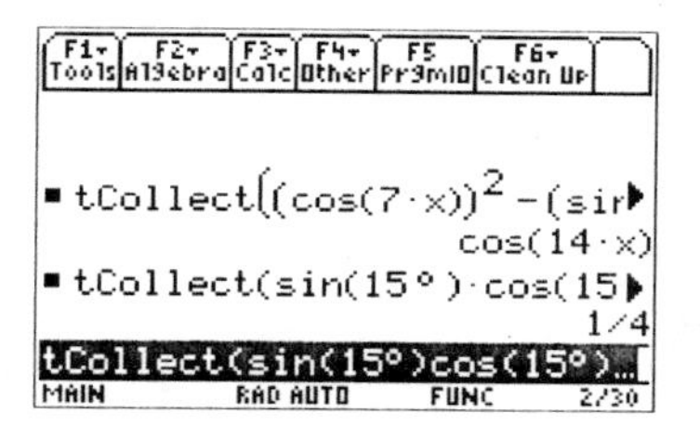

Note, however, that there are often many equivalent ways to re-write a trigonometric expression. For Example 4, `tExpand(sin(3x))` does not give the same answer as the text.

Section 9.4 Example 1 (page 706) — Finding Inverse Sine Values

Section 9.4 Example 2 (page 707) — Finding Inverse Cosine Values

Of course, it is not necessary to graph $y = \sin^{-1} x$ to find these values; one can simply enter, e.g., `sin-1(1/2)`. Angles are given in radians or degrees, depending on the TI-89's mode setting.

Note that if the TI-89's `Complex Format` is RECTANGULAR or POLAR, it does *not* give an error for the input `sin-1 (-2)`, but instead gives a complex result. (A `Non-real result` error occurs if `Complex Format` is set to REAL.) This is technically a correct result, but is not appropriate for the problems in this text.

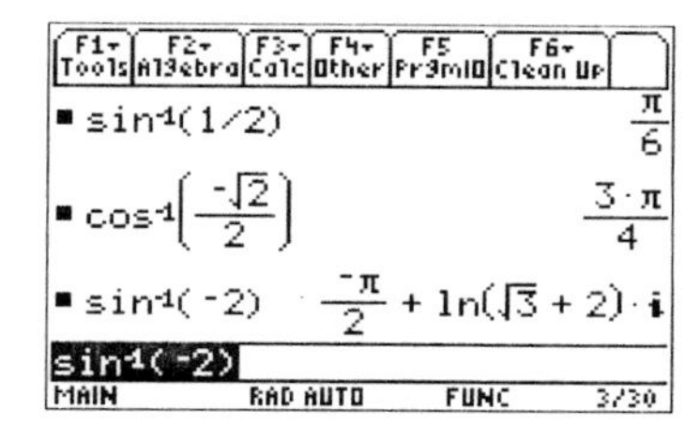

Section 9.4 Example 4 (page 710) — Finding Inverse Function Values with a Calculator

Note that the answer given for (b), 109.499054°, overrepresents the accuracy of that value. A typical rule for doing computations involving decimal values (like −0.3541) is to report only as many digits in the result as were present in the original number—in this case, four. This means the reported answer should be "about 109.5°," and in fact, any angle θ between about 109.496° and 109.501° has a cotangent which rounds to −0.3541. (See also the discussion in the text on page 608.)

Section 9.4 Example 7 (page 712) Writing Function Values in Terms of *u*

The TI-89 handles both of these expressions quite nicely, although it does not rationalize the denominator for (a), and only changes the second expression if the `tExpand` command is used.

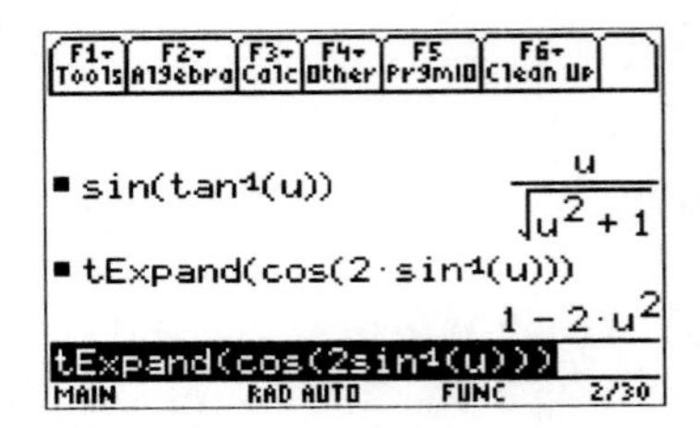

Section 9.5 Example 1 (page 718) Solving a Trigonometric Equation by a Linear Method

See page 111 for extensive information about solving equations with the TI-89.

The text gives the answers $x = \frac{\pi}{6}$ and $x = \frac{5\pi}{6}$. The TI-89's `solve` command (from the Algebra menu) will give those same answers, plus a little more information. Shown is the output of this command; the part of the solution that is not visible in the history area is "x=2·@n1·π+$\frac{\pi}{6}$." On the TI-89, the symbols `@n1`, `@n2`, ... represent arbitrary integers. (The TI-89 will cycle through the numbers 0 to 255 in the symbol "`@n_`" as it solves equations whose solutions are expressed using arbitrary integers.) In other words, the TI-89 is telling you that, by adding any multiple of 2π to $\frac{\pi}{6}$ and $\frac{5\pi}{6}$, you can find other solutions to this equation. (This example only asked for solutions in the interval $[0, 2\pi)$, so we do not want those other solutions.)

Section 9.6 Example 8 (page 727) Describing a Musical Tone from a Graph

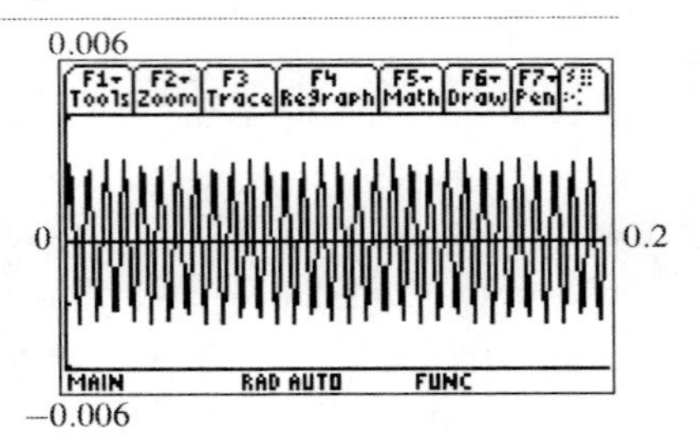

Note that the calculator screen shown in Figure 40 illustrates the importance of choosing a "good" viewing window. If we choose the wrong vertical scale (`ymin` and `ymax`), we might not be able to see the graph at all—it might be squashed against the x-axis. If we make the window too wide—that is, if `xmax` minus `xmin` is too large—we might see the "wrong" picture, like the one on the right: We see a periodic function in this view, but not the one we want. There are actually 30 periods in this window, but the TI-89's limited resolution cannot show them all.

This observation—that a periodic function, viewed at fixed intervals, can appear to be a *different* periodic function—is the same effect that causes wagon wheels to appear to run backwards in old movies.

Section 10.1 Technology Note (page 743) Programs to solve triangles
Section 10.2 Technology Note (page 756)

See section 13 of the introduction (page 107) for information about installing and running programs on the TI-89. If you wish to type these in on your own, here are two programs that give the same output as those shown in the text. (The command `setMode...` in the second program ensures that the TI-89 is in Degree mode. The first program will operate correctly regardless of the mode.)

```
asa(a,b,d)
Prgm
Local c,e,f
180-a-b→c
d*sin(a°)/(sin(c°))→e
d*sin(b°)/(sin(c°))→f
Disp "Other angle",c
Disp "Other sides",e,f
EndPrgm
```

```
sss(d,e,f)
Prgm
Local a,b,c
setMode("Angle","DEGREE")
cos⁻¹((d^2+e^2-f^2)/(2*d*e))→c
cos⁻¹((d^2+f^2-e^2)/(2*d*f))→b
180-b-c→a
Disp "Angles are",a,b,c
EndPrgm
```

To use these programs, type (for example) `asa(112.9,31.1,347.6)` or `sss(9.47,15.9,21.1)`. Upon pressing [ENTER], the other parts of the triangle will be given. (When finished reading those other parts, press [ESC] to return to the home screen.)

Section 10.3 Example 1 (page 766) Finding Magnitude and Direction Angle

Section 10.3 Example 2 (page 767) Finding Vertical and Horizontal Components

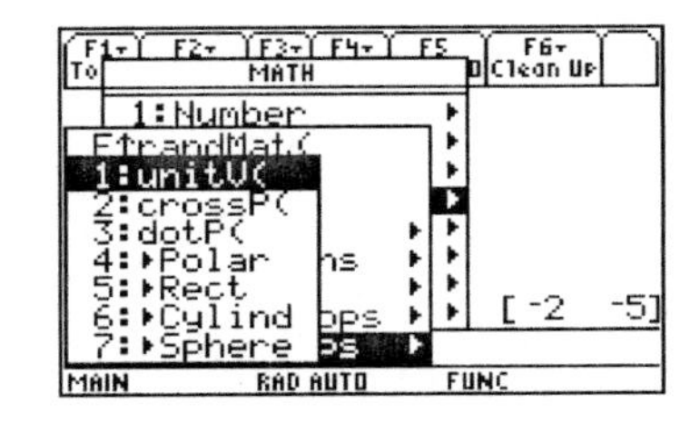

The TI-89 recognizes vectors entered in either of two formats:

[*horizontal component*, *vertical component*] —rectangular format, or
[*magnitude*, ∠ *direction angle*] —polar format.

(The square brackets are [2nd][,] and [2nd][÷], and "∠" is [2nd][EE].) Regardless of how the vector is entered, the TI-89 displays it according to the `Vector Format` mode setting (see page 101); specifically, it displays the vector in component form in RECTANGULAR mode, and in magnitude/direction form for either of the other two modes. (There are, however, commands to override how the vector is displayed.)

Commands for manipulating vectors are buried several levels deep in the MATH:Matrix:Vector ops menu— [2nd][5][4][alpha][4].

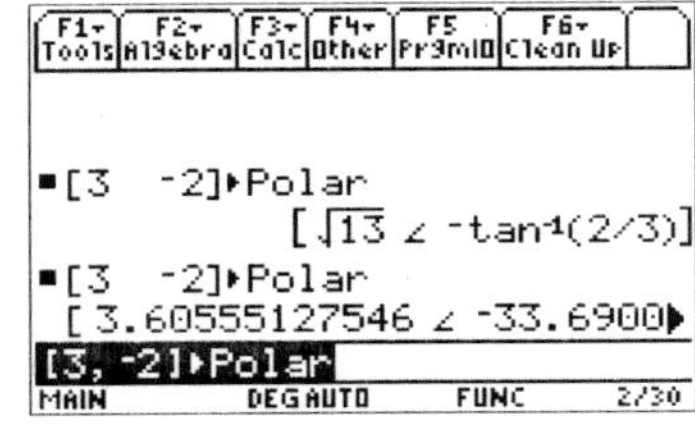

For Example 1, the vector $\mathbf{u} = \langle 3, -2 \rangle$ is entered on the TI-89 as `[3,-2]`. To convert to magnitude/direction form, either put the TI-89 in CYLINDRICAL or SPHERICAL mode, or use the `▸Polar` command (either from the menu referred to above, or selected from the [CATALOG]), as the screen on the right illustrates. The second entry shows the decimal approximation of the magnitude and direction. (With the TI-89 in Degree mode, the angle returned is in degrees.) Also note that the vector is *entered* with a comma between the components, but *displayed* (in the history area) without the comma.

For Example 2, one can find both the horizontal and vertical components at once, as the screen on the right shows. These computations were done with the TI-89 in Degree mode, so it was not necessary to include the degree symbol on the angle. Additionally, the TI-89 was in RECTANGULAR vector display mode; if it had not been, the `▸Rect` command (found in the MATH:Matrix:Vector ops menu) could be used to force conversion to rectangular format. Note that the vector is *entered* with a comma before the ∠ symbol, but *displayed* (in the history area) without the comma.

Alternatively, the TI-89 has `R▸Pr`, `R▸Pθ`, `P▸Rx`, and `P▸Ry` conversion functions, found in the MATH:Angle menu ([2nd][5][2]) as options 3 through 6. As shown in text Figures 28 and 30, these can be used to convert between the vector formats. "R" and "P" stand for "rectangular" and "polar" coordinates. Rectangular coordinates are the familiar x and y values. Polar coordinates, described in Section 10.6 of the text, are r (which corresponds to the magnitude of a vector) and θ (the direction angle).

Finally, observe that the conversion for Example 2 could be done by typing `25cos(41.7)` and `25sin(41.7)`; which approach to use is a matter of personal preference.

Section 10.3 Example 3 (page 767) Writing Vectors in the Form ⟨*a*, *b*⟩

Note how easily these conversions are done in the TI-89's polar vector format. (The TI-89 was in Degree mode.)

```
■[5 ∠ 60]                 [5/2  5·√3/2]
■[2 ∠ 180]                      [-2  0]
■[6 ∠ 280]
 [1.041889066    -5.90884651▸
[6,∠280]
MAIN        DEG AUTO      FUNC     3/30
```

Section 10.3 Example 5 (page 768) Performing Vector Operations

Section 10.3 Example 6 (page 769) Finding the Dot Product

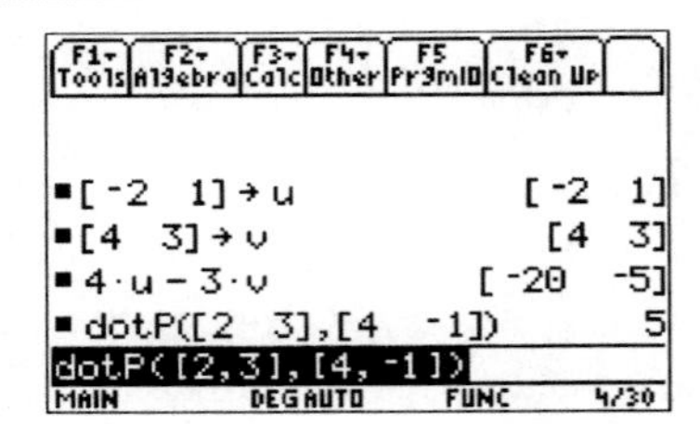

Figure 35 shows vectors being added, subtracted, and multiplied by constants using lists. This method can also be used with the TI-89, but the built-in support for vectors is a better approach. Using the TI-89's vector notation, the vectors can be stored in calculator variables (see page 101) which can then be used to do the desired operations. Shown are the commands to compute 5(c): $4\mathbf{u} - 3\mathbf{v}$, and 6(a): $\langle 2, 3\rangle \cdot \langle 4, -1\rangle$. The `dotP` function is in the [CATALOG] or the MATH:Matrix:Vector ops menu—[2nd][5][4][alpha][4][3].

Section 10.4 Example 2 (page 779) Converting from Trigonometric Form to Rectangular Form

```
■(2 ∠ 300°)                  1 - √3·i
■2·e^(i·300°)                1 - √3·i
2e^(i*300°)
MAIN        RAD AUTO      FUNC     2/30
```

With the TI-89 set to `Complex format:RECTANGULAR` (see page 102), or by adding the `▸Rect` command to the end of the computation, the conversion to the format $a + bi$ is done automatically when a complex number is entered. Complex numbers given in trigonometric format (like those in this example) can be entered much more conveniently using either of the other two complex formats: $r \cdot e^{i\theta}$ or $(r\angle\theta)$. The entries on the right illustrate this. These computations were done in Radian mode; note the degrees symbol. The $(r\angle\theta)$ format is similar to the polar format for vectors (see page 141; "∠" is [2nd][EE]).

This conversion can also be done using the `real` and `imag` functions from the MATH:Complex menu, or the `P▸Rx` and `P▸Ry` conversion functions from the MATH:Angle menu. These approaches are not illustrated here, but are analogous to those discussed for vector conversion above.

Section 10.4 Example 3 (page 780) Converting from Rectangular Form to Trigonometric Form

Section 10.4 Technology Note (page 782) The angle and abs commands

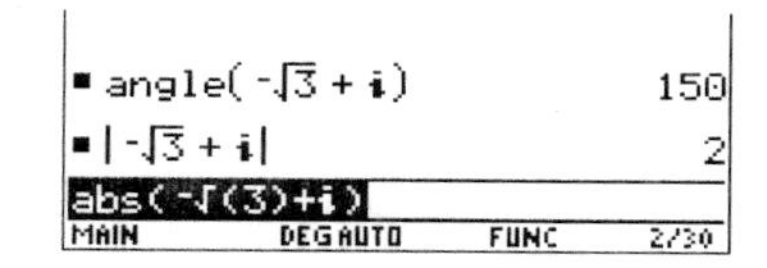

The abs and angle commands, found in the TI-89's MATH:Complex menu ([2nd][5][5]), can be used to compute the modulus r and the argument θ (respectively). The angle command gives θ in radians or degrees, depending on the mode setting. On the screen on the right, note that what is displayed as "$|-\sqrt{3}+$ i|" was entered as "abs(-√(3)+i)." Note that angle(-3i) and R▸Pθ(-3i) give the result -90°, which is coterminal with the answer given in the text, 270°.

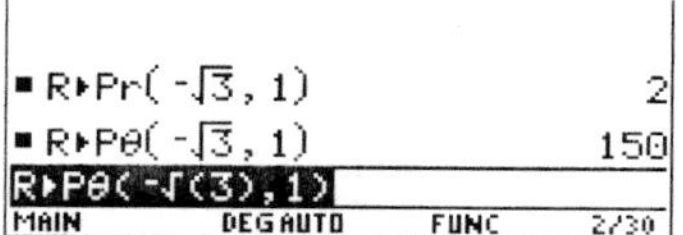

Another approach is to use R▸Pr and R▸Pθ from the MATH:Angle menu. To do this, enter the complex number $a+bi$ as an ordered pair (a,b), as is illustrated in the screen on the right.

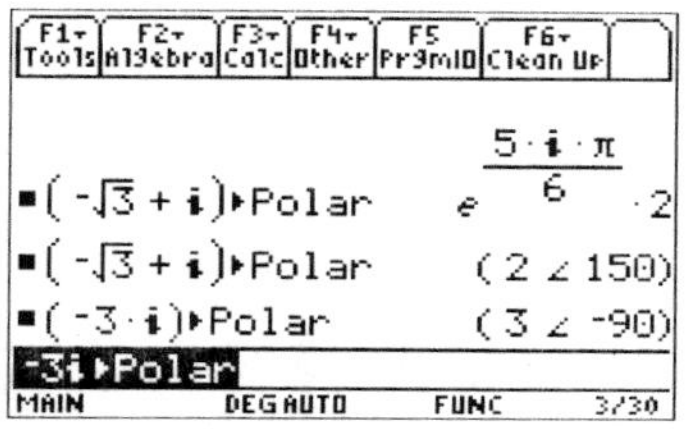

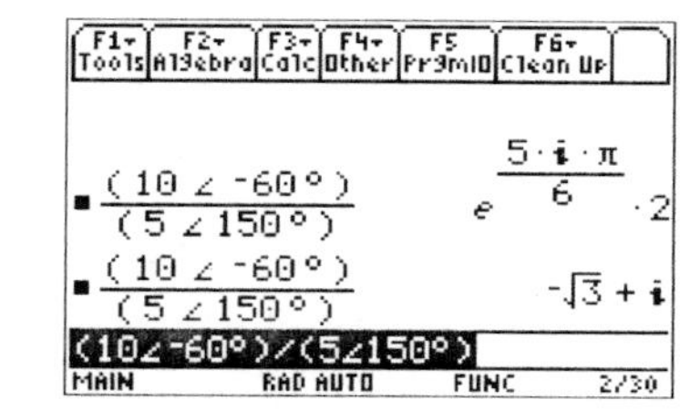

More convenient is the TI-89's ▸Polar command, found in the [CATALOG] or in the MATH:Matrix:Vector ops menu. This gives both the magnitude and angle at the same time, although the format in which this is displayed depends on whether the TI-89 is in Radian or Degree mode. In Radian mode (the first entry in the history area on the right), the result is displayed in the format $r \cdot e^{i\theta}$—hence, $r=2$ and $\theta = 5\pi/6$. When in Degree mode (the second and third entries on the right), the result is displayed as $(r\angle\theta)$—so $r=2$ (as before), $\theta = 150^\circ$ for (a), and $r=3, \theta=-90^\circ$ for (b). Note that -90° is coterminal with 270°, given in the text.

Furthermore, with the calculator's Complex format setting as POLAR (see page 102), the conversion can be done all at once, without the ▸Polar command. (As with the ▸Polar command, the format of the output depends on whether the TI-89 is in Radian or Degree mode.)

Section 10.4 Example 5 (page 783) Using the Product Theorem

Section 10.4 Example 6 (page 784) Using the Quotient Theorem

If doing such computations on a calculator, the TI-89's polar format can save a lot of typing. The expression in Example 4, e.g., can be entered as (3∠45°)(2∠135°). For Example 5, enter the expression as shown on the right; note that the output format can be affected both by the Complex Format setting and whether the TI-89 is in Radian or Degree mode.

Section 10.5 Example 1 (page 787) Finding a Power of a Complex Number

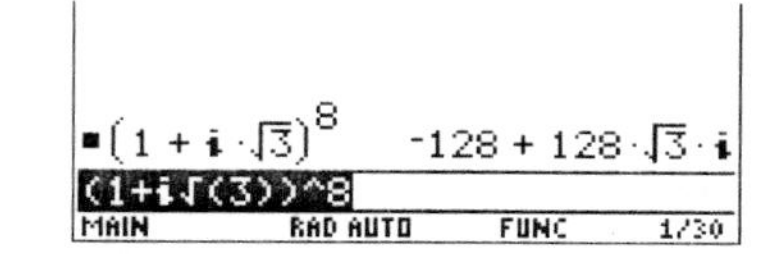

While the TI-89 could be used to perform the various steps illustrated in the text (conversion to trigonometric form, etc.), note that it will compute the exact value of $(1+i\sqrt{3})^8$ directly.

Section 10.5 Example 4 (page 790) Solving an Equation by Finding Complex Roots

The TI-89's `cSolve` (complex solve) function can find all complex roots of a polynomial equation. The computation on the right was done in Degree and `POLAR` complex display mode.

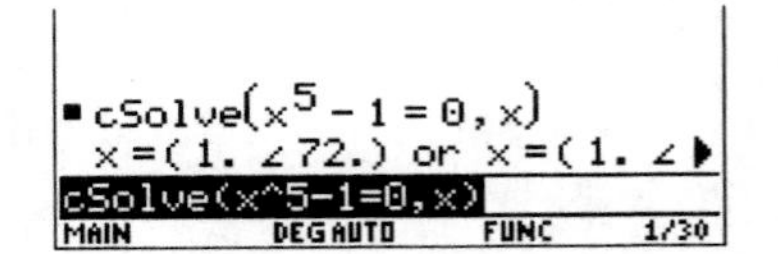

Section 10.6 Example 1 (page 794) Plotting Points with Polar Coordinates

Section 10.6 Example 2 (page 795) Giving Alternative Forms for Coordinates of a Point

Converting between polar and rectangular coordinates can be done using any of the conversion methods for vectors and complex numbers (covered beginning on page 141).

Section 10.6 Example 3 (page 796) Examining Polar and Rectangular Equations of Lines and Circles

Section 10.6 Example 4 (page 796) Graphing a Polar Equation (Cardioid)

Section 10.6 Example 5 (page 797) Graphing a Polar Equation (Rose)

Section 10.6 Example 6 (page 798) Graphing a Polar Equation (Lemniscate)

To produce these polar graphs, the TI-89 should be set to Degree and Polar modes (see the screen on the right). In this mode, the Y= screen is really the "r=" screen; the TI-89 allows entry of up to 99 polar equations (r as a function of θ). One could also use Radian mode, adjusting the values of `θmin`, `θmax`, and `θstep` accordingly (e.g., use 0, 2π, and $\pi/30$ instead of 0, 360, and 5).

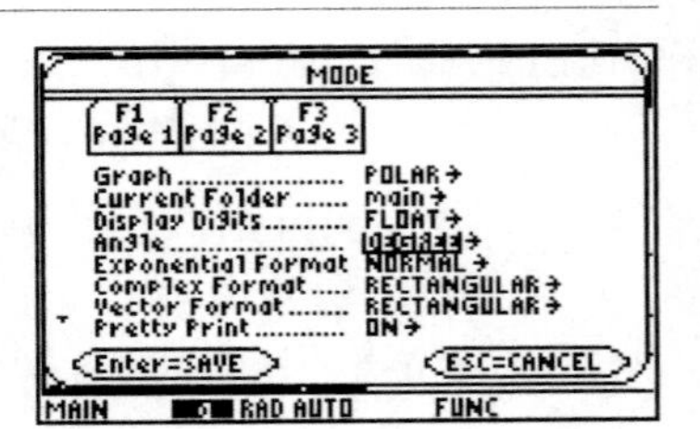

Press [◆][F1] and set `r1` to the desired expression (use [◆][^] to type θ). For the cardioid, rose, and lemniscate, the window settings shown in the text show these graphs on "square" windows (see section 11 of the introduction, page 104), so one can see how their proportions compare to those of a circle. Note, however, that on the TI-89, these window are not square; the Zoom:ZoomSqr option can be used to adjust the window dimensions to make it square.

For the cardioid, the value of `θstep` does not need to be 5, although that choice works well for this graph. Too large a choice of `θstep` produces a graph with lots of sharp "corners," like the one shown on the right (drawn with `θstep=30`). Setting `θstep` too small, on the other hand, produces a smooth graph, but it is drawn very slowly. Sometimes it may be necessary to try different values of `θstep` to choose a good one.

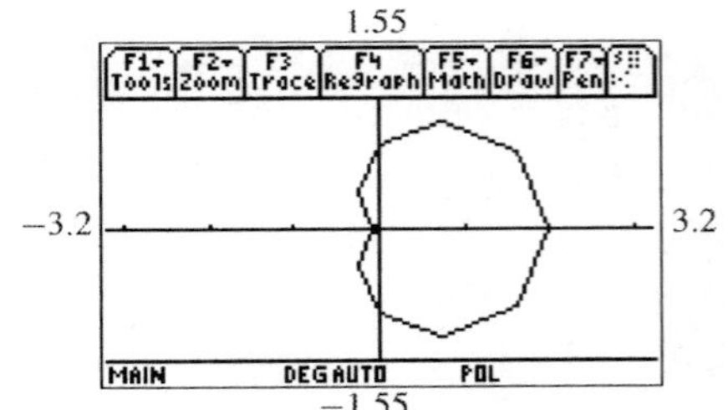

The lemniscate can be drawn by setting `θmin=0` and `θmax=180`, or `θmin=-45` and `θmax=45`. In fact, with θ ranging from -45 to 225, the graph of `r1=√(cos(2θ))` (alone) will produce the entire lemniscate. (`θstep` should be about 5.) The rose can be produced by setting `θmin=0` and `θmax=360`, or using any $360°$-range of θ values (with `θstep` about 5).

Section 10.6 Example 7 (page 798) Graphing a Polar Equation (Spiral of Archimedes)

To produce this graph on the viewing window shown in the text, the TI-89 must be in Radian mode. (In Degree mode, it produces the same shape, but magnified by a factor of $180/\pi$ — meaning that the viewing window needs to be larger by that same factor.)

Section 10.7 Example 2 (page 805) Graphing an Ellipse with Parametric Equations

See page 124 for information about using parametric mode. This curve can be graphed in Degree mode with `tmin=0` and `tmax=360`, or in Radian mode with `tmax=2π`. In order to see the proportions of this ellipse, it might be good to graph it on a square window. This can be done most easily with the ZOOM:ZSQR option ([◆][F3][F3][MORE][F2]). On a TI-89, initially with the window settings shown in the text, this would result in the window $[-8.3, 8.3] \times [-4, 4]$.

Section 10.7 Example 3 (page 805) Graphing a Cycloid

The TI-89 *must* be in Radian mode in order to produce this graph.

Section 10.7 Example 5 (page 807) Simulating Motion With Parametric Equations

Section 10.7 Example 7 (page 808) Analyzing the Path of a Projectile

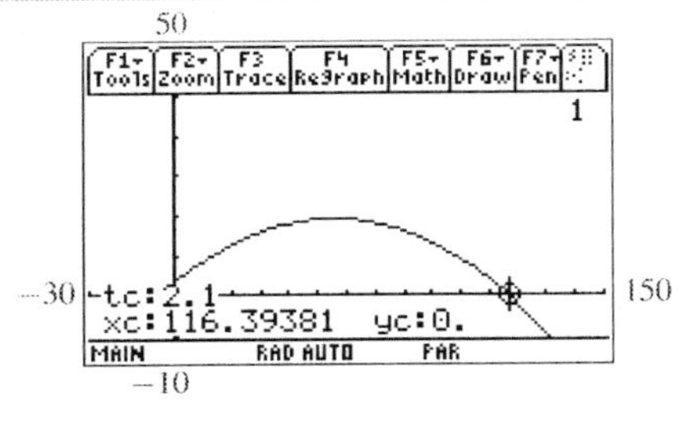

Parametric mode is particularly nice for analyzing motion, because one can picture the motion by watching the calculator create the graph, or by using TRACE ([◆][F3][F3]) and watching the motion of the trace cursor. (When tracing in parametric mode, the ▶ and ◀ keys increase and decrease the value of t, and the trace cursor shows the location (x, y) at time t.) The screen on the right is essentially the same as Figure 82, and illustrates tracing on the projectile path in Example 7. Note that the value of t changes by ±`tstep` each time ▶ or ◀ is pressed, so obviously the choice of `tstep` affects which points can be traced. This graph was produced by setting `tMin=0`, `tMax=3`, and `tStep=0.1`.

Rather than entering three separate pairs of equations for Example 5, the TI-89's list features can be used to graph all three curves with a single pair of equations: Define

`xt1=132cos({30,50,70})*t` and `yt1=132sin({30,50,70})*t-16t^2`

with the calculator in Degree mode.

Section 11.1 Example 1 (page 821)

Finding Terms of Sequences

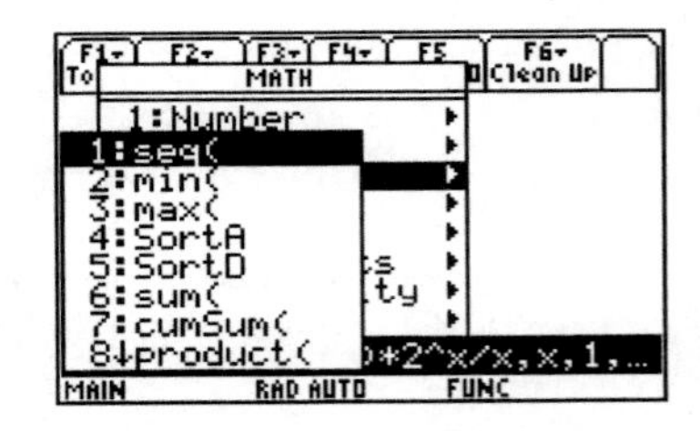

The seq(command can be found in the MATH:List menu ([2nd][5][3]). Given a formula a_n for the nth term in a sequence, the command

seq(*formula*, *variable*, *start*, *end*, *step*)

produces the list $\{a_{start}, a_{start+step}, \ldots, a_{end}\}$. If *step* is omitted, a value of 1 is assumed. The size of the resulting list is limited by available memory.

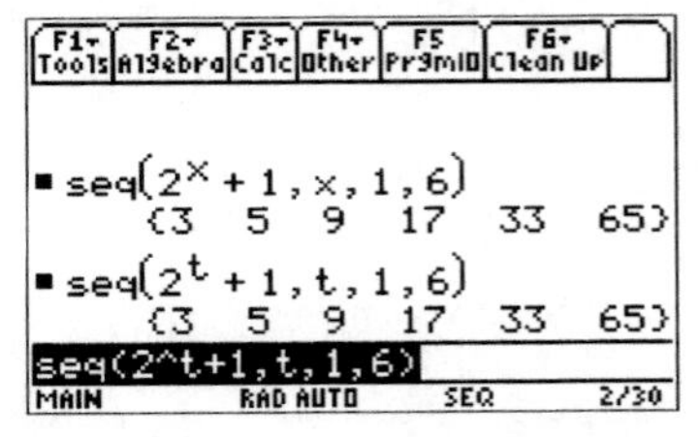

Note that *variable* can be any letter (or letters). The text uses the character n, which is not available on the TI-89 (although the non-italicized n works), but x, y, z, or t would be more convenient (since they can be typed with a single key).

Section 11.1 Technology Note (page 822)

Sequence mode

The TI-89 has a Sequence mode setting; information about mode settings can be found in section 9 of the introduction (page 101). Note that it is not necessary to use Sequence mode to use the seq command.

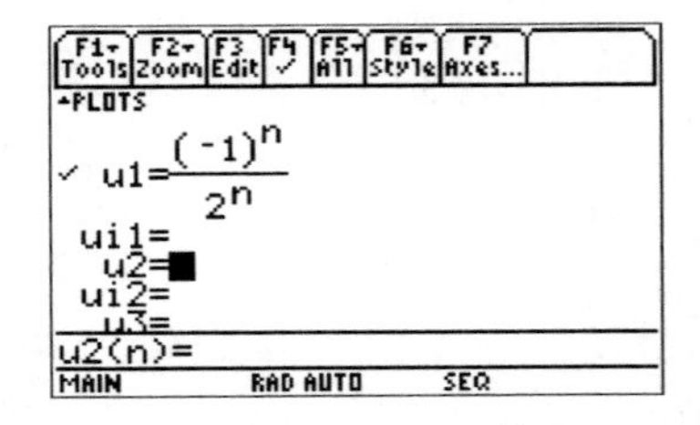

With the TI-89 in Sequence mode, [◆][F1] accesses the screen shown on the right, which allows the definition of up to 99 sequences, called u1, u2, *etc*. Note that the TI-89 represents terms of the sequence as u1(n) rather than a_n; otherwise, this is very similar to the notation used in the text. The method of defining a sequence is to first give a formula for the nth term in the sequence, and then (if necessary) to specify the initial values for the sequence (the "i" in ui1, for example, stands for "initial"). (The TI-83, used for the screen shown in the text, allows a sequence to begin at any index, as specified by nMin; on the TI-89, all recursively defined sequences must begin at index 1.)

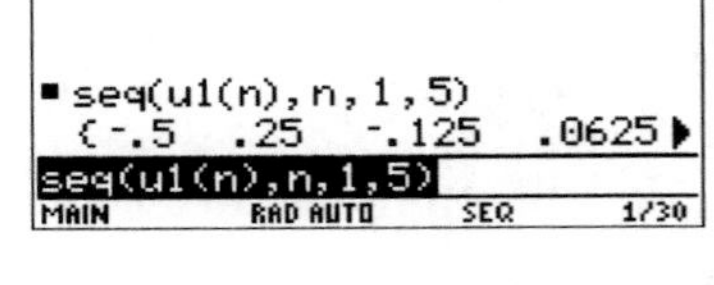

Once a sequence formula has been entered, it can be referenced from the home screen, as the screen on the right illustrates. Note that while sequences must be defined (on the Y= screen) in terms of the variable n, they can be referenced on the home screen using any variable.

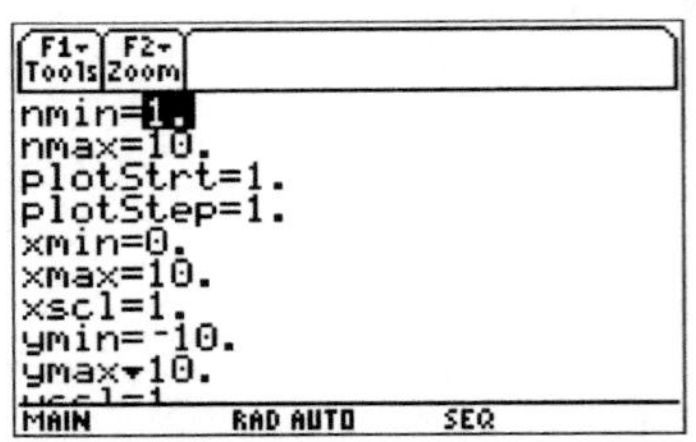

Scatter diagrams of sequences (like those shown in the text) can be produced fairly easily once a formula has been entered. For example, here are the appropriate window settings for producing a plot of the first 10 terms of a sequence. (The window variable not visible off the bottom of the screen is the usual yscl.) Pressing [◆][F3] should produce the correct plot; if it does not, press [◆][F1][2nd][F2] and check that Axes is set to TIME.

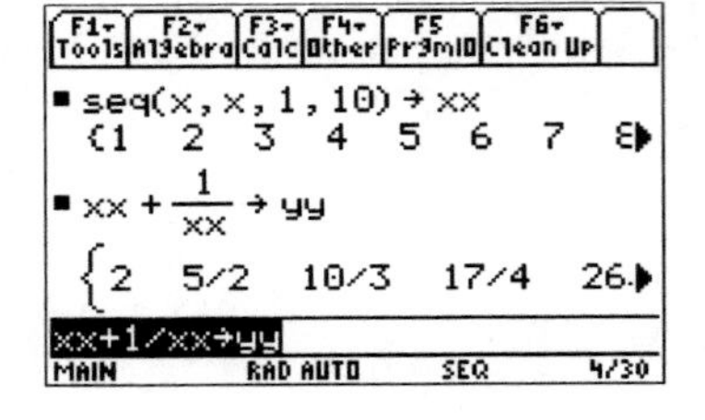

A second method to produce a scatterplot begins by storing lists in two variables (we have used xx and yy) as shown on the right. Note that this approach is less appealing for recursively defined sequences, but it does not require that the TI-89 be placed in Sequence mode.

Next press [◆][F1][▲][ENTER] and make the settings for a statistics plot as shown on the right. Finally, set up the viewing window, and check that nothing else will be plotted; that is, go to the Y= screen and make sure that the only `Plot1` has a check mark next to it. Pressing [◆][F3] should produce a plot like that shown in the text.

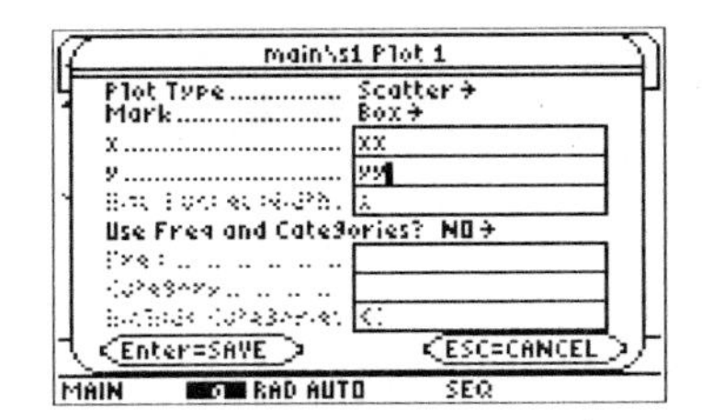

Note: When finished with a statistics plot like this one, it is a good idea to turn it off so that the TI-89 will not attempt to display it the next time [◆][F3] (GRAPH) is pushed. This can be done from the Y= screen using [F4] to un-check the plot, or by pressing [F5][5] (All:Data Plots Off).

Section 11.1 Example 2 (page 822)

Using a Recursion Formula

In Sequence mode, recursion formulas can be entered as shown on the screen on the right. The initial term of the sequence $a_1 = u1(1)$ is entered on the `ui1=` line. (The "`i`" stands for "initial.")

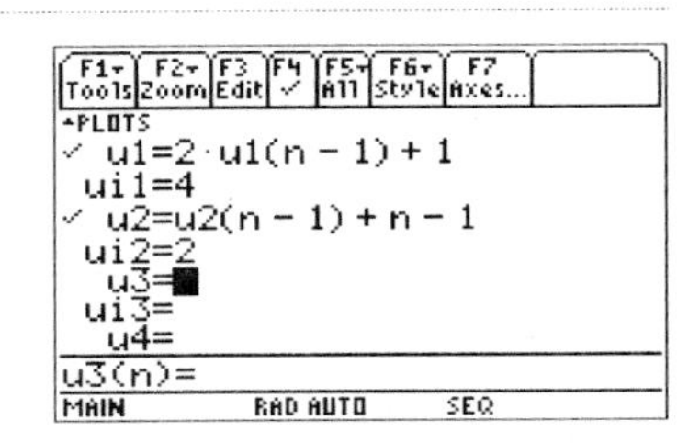

The TI-89 can also compute terms of such recursive sequences using the `ans(1)` storage variable. The screen on the right illustrates this approach: After entering the number 4, followed by the formula `2ans(1)+1`, pressing [ENTER] repeatedly computes successive terms of the sequence. Note that this approach does not lend itself to recursion formulas involving more than previous term (see the next paragraph), nor is it as useful for summations (see the next example).

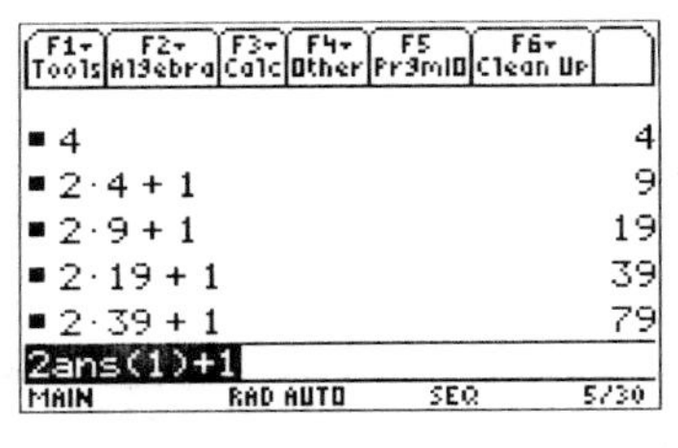

The Fibonacci sequence is described in the "For Discussion" section following this example in the text. It is typically defined recursively by $f_0 = f_1 = 1$, and $f_n = f_{n-1} + f_{n-2}$ for $n \geq 2$. In Sequence mode, the two screens below on the left show how to easily compute terms of this sequence (allowing that `u1(1)` represents f_0, and in general `u1(n)` is f_{n-1}). The `ans(1)` method also works here, since `ans(2)` references the *second-to-last* computed value (and `ans(3)` is the third result from the end, etc.). Therefore, as the third screen shows, `ans(1)+ans(2)` adds the previous two results, and can therefore be used to generate terms in the Fibonacci sequence.

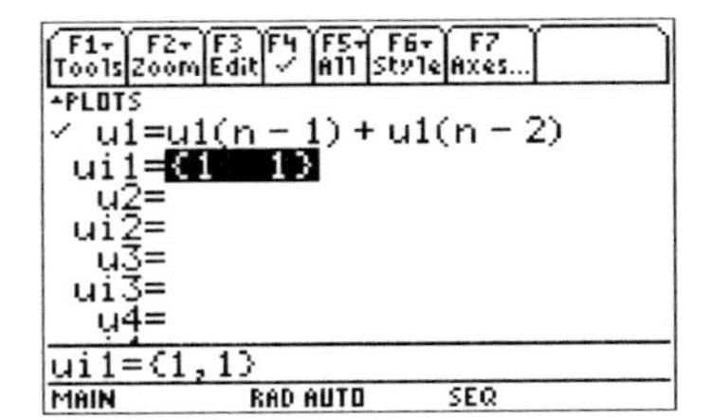

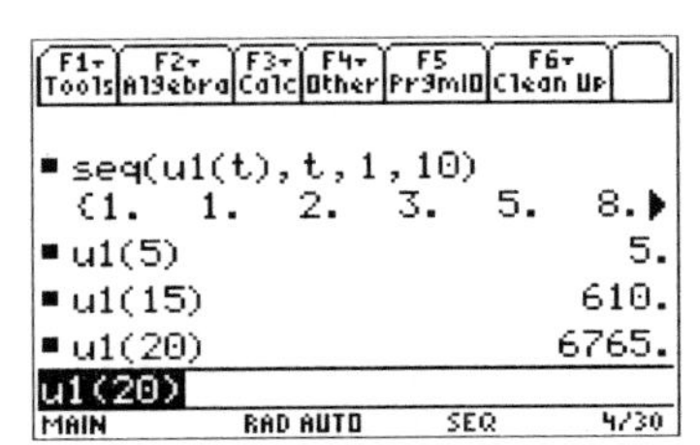

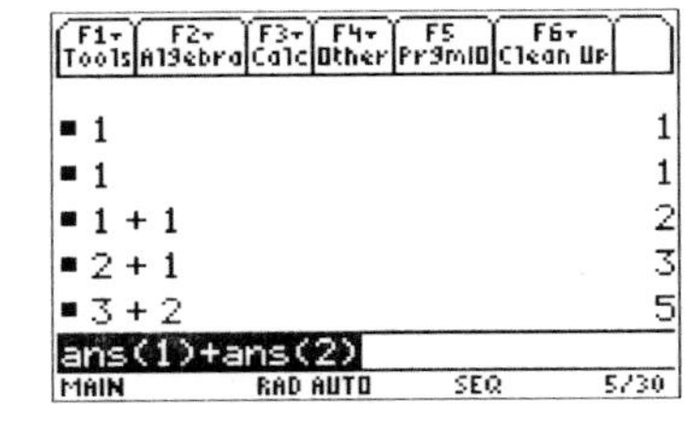

Section 11.1 Example 4 (page 825) — Using Summation Notation

The `sum(` command is also found in the MATH:List menu ([2nd][5][3]). `sum` can be applied to any list—either to a list variable, or directly to a list created with the `seq(` command. Evaluating summations on the TI-89 requires first generating the sequence as a list, then summing the list. The summation in Example 4 can be performed as shown on the right, using the `ans(1)` storage variable.

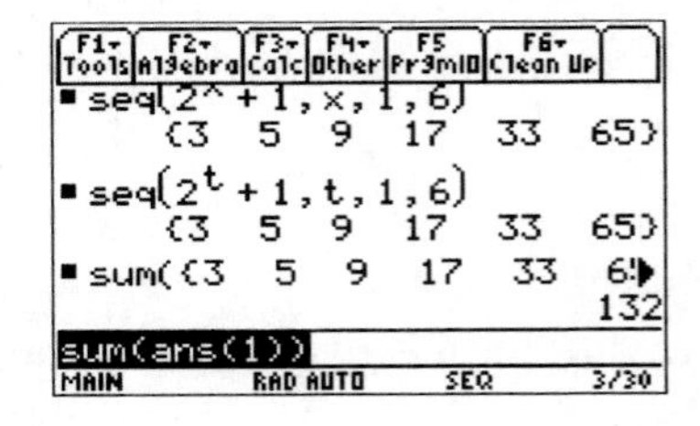

Alternatively, the `sum(` and `seq(` commands can be combined on a single line. The screen on the right uses this approach to find $\sum_{k=1}^{10}(4k+8)$.

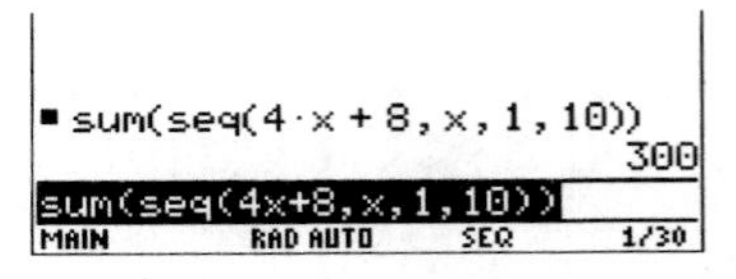

The $\sum$ function, found in the Calc menu ([F3]), essentially combines the actions of the `sum` and `seq` commands. It has the added benefit that the entry in the history area looks the same as in the text.

$\sum_{k=1}^{6}(2^k+1)$ 132
Σ(2^k+1,k,1,6)
MAIN RAD AUTO SEQ 1/30

Section 11.3 Example 7 (page 844) — Summing the Terms of an Infinite Geometric Series

Using the $\sum$ function (mentioned above), the TI-89 can give exact values for some kinds of infinite series, including this geometric series.

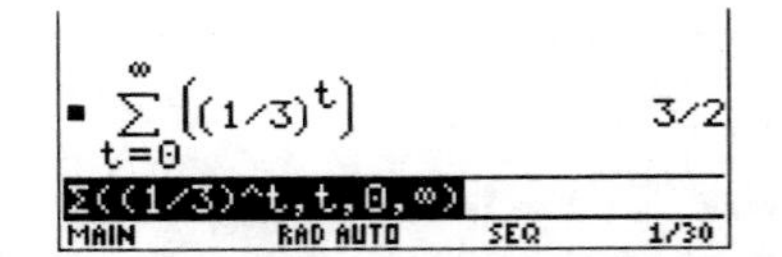

Section 11.4 Technology Note (page 851) — Computing Factorials

Section 11.4 Example 1 (page 852) — Evaluating Binomial Coefficients

Section 11.6 Example 4 (page 865) — Using the Permutations Formula

The `nCr(` and `nPr(` functions and the factorial operator "`!`" are found in the MATH:Probability menu ([2nd][5][7]). Note, though, that "`!`" is much more easily typed as [◆][÷]—this is one of those "hidden" key combinations that is revealed by [◆][EE]. The TI-89's format for `nCr` and `nPr` is different from the TI-83's (shown in the text); instead, enter these functions as shown on the right.

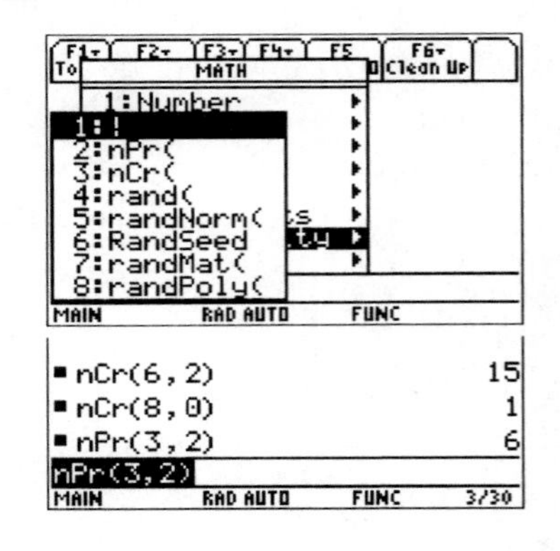

Section 11.7 Example 6 (page 878) Using a Binomial Experiment to Find Probabilities

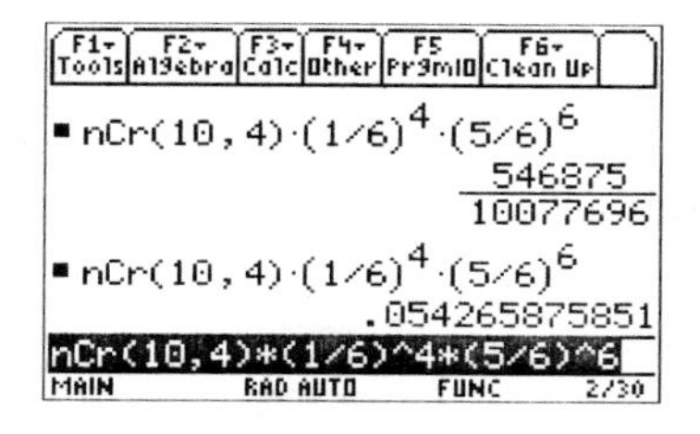

The TI-89 does not have statistical distribution functions like `binompdf`. The computations shown in this example must be done by manually entering the entire binomial probability formula (or by obtaining a program to automate such computations.) Note that if the TI-89 is in `AUTO` or `EXACT` mode, the results can be more detailed than is useful.

Section R.1 Example 4 (page 896) Adding and Subtracting Polynomials

Section R.1 Example 5 (page 897) Multipliying Polynomials

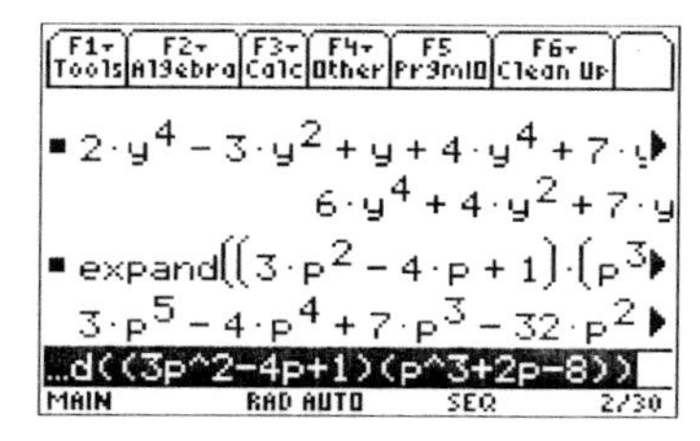

The TI-89 will automatically simplify sums and differences of polynomials, and will expand products with the `expand` command (in the Algebra menu, [F2][3]). Make sure that the variables used (`y` and `p` in this screen) are undefined. For example, if `y` had been assigned the value 0, the TI-89 would report 0 for the first computation. The simplest way to fix this is to type `DelVar y`, or press [2nd][F1][1][ENTER] (Clean up:Clear a-z).

Section R.2 Example 1 (page 900) Factoring Out the Greatest Common Factor

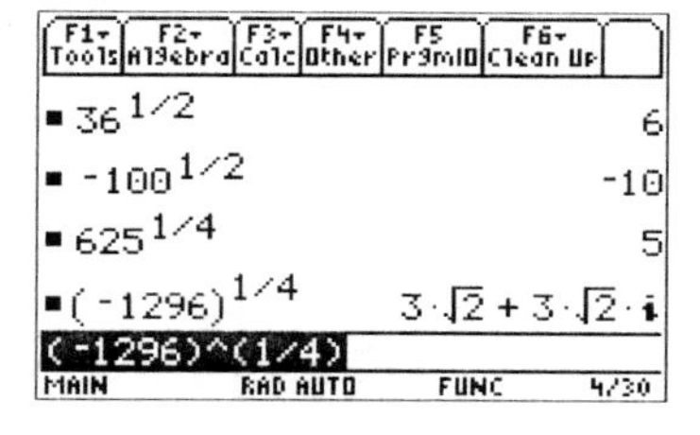

The `factor` command (in the Algebra menu, [F2][2]) will factor expressions like these, provided the variables are undefined (see the comments above.)

Section R.4 Example 4 (page 916) Using the Definition of $a^{1/n}$

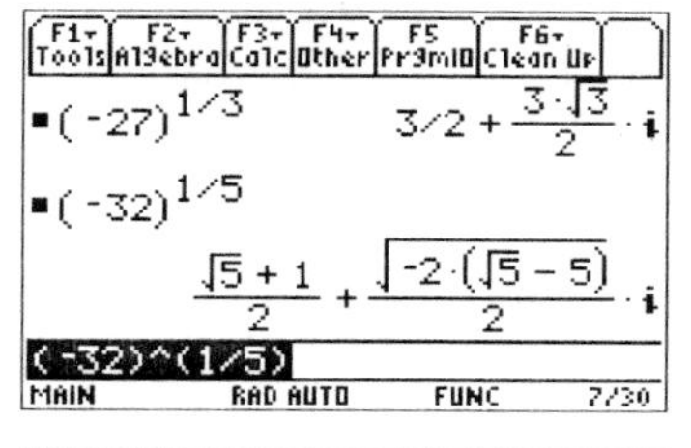

In evaluating these fractional exponents with a calculator, be sure to put parentheses around the fractions. Note for (d), where the text notes that $(-1296)^{1/4}$ is not a real number, the TI-89 gives a complex result if the complex format setting (see page 102) is something other than `REAL`.

In fact, if the complex format setting is `RECTANGULAR` or `POLAR`, the TI-89 is a bit too clever to be useful! Observe the results reported for (e) and (f) in the screen shown here. These *are* technically correct; for example, the cube of $\frac{3}{2} + \frac{3\sqrt{3}}{2}i$ really is -27, but they are not the results that we are looking for.

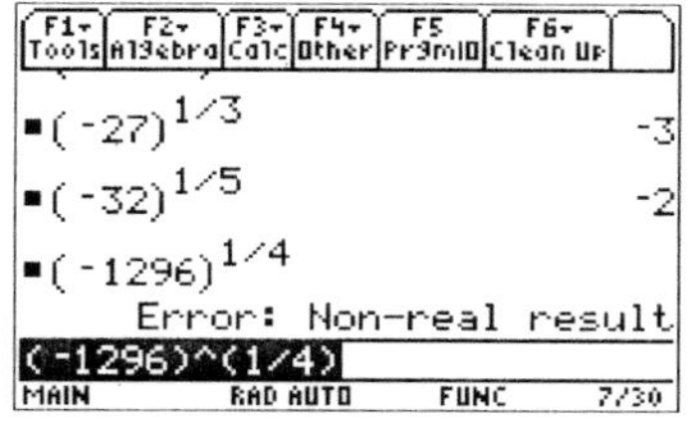

If we set the complex format to `REAL`, we get the desired results.

Section R.4 Example 5 (page 917)

Using the Definition of $a^{m/n}$

For (f), attempting to evaluate $(-4)^{5/2}$ will produce either a complex result or the error message shown, depending on the complex format setting (see the previous example).

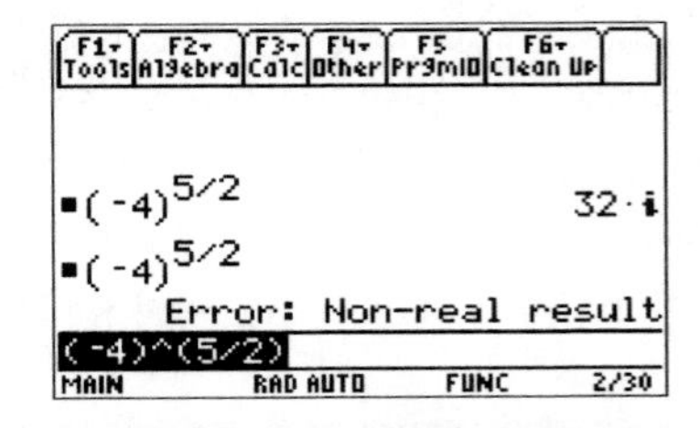

Appendix B Example 3 (page 934)

Using the SSE Program

See section 13 of the introduction (page 107) for information about installing and running programs on the TI-89. and see page 110 for a description of how to enter data into the calculator.

Here is the program shown in Figure 2, translated for the TI-89 and TI-92:

```
sse(xval,yval)
Func
Return sum((yval-y1(xval))^2)
EndFunc
```

As in the text, we assume that `y1` contains the formula for the model. Unlike the TI-83 (used for the calculator screens in the text), the TI-89 stores regression data in *data* variables rather than *lists*. If, for example, we have entered x and y values into a data variable called `wages`, with x in the first column (usually called `c1`) and y in the second column (`c2`), then the command

```
sse(wages[1],wages[2])
```

would compute the SSE using those x and y values.

Alternatively, we could store the data in list variables instead of in a data variable. If the lists `xx` and `yy` contain the data, then `sse(xc,yy)` would compute the SSE.

Appendix C Example 1 (page 937)

Finding the Distance between Two Points in Space

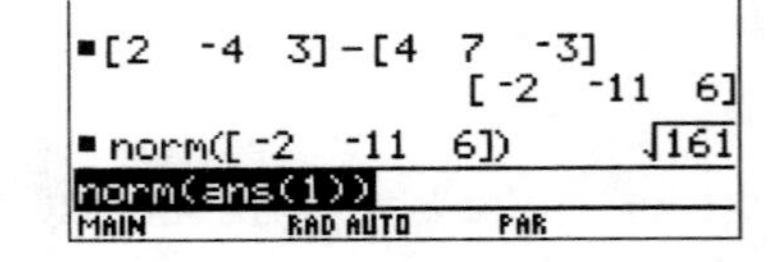

The TI-89's built-in support for vectors (previously described for two-component vectors on page 141) can be used to compute the distance between two points. Coordinates of points can be entered as vectors in the form `[x,y,z]` (the square brackets are [2nd][,] and [2nd][÷]), and then the `norm` function, found in the MATH:Matrix:Norms menu, can be used as shown on the right. The `norm` function adds the squares of each number in the vector, then takes the square root—in other words, it performs the computations of the distance function.

Appendix C Example 4 (page 938) Performing Vector Operations

The TI-89's built-in support for vectors extends to three-dimensional vectors. As before, the TI-89 uses square brackets rather than angle brackets.

The screen on the right shows three of the computations for this example.

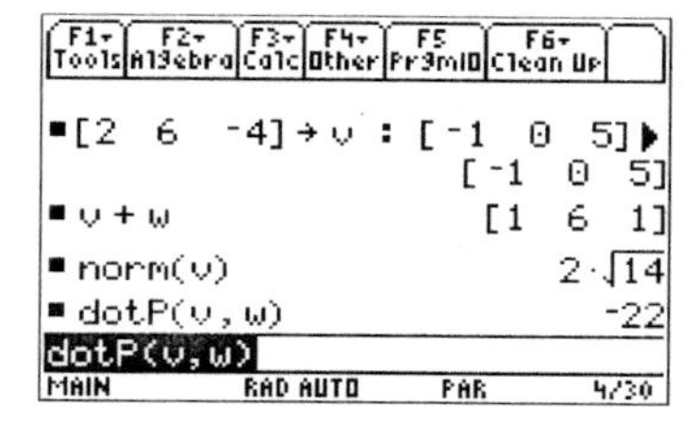

(a) $\mathbf{v} + \mathbf{w} = \langle 1, 6, 1 \rangle$

(d) $|\mathbf{v}| = 2\sqrt{14}$ `norm`* is in the MATH:Matrix:Norms menu

(e) $\mathbf{v} \cdot \mathbf{w} = -22$ `dotP` is in the MATH:Vector ops menu.

*For (d), we use `norm` rather than `abs`—even though `abs(v)` shows up in the history area as |v| in pretty print mode. `abs(v)` simply returns the vector `[2,6,4]`—it takes the absolute value of each component of the vector **v**.